Integrative Multi-Omics in Biomedical Research

Integrative Multi-Omics in Biomedical Research

Editors

Michelle Hill
Christopher Gerner

MDPI • Basel • Beijing • Wuhan • Barcelona • Belgrade • Manchester • Tokyo • Cluj • Tianjin

Editors
Michelle Hill
Cell Molecular Biology
QIMR Berghofer Medical
Research Institute
Australia

Christopher Gerner
Department of Analytical
Chemistry
University of Vienna
Austria

Editorial Office
MDPI
St. Alban-Anlage 66
4052 Basel, Switzerland

This is a reprint of articles from the Special Issue published online in the open access journal *Biomolecules* (ISSN 2218-273X) (available at: http://www.mdpi.com).

For citation purposes, cite each article independently as indicated on the article page online and as indicated below:

LastName, A.A.; LastName, B.B.; LastName, C.C. Article Title. *Journal Name* **Year**, *Volume Number*, Page Range.

ISBN 978-3-0365-2582-2 (Hbk)
ISBN 978-3-0365-2583-9 (PDF)

Contents

About the Editors

Michelle Hill obtained her PhD at the University of Queensland, Brisbane, Australia. After postdoctoral fellowships with Brian Hemmings (Friedrich Miescher Institute for Biomedical Research, Basel, Switzerland) and Seamus Martin (Trinity College Dublin, Dublin, Ireland), she worked as Research Officer with Robert Parton and John Hancock at Institute for Molecular Bioscience, The University of Queensland. In 2009, Michelle established the Cancer Proteomics Group at the University of Queensland Diamantina Institute at the Princess Alexandra Hospital campus, which relocated to QIMR Berghofer Medical Research Institute as Precision and Systems Biomedicine Laboratory. Her research aims to improve modifiable health outcomes by developing better diagnostics and integrative study of cell membrane disturbance that lead to diseases such as cancers, chronic diseases and infection. The program focus on early disease detection and non-drug prevention approaches, to provide tools for actively maintaining health. Their research combines cell biology and biochemistry knowledge with cutting edge multi-omics technologies and computational analyses.

Christopher Gerner studied biochemistry at the University of Vienna. After his post-doc in the labs of Rolf Schulte-Hermann, University of Vienna and Seamus Martin, Trinity College Dublin, he became Associate Professor at the Medical University of Vienna in 2003, then heading a clinical proteomics laboratory. 2012 he became full professor for bioanalytics at the University of Vienna. He is now heading the Joint Metabolome Facility, focusing on post-genomic analysis of clinical samples. Thus, he is characterizing patho-mechanisms mainly governed by biochemical processes such as hypoxia, oxidative stress or metabolic shortcomings. In his lab, several methods supporting analysis of humans in vivo (e.g., metabolomics based on finger sweat analysis) and the analysis of human or animal model-derived materials with regard to proteins, lipids and metabolites are well established. Corroborated by phospho-proteomics analyses focusing on short-term effects, this toolbox proves highly versatile for the investigation of drug effects and disease mechanisms.

 MDPI

Editorial

Integrative Multi-Omics in Biomedical Research

Michelle M. Hill [1,2,*] and Christopher Gerner [3]

1 QIMR Berghofer Medical Research Institute, Herston, Brisbane, QLD 4006, Australia
2 UQ Centre for Clinical Research, Faculty of Medicine, The University of Queensland, Herston, Brisbane, QLD 4006, Australia
3 Department of Analytical Chemistry, University of Vienna, 1090 Vienna, Austria; christopher.gerner@univie.ac.at
* Correspondence: michelle.hill@qimrberghofer.edu.au

Genome technologies have revolutionized biomedicine, but the complexity of biological systems cannot be explained by genomics alone. Advances in sequencing and mass spectrometry technologies coupled with methodological and computational innovations are essential in driving multidimensional omics applications.

This Special Issue covers the latest methods and novel findings from integrative analysis of multiple omics datasets to address diverse questions in biology and pathology.

The scene is set with a review article by Lancaster et al. [1], which introduces six players (genome, epigenome, transcriptome, metagenome, proteome and metabolome) that use two different technologies (sequencing and mass spectrometry). After characterizing individual omics data and analytical approaches, considerations for multi-omic study design and data integration methods are discussed.

The contributed research papers span a broad range of studies from clinical cohorts and mouse models to cell-based investigations, thus illustrating the diverse applications of multi-omics.

Two papers applied multi-omics to investigate physiological interventions.

Odenkirk et al. [2] compared the blood lipidome and metabolome in two cohorts of patients undergoing exercise and planned myocardial infarction, respectively, to gain insight on the metabolic pathways underlying the disease and its prevention.

Molendijk et al. [3] applied lipidomic and metagenomic profiling in a dietary model of gastro-esophageal reflux disease and associated esophageal pathology in mice, revealing increased microbiome diversity and a lipidomics signature associated with esophageal inflammation and metaplasia.

Five papers applied multi-omics to diverse cell models, with a study by Niederstaetter et al. [4] highlighting the variability and influence of fetal calf serum (used in culture media)-contained eicosanoids on cellular function, evaluated via proteomics and lipidomics. Neuditschko et al. [5] investigated endometrial pain mechanisms by applying proteomics, metabolomics and eicosanoid profiling to cells derived from endometriotic lesions.

Gillen et al. [6] applied metabolic measurements with secretome profiling to assess the impact of endotoxin (LPS) on macrophages, while Novikova et al. [7] combined transcriptome and proteomic profiling to investigate granulocyte differentiation and discovered HIC1, CEBPB, LYN and PARP1 as potential therapeutic targets in acute myeloid leukemia.

Finally, the paper by Kim et al. [8] illustrates the standardized application of combining drug affinity responsive target stability (DARTS) and mass spectrometry imaging (MSI) to facilitate target protein identification for other existing natural therapeutic compounds.

To wrap up this Special Issue, the comprehensive review article by Howard and Cristea [9] highlights the role of integrative multi-omics in deciphering system-level mechanisms of DNA sensing during viral infections. Following viral infection, protein–protein interactome and protein post-translational modifications drive the remodeling of the cellular transcriptome, proteome and secretome; hence, multi-omic investigations should also include interactome and modification analyses such as phosphoproteome.

Citation: Hill, M.M.; Gerner, C. Integrative Multi-Omics in Biomedical Research. *Biomolecules* **2021**, *11*, 1527. https://doi.org/10.3390/biom11101527

Received: 6 October 2021
Accepted: 8 October 2021
Published: 16 October 2021

In conclusion, multi-omic investigation has become a central technique for deciphering complex biological systems. Continued innovations in technologies, methodologies and applications will enable and support further expansion and integration of multi-omics in future biomedical research.

Conflicts of Interest: The authors declare no conflict of interest.

References

1. Lancaster, S.M.; Sanghi, A.; Wu, S.; Snyder, M.P. A Customizable Analysis Flow in Integrative Multi-Omics. *Biomolecules* **2020**, *10*, 1606. [CrossRef] [PubMed]
2. Odenkirk, M.T.; Stratton, K.G.; Bramer, L.M.; Webb-Robertson, B.M.; Bloodsworth, K.J.; Monroe, M.E.; Burnum-Johnson, K.E.; Baker, E.S. From Prevention to Disease Perturbations: A Multi-Omic Assessment of Exercise and Myocardial Infarctions. *Biomolecules* **2021**, *11*, 40. [CrossRef] [PubMed]
3. Molendijk, J.; Nguyen, T.M.; Brown, I.; Mohamed, A.; Lim, Y.; Barclay, J.; Hodson, M.P.; Hennessy, T.P.; Krause, L.; Morrison, M.; et al. Chronic High-Fat Diet Induces Early Barrett's Esophagus in Mice through Lipidome Remodeling. *Biomolecules* **2020**, *10*, 776. [CrossRef] [PubMed]
4. Niederstaetter, L.; Neuditschko, B.; Brunmair, J.; Janker, L.; Bileck, A.; Del Favero, G.; Gerner, C. Eicosanoid Content in Fetal Calf Serum Accounts for Reproducibility Challenges in Cell Culture. *Biomolecules* **2021**, *11*, 113. [CrossRef] [PubMed]
5. Neuditschko, B.; Leibetseder, M.; Brunmair, J.; Hagn, G.; Skos, L.; Gerner, M.C.; Meier-Menches, S.M.; Yotova, I.; Gerner, C. Epithelial Cell Line Derived from Endometriotic Lesion Mimics Macrophage Nervous Mechanism of Pain Generation on Proteome and Metabolome Levels. *Biomolecules* **2021**, *11*, 1230. [CrossRef] [PubMed]
6. Gillen, J.; Ondee, T.; Gurusamy, D.; Issara-Amphorn, J.; Manes, N.P.; Yoon, S.H.; Leelahavanichkul, A.; Nita-Lazar, A. LPS Tolerance Inhibits Cellular Respiration and Induces Global Changes in the Macrophage Secretome. *Biomolecules* **2021**, *11*, 164. [CrossRef] [PubMed]
7. Novikova, S.; Tikhonova, O.; Kurbatov, L.; Farafonova, T.; Vakhrushev, I.; Lupatov, A.; Yarygin, K.; Zgoda, V. Omics Technologies to Decipher Regulatory Networks in Granulocytic Cell Differentiation. *Biomolecules* **2021**, *11*, 907. [CrossRef] [PubMed]
8. Kim, Y.; Sugihara, Y.; Kim, T.Y.; Cho, S.M.; Kim, J.Y.; Lee, J.Y.; Yoo, J.S.; Song, D.; Han, G.; Rezeli, M.; et al. Identification and Validation of VEGFR2 Kinase as a Target of Voacangine by a Systematic Combination of DARTS and MSI. *Biomolecules* **2020**, *10*, 508. [CrossRef] [PubMed]
9. Howard, T.R.; Cristea, I.M. Interrogating Host Antiviral Environments Driven by Nuclear DNA Sensing: A Multiomic Perspective. *Biomolecules* **2020**, *10*, 1591. [CrossRef] [PubMed]

Review

A Customizable Analysis Flow in Integrative Multi-Omics

Samuel M. Lancaster [1,2,*], Akshay Sanghi [1,2,*], Si Wu [1,2,*] and Michael P. Snyder [1,2,*]

1 Department of Genetics, Stanford School of Medicine, Stanford, CA 94305, USA
2 Cardiovascular Institute, Stanford School of Medicine, Stanford, CA 94305, USA
* Correspondence: slancast@stanford.edu (S.M.L.); asanghi@stanford.edu (A.S.); siwu@stanford.edu (S.W.); mpsnyder@stanford.edu (M.P.S.); Tel.: +1-612-600-4033 (S.M.L.); +1-650-723-4668 (M.P.S.)

Received: 30 September 2020; Accepted: 23 November 2020; Published: 27 November 2020

Abstract: The number of researchers using multi-omics is growing. Though still expensive, every year it is cheaper to perform multi-omic studies, often exponentially so. In addition to its increasing accessibility, multi-omics reveals a view of systems biology to an unprecedented depth. Thus, multi-omics can be used to answer a broad range of biological questions in finer resolution than previous methods. We used six omic measurements—four nucleic acid (i.e., genomic, epigenomic, transcriptomics, and metagenomic) and two mass spectrometry (proteomics and metabolomics) based—to highlight an analysis workflow on this type of data, which is often vast. This workflow is not exhaustive of all the omic measurements or analysis methods, but it will provide an experienced or even a novice multi-omic researcher with the tools necessary to analyze their data. This review begins with analyzing a single ome and study design, and then synthesizes best practices in data integration techniques that include machine learning. Furthermore, we delineate methods to validate findings from multi-omic integration. Ultimately, multi-omic integration offers a window into the complexity of molecular interactions and a comprehensive view of systems biology.

Keywords: multi-omics; multi-omics analysis; study design; bioinformatics; machine learning; analysis flow

---

1. Introduction

Omics measurements are unbiased samples of molecules from a biological specimen. The genome was the first ome studied [1,2], and subsequent omes followed, building off DNA sequencing technology. Transcriptomics sequences the RNA content in cells, and metagenomics sequences all the genetic material from a group of organisms, usually microbial populations. Chromatin accessibility measurements select for sections of DNA to sequence that are differentially bound by chromatin—believed to affect transcription.

Omic measurements are not limited to nucleic acid sequencing. The most common omics methods orthologous to nucleotide sequencing involve mass spectrometry (MS). These include proteomics, metabolomics, and lipidomics, which are all vitally important and account for innumerable actionable discoveries. There are many other omic measurements, which all work together to improve understanding of systems biology.

Understanding each of these omes is vitally important and integrating them provides a more comprehensive picture of biology. For example, to understand the biochemical effects of changes in transcription, one must understand the metabolome and proteome as well. However, with the different natures of omic measurements, and the fact that they are best modeled by different statistical distributions, integrating this vast information in these distinct biological layers is challenging to non-experts. Using these omic measurements as examples, we will highlight potential integration methods that will reveal trends in multi-omics data.

2. Analysis of Single Omics Prior to Integration

Each of these omic methods is analyzed differently, with similar analyses shared between the more similar methods. One cannot discuss multi-omic integration without first having a shared understanding of how to analyze the individual omic measurements.

2.1. Genome Analysis

The genome is the core ome, and it codes for the basic information that inevitably is pushed into the other omes. For example, the transcriptome is aligned with the genome. This task is complicated because of the numerous mRNA isoforms, and the non-normal distribution of reads, which can be modeled using a negative binomial distribution [3]. After alignment and normalization, the read depth is used as a measurement of expression, reviewed below. Similarly, in the metagenome data, reads are aligned with the set of known microbiome data and read depth is assumed to be an abundance of each microorganism [4]. Chromatin accessibility measurements, such as the assay for transposase-accessible chromatin using sequencing (ATACseq), follow a similar principle. In this case read depth is a measure for how open the chromatin is.

Most genomes are sequenced on an Illumina platform, generating short reads. First, the quality of these reads must be determined, which informs one how well the sequencing was performed. Generally speaking a PHRED score of 30 is used as a threshold for keeping a read, although it may be altered depending on the needs of a study [5]. These scores are saved in FASTQ files as one of the four rows for each read, and they may be pulled out using several different programs. Another main sequencing type, long read sequencing, usually allows for the retrieval of much longer (>10,000 bp) sequencing reads (e.g., PacBio) and may be used to better capture repetitive regions or insertions or deletions, but it is often more expensive per base.

The reads that pass quality controls must be aligned with a known genome. For organisms without assembled reference genomes, which are increasingly rare, such reads must first be assembled into a genome with large contiguous chunks of DNA, or contigs (reviewed in [6]). Alignment tools such as BWA and Bowtie allow alignment of reads with a given number of mismatches, because no genome will be identical to the reference genome [7,8]. These alignments generate a sequence alignment map (SAM) file and their more compressed binary format BAM file [9]. From these files, variants between the sequenced genome and referenced genome can be determined using Samtools or other software and saved as a variance call format (VCF) file [10]. These may be DNA insertions, deletions, or nucleotide variations. From these files, biologically relevant genetic differences, for example, those that affect protein translations, may be determined. In some cases, single nucleotide polymophisms (SNPs) can be associated with known phenotypes or may even be proved causative for a disease.

2.2. Epigenomic Analysis

Epigenomic analysis aims to understand the functional context of the genome. For example, an animal has multiple organs with the same genome, but the genes expressed vary between organs depending on the tissue's epigenetic state. The genome is contained within a larger chromatin context that regulates which genes have access to transcriptional machinery and which are insulated from active machinery.

Various technologies have been developed to profile the epigenetic landscape, and particularly in the last decade, next-generation technologies have been applied to comprehensively map the epigenetic patterns in mammalian species [11,12]. One of the newest technologies in epigenetic analysis is assay transposase-accessible chromatin sequencing (ATAC-seq) [13]. The benefits of the ATAC-seq are (1) it provides direct evidence of genomic positions of nucleosome-depleted chromatin, which are permissible to transcriptional machinery binding, and (2) the assay only requires 10,000–50,000 cells as input, so it is particularly useful for animal tissue and limited specimens [14].

Similarly to whole-genome sequencing, ATAC-seq data are generated on the Illumina platform, giving high resolution information of open chromatin regions throughout the entire genome. After alignment with the same genome aligners, such as Bowtie, a critical step is filtering out low-quality and insignificant reads. This especially entails removing the high fraction of mitochondrial reads, which because of their high degree of accessibility are preferentially ligated with sequencing adapters. The sequencing reads are then analyzed for their pile-ups in peaks. The general purpose of peak calling is to find regions (on the order of hundreds of base pairs) that have significantly more reads piled up compared to the background reads across the genome [15]. ATAC-seq peaks represent the functional output of the data, and peaks are used in several types of analyses [16]. One very interesting analysis is transcription factor footprinting, which predicts which transcription factors are actively bound to chromatin and where the transcription factors activate transcription, giving insights into the regulatory pathways that affect gene expression [15].

2.3. Transcriptome Analysis

Transcriptomic data are generated in a similar way to genome sequencing libraries, except cDNA from reverse transcription is sequenced rather than genomic DNA. Aligning these reads to a transcriptome is a more complicated problem than aligning to a genome because of RNA variants, splicing, and otherwise uneven transcription of the genome. Transcriptome alignment tools require aligners such as Bowtie or BWA but require different information to annotate the transcription of the genome. The most commonly used program for transcriptome analysis is Spliced Transcripts Alignment to a Reference (STAR) [17]. This program is what is used by The Encyclopedia of DNA Elements (ENCODE), so should be used if someone wants to directly compare their results to most other experiments [18]. A newer program that is even faster than STAR is Kallisto, which is beneficial because it reduces computational expenses for very large experiments [19]. Salmon is another reputable transcriptomic software as well, among others [20]. Any of these different software algorithms will produce useful results for your experiment that may be later used for multi-omic integration.

Once this software has been run, several metrics will be generated for every transcript in each sample, including transcripts per million (TPM) and reads per kilobase of transcript, per million mapped reads (RPKM). To begin your analysis, several steps need to be taken. One analysis program in particular is used because of its end-to-end capabilities: the R package DESeq [3]. Similar packages include edgeR and limma [21,22]. The normalized reads can then be used for downstream analyses listed below. To perform custom analyses, the data should be read into a data matrix, which is helped by a program such as the R program tximport [23]. In this way TPM or RPKM can be pulled out for every sample. These should then be corrected for batch effects, for which RNAseq is particularly sensitive. The program sva::COMBAT() from R is excellently suited for batch correction [24]. Once corrected, the data are ready for downstream data analysis as illustrated below.

The first step in differential analysis workflow is data normalization, in order to guarantee the accurate comparisons of gene expression between and/or within samples. Proper normalization is essential not only for differential analysis, but also for exploratory analysis and visualization of data. The main factors that we often need to consider during count normalization are sequencing depth, gene length, and RNA composition. There are several common normalization methods to account for the "unwanted" variates, including counts per million (CPM), TPM, reads/fragments per kilobase of exon per million reads/fragments mapped (RPKM/FPKM), and DESeq2's median of ratio trimmed mean of M values (TMM) [3,25].

CPM, TPM, and RPKM/FPKM are the traditional normalization methods for sequencing data, but they are not suitable for differential analysis due to the fact that they only account for sequencing depth and gene length, but not RNA composition. Accounting for RNA composition is especially crucial for the scenario with a few highly differentially expressed genes between samples—big differences in the number of genes expressed between samples, which can skew the traditional types of normalization methods. It is highly recommended to account for RNA composition, especially for differential analysis [3]. Due to this, TMM normalization was developed and can be conducted in the edgeR package [22]. DESeq2 package implements the normalization method of median of ratio [3]. The DESeq2 package implements transformations by computing a variance stabilizing transformation which is roughly similar to log2 transformation of data, but also deals with the sample variability of low counts, generating vst and rlog formats of data. However, both formats are designed for applications other than differential analysis, such as sample clustering and other machine learning applications.

From these data, transcript enrichment can be performed using gene ontology (GO) or another such categorization method. GO involves assigning one or more functions to each gene based on its experimental function or categorization, and these categories are assigned to several genes. Then between the cases and controls one may see whether the GO categories are significantly enriched. DAVID bioinformatics offer a wide range of enrichment methods, including GO enrichments [26], although there are many similar GO algorithms such as GOrilla [27]. Another powerful pathway analysis and mechanism elucidation tool is ingenuity pathway analysis (IPA). It is an all-in-one, Web-based software application, enabling analysis, integration, and understanding of omics data, including gene expression, miRNA, SNP microarray, proteomics, metabolomics, etc. However, one of its downsides is that it is only commercially available. Nguyen et al. [28] systematically investigated and summarized the comprehensive pathway enrichment analysis tools, and concluded that topology-based methods outperform other methods, given the fact that topology-based methods take into account the structures of the pathways and the positions of each molecule in the biological system map. The best topology-based approaches include SPIA (signaling pathway impact analysis) and the ROntoTool R package.

Deeper analyses may be performed as well. For example, from peripheral blood mononuclear cells (PBMCs) the composition of the white blood cells may be estimated from expression of marker genes using software such as immunoStates [29], although others are effective as well [30]. These data may complement integrative analyses, integrating enrichment software from various omes.

2.4. Metagenomic Analysis

Metagenomic analysis also is similar to other nucleic acid omes. All the genetic material from a microbiome sample, often from stool, is sequenced. This review will discuss sequencing on an Illumina platform; however, other sequencing platforms are appropriate as well. This is all the genetic material from multiple organisms, hence the metagenome. These reads must be queried against a database, similarly to the previous methods. For example, the pipeline can query the human microbiome project database not just for different taxa, but also for biochemical pathways and even related individual genes [31]. This is important because taxa alone do not provide all the functional biological data about a microbial population. Such data provide a wealth of information about several levels of the microbiome. A fast, highly sensitive, although less specific method is querying chunks of the reads, or kmers, against a database, as used in Kraken. To aid with the problems presented by genomic flexibility in microorganisms, a kmer approach is increasingly being utilized, which requires only aligning part of the read, not the entire one [32].

These methods require very deep sequencing, and a more cost-effective method may be to sequence the 16s rDNA gene from bacteria [33]. This gene acts like a molecular clock, determining which taxa the read is from just like a clock determines time, with different parts of the gene highlighting different granularity in the taxonomic tree [34]. This method and metagenomic methods return count data, where read depth is used as a measurement of how abundant that particular part of the microbiome is. Methods to determine absolute abundance, not just relative abundance, should be used as well—for example, spiking a known amount of a microbe or DNA into the sample. Once the count data has been determined, it may as well be batch corrected. From these data, microbes with known importance or microbial pathways with known biological relevance to the host may be determined using methods described below.

2.5. Mass Spectrometry for Biomolecules

Like for the nucleic acid methods, mass spectrometry (MS)-based methods share some similarity, but also have their unique properties. Each ome is first fractionated through liquid chromatography (LC), the parameters of which are determined according to its own unique biochemistry properties. In proteomics, the proteins are typically digested into shorter peptides first. After LC, proteins are sent through data dependent MS/MS or data independent acquisition. The software for calling peaks depends on the platform used, with the most popular being Skyline and Perseus [35,36]. Data-dependent acquisition generates more identified proteins but is less comparable between samples. Each method requires its own data analysis software to call peaks, for example, openSWATH for the data independent acquisition [37]. Furthermore, even with data from a single piece of software, the library compared against is absolutely essential for data quality. For example, the TWIN library may produce particularly good data with the openSWATH platform [38].

Metabolomics data can be generated in different platforms, such as reversed RPLC-MS (reversed phase liquid chromatography-mass spectrometry), HILIC-MS (hydrophilic interaction chromatography-mass spectrometry), and so on. These are then imported into Progenesis QI 2.3 software which is able to convert spectra data to data matrix for further downstream analysis. Further data preprocessing steps include but not limited to filtering noise signals, data imputation, retention time adjustment, and data normalization [39]. Removing batch effects is one of the most crucial tasks in metabolomics, and various classic and advanced methods have been developed. Recently, Feihn et al. published a random forest model-based normalization method SERRF (systematic error removal using random forest), and the authors claimed that this method outperforms other normalization methods, including median, PQN, linear method, and LOESS [40]. This normalization can be applied on both untargeted metabolomic and lipidomic datasets. After data cleaning, one can use either an in-house metabolite library or public databases (HMDB, Metlin, MassBank, NIST, etc.) for metabolite annotation. With different available data, the annotation needs to be defined clearly with confidence levels. Our laboratory uses a Lipidyzer, a semi-targeted lipidomics platform, to determine the lipid absolute abundance by using lipid chemical standards [41]. The software that calls the lipid species is LWM (Lipidomics Workflow Manager), although this area is a fertile one for growth.

From these methods, individual molecules, proteins, and lipid species may be associated with biological questions of interest. Furthermore, in the lipidomic data classes of molecules may be enriched from their individual species. For example, triacylglycerols as a whole, not just individual triglyceride species, may be associated with the biological question of interest. Proteomic and metabolomic data enrichments may be performed with DAVID or MetaboAnalyst [42]. Ingenuity pathway analysis (IPA) from Qiagen may also be used for enrichments of the proteomic and metabolomic data, and it may also be used to integrate the two together. The Kolmogorov–Smirnov method is an alternative approach for pathway/chemical class enrichment analysis in the metabolomics and lipidomics field, which is able to use ranked significance levels as input. There are many other pre-written computer programs available, such as IPA, to analyze multi-omics data (reviewed in [43]), but we will focus on methods for developing your own customized pipeline, rather than pre-built Web-based software.

From all these individual methods, information is gleaned about that particular omic measurement. Furthermore, these methods all generate data structured similarly that facilitate omic integration. They all generate a list of analytes for every sample, be it a transcriptome, microbiome, proteome, lipidome, or metabolome. These analytes are then associated with a particular intensity (Figure 1). There are many differences between these omic measurements, but this similarity in data structure facilitates downstream analysis (Figure 1). There are many other omics measurements that share similarities with those mentioned, and there are numerous databases containing already-generated datasets, which may also be used for integrative multiomics rather than generating new data [44]. For example, http://educationknowengorg/sequenceng/, mentions 68 different next-generation sequencing technologies, most of which are omics measurements. Nonetheless, most share similarities with those already discussed here.

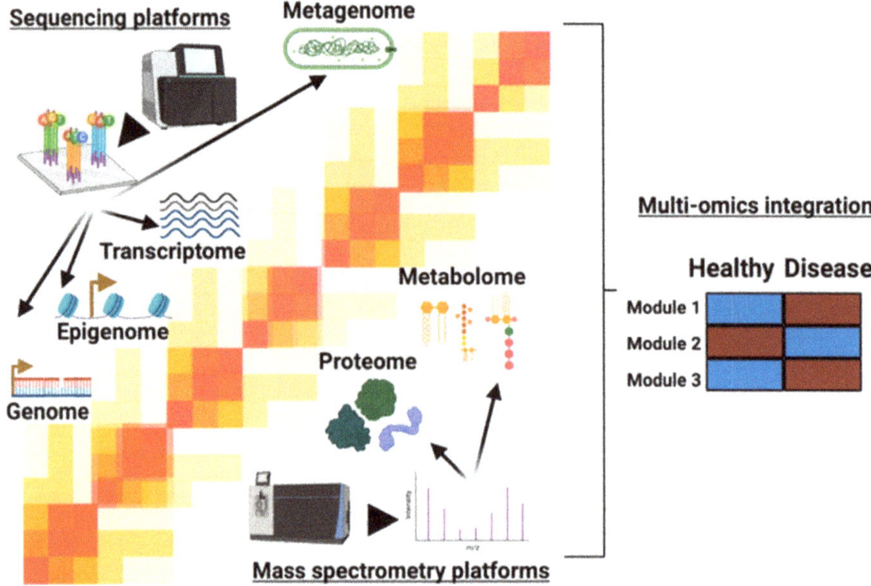

Figure 1. The molecules profiled in multi-omics studies. We describe 6 levels of information, starting from the bottom to the top: genome, epigenome, transcriptome, proteome, metabolome, and metagenome. The genome, epigenome, transcriptome, and metagenome are profiled by sequencing-based technologies such as sequencing by synthesis, depicted here, to profile a comprehensive set of nucleic acid molecules. On the other hand, mass spectrometers generate proteome and metabolome profiles as depicted here through measurements of biomolecules' masses and charges. For overlapping technologies, each omic level provides unique information and insights into cellular activity present in conditions being studied. By leveraging the layers of information, longitudinal and cross-sectional multi-omics studies find modules (e.g., cell signaling pathways) that are differential between healthy and disease states. These modules represent complex system biology networks that give precise insights into the molecular dysregulation in disease states.

3. Designing a Quality Study

The first step in understanding an analysis flow for integrative multi-omics is determined by your study design. Cross-sectional and association studies are beneficial in their relative ease to implement, and their ability to generate large amounts of data (Figure 2a). Typically, cross-sectional studies do not involve a randomized intervention, precluding causal inference. They involve taking a population split between cases and controls, and then sampling them evenly, and are excellent methods for determining associations.

Conversely, longitudinal studies are relatively difficult to recruit large numbers of participants to because they generate large numbers of time points and become expensive. However, the longitudinal nature increases the statistical power of a relatively small number of participants [45] (Figure 2b). Longitudinal studies further facilitate making causal inferences and allow for more accurate predictions. Each study design, with its strengths and weaknesses, has a slightly different flow of analysis. Wherever possible, the multiple omic measurements should be selected not staggered in time. For example, if the treatment course is seven days, all the participants should be sampled on the same days during treatment. This will greatly facilitate the analysis methods.

Some advantages of longitudinal studies include the ability to associate events chronologically with particular interventions or exposures. They allow a study of change over time, or a delta measurement from baseline, as discussed below, which can be more powerful than studying a single point in time for the effects of an exposure or intervention. They also allow for establishing the chronological order of events, which is essential for establishing causation—again, something that is precluded in cross-sectional association studies. There are relatively few negative effects other than the difficulty of recruiting large numbers of participants, but they may also include loss of individuals over time, confounding results [46].

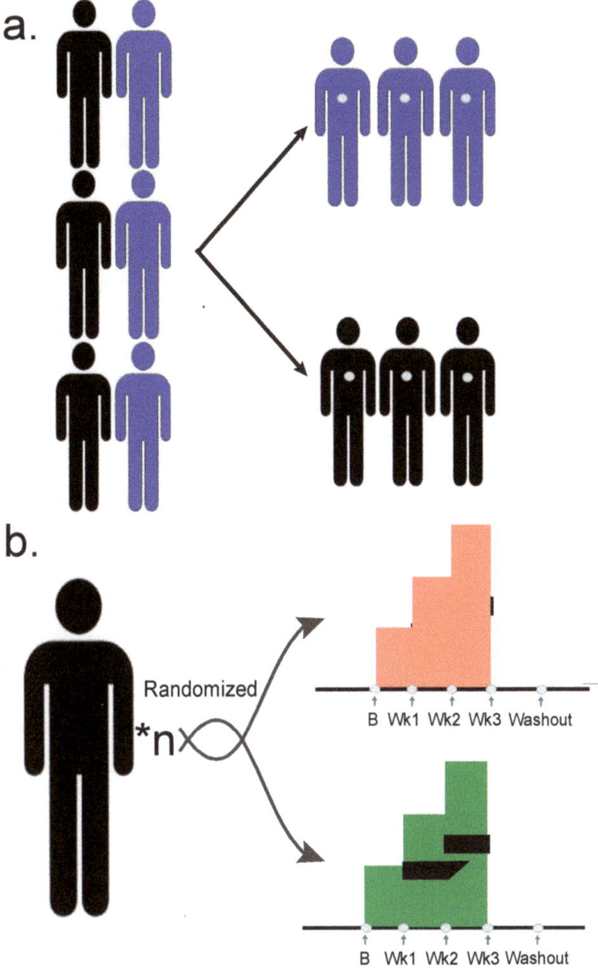

Figure 2. Typical multiomic study designs. Gray dots represent samples taken. (**a**) A case control observational study. A population is taken with participants that have the phenotype of interest (cases) and those without (controls). Cases and controls are sampled in even amounts. (**b**) A randomized longitudinal study where n participants are randomized into two arms of a study. In this case an increasing treatment dose is administered, and samples are taken every week.

In each individual, there are apparent biases in the technologies and analytical methods, which limit insights into biology. Often signals from individual omes are difficult to label as accurate or relevant because the information does not connect to the broader context of the system. Multi-omic integration offers an opportunity to use orthogonal methods to measure the same molecular pathways and processes. Such methods partially mitigate the inherent false positives and false negative rates in the single omes, as finding the similarities and biological connections supports the truly biologically-relevant information [47].

4. Analysis Methods for Multi-Omic Integration

4.1. Dimensionality Reduction

The first step in an omics study is to reduce the dimensionality of your data so they can be visualized. In a metagenome, for example, there may be hundreds of microbial species. This means that every sample is a data point with hundreds of dimensions. Dimensionality reduction techniques will take the data and reduce them to fewer dimensions, often as few as two or three, that represent most of the variation in the data. Then it is easier to visualize and use statistics that require fewer dimensions.

The first dimensionality reduction technique invented is principle component analysis, which is a widely used unsupervised method. This method, though yielding valuable results, is not the most statistically precise because it assumes normally distributed data. Anyone who works in omics will testify that the data are never normally distributed, although transformations can make the data approximately normal. One superior method is non-metric multidimensional scaling. This method is iterative and nonparametric, avoiding problems with unusual distributions, and it handles zero-truncated data well—a phenomenon in which in some samples a particular analyte is undetectable and in others it exists at a high level. Another method, tSNE, is particularly well designed to separate well defined groups. Besides t-SNE, UMAP (uniform manifold approximation and projection) [48] is a newly developed dimension reduction technique for non-linear relations. It usually implements faster than t-SNE, especially when it concerns large number of data points or a number of embedding dimensions greater than 2 or 3. It has many applications in single-cell sequencing data. Other methods include principal coordinate analysis and multidimensional scaling. Every method is capable of providing useful information; however, properly selecting a method can increase your statistical power.

The information gleaned from dimensionality reduction is similar across omic techniques. It can discover batch effects, particularly in mass spectral data. If two batches do not overlap, then additional correction techniques need to be applied. This method can find samples that failed, which would be represented as outliers in the data. Once data quality has been established, these methods can find any structure in the data that might be associated with biologically relevant variants. This is the most basic example where a metadatum, participant ID, may be grouped together. However, there are many more—sex, insulin resistant status, etc. In the case of the microbiome, it can also be used to measure beta diversity, as outlying samples will have different microbial compositions than the rest of the cohort.

4.2. Normalizing the Data

Once the structure of the data has been determined, omics measurements can be grouped together for integration. Usually they are done so after log, log2, or other transformations to facilitate downstream statistics [49,50]. The log transformation is normally used to make highly skewed data approximately approach a normal distribution. This can be useful both for facilitating the data to meet the assumption of statistic models and for making patterns in the data more interpretable. Microbiome data are so unusually distributed, other transformations may be applied, such as arcsin. With certain longitudinal designs they can be normalized to the baseline measurements to only measure the deltas from the baseline, reducing the effects of inter-individual variability. This is absolutely essential in longitudinal data to reduce the effects multiple individuals would have on biasing a sample, and is one of several strengths of that study design.

A z-score is another normalization method that standardizes all the analytes to the same range. This alleviates the problem of vastly different expression levels, facilitating grouping several different omes together for integration. For example, if one wanted to integrate the metabolome and gut microbiome, the values for the metabolome may be in the tens of millions, while analytes in the microbiome may be zero truncated, with most values being 0. To compare these two, particularly visually, they must be on a similar scale. Z-scoring makes the average value for every analyte 0, and then one standard deviation above that 1, etc.

4.3. Correlation Networks Analyses

Once these normalizations and transformations are performed, correlation metrics can inform one about the most basic relationships between the analytes. Pearson correlation coefficients (PCC) and spearman correlation coefficients (SCC) are the two most typical types of correlation metrics. The PCC is a parametric metric with more accuracy, whereas the SCC is more robust if outlier samples are present. One should target analytes of the most interest (e.g., only the significant molecules) if possible, because with too many analytes in networks, it is difficult to capture the most useful biological information and it is inclined to be masked by the underlying noise. Correlation networks are much more effective when dealing with deltas in longitudinal data that reduce interindividual variability. If more than one sample, not corrected to baseline, is from a single individual, such an analysis will be overfit and produce false positives. Additionally, one must always correct for multiple hypotheses during these projects to reduce false positives. In these data a Benjamini–Hochberg correction is appropriate. One may also use a Bonferroni correction, but in some omic studies that may overcorrect, losing true positives. Both will have their uses and may be differentially used in longitudinal baseline normalized vs. unnormalized data. This correlation analysis can be plotted as a network diagram, which is a fantastic visualization tool for this type of data. Though high-level visualizations, network diagrams offer compelling, informative overviews of interactions in biological systems [51].

When comparing interaction networks across different conditions, disease states, or interactions, a network analysis may provide you appropriate information about how the two states differ. A network analysis will provide one with total nodes (analytes) that are connected in the network, the total number total edges between the networks (significant correlations), and many other important relationships, such as the numbers of positive and negative correlations. Complementing visualizations, these summary statistics provide an excellent overall view of the co-correlations occurring in any multi-omic project. This type of topological analysis is not only able to provide practitioners straightforward and clear ideas when comparing multiple networks, but also provide insights into network hubs and centers, which may have many applications in drug target selection and identification of key regulators. There are several packages for R—igraph, statnet, ggnetwork, ggnet, ggraph, etc.—with highly related functionalities that perform these analyses [52,53]. R packages "igraph" and "statnet" are able to provide quick visualizations, which are good for a quick exploration about the network structure but are not necessarily the most efficient ways for aesthetically perfect visualization. R packages "ggnet" and "ggnetwork" are very similar packages, and both seem to use a variant of the ggplot syntax, meaning that they would be advantageous if you are familiar with the ggplot system.

4.4. Cross-Sectional Analyses and Testing Categorical Variables

In a cross-sectional study, when testing a single analyte between two sets of samples, the nonparametric version of the student's t-test, Wilcoxon rank sum test, is appropriate. A t-test assumes a normal-like distribution and should be used with care, as omics measurements are extremely rarely Gaussian. If confident that prior information will be obtained before the test, Bayesian counterparts to these tests will provide more power. These are not necessary, and should only be used by an expert. In a cross-sectional study where two categories are being tested against, one may further use logistic regression as a means of regressing between these categories. This regression fits a curve to binary data, generating an odds ratio and p-value.

When analyzing across more than two categories of data, one should use the non-parametric analysis of variance (ANOVA), Kruskal–Wallis. This method may be used to test a trend in your data over categorical variables. When correcting for multiple variables, one may use a multivariate ANOVA (MANOVA), but this should be used with care because ANOVAs assume a normal distribution. To avoid these assumptions about distributions, a permutational multivariate analysis of variance (PERMANOVA) should be used.

4.5. Testing along Continuous Variables

Another method determining trends over categorical variables is multiple linear regression. Like ANOVA, this may be used to find trends in one or more categorical variables. However, multiple linear regression can find trends over continuous variables as well, or any combination thereof. Although multiple linear regression also assumes a normal distribution, it can still be a valuable tool for detecting trends in data and is widely used by multi-omic researchers. In cases like this, where the statistical assumptions do not perfectly match data distributions, orthologous methods should be used for confident assessments.

Even more sophisticated than multiple linear regression is a mixed model. These can find trends in data and can also find the variance in data for random variables. Random variables are those that are randomly distributed in your data, say a random assignment of sex, so they are not associated with the outcome variable. Nonetheless, these variables can add variance, making the data noisier. Further, these mixed models can select other distributions than Gaussian, such as Poisson, so variables that violate normality may be modeled better. Mixed models are appropriate to account for complicated and heterogeneous datasets with confounders—gender, race, age, BMI, etc. Mixed models are particularly well suited for tracking longitudinal data [45]. Together, these methods are powerful for detecting trends in the data.

4.6. Clustering Algorithms

Clustering algorithms group similar samples or analytes together. Two primary clustering algorithms are hierarchical and k-means clustering. These are "hard" clustering algorithms which force analytes or samples into particular groups. This may be useful to determine whether samples cluster by individual, batch, or some other biological measurement, for dimensionality reduction techniques. They can also be used to determine outliers in the data, which may be of special interest to the researcher.

To find trends in longitudinal data, fuzzy c-means clustering is a powerful tool. The R mfuzz package provides tools for this analysis [54]. This is a "soft" clustering algorithm, giving analytes a score known as membership in every cluster, rather than forcing them into a single cluster. However, like other clustering algorithms, it still finds analytes with similar expression profiles. Using the previously mentioned z-scores, c-means clustering finds longitudinal trends in data for multiple omic measurements. These trends are powerful if one wants to find dose, temporal, delayed, or other response patterns in multi-omics data [55,56].

One of the most critical and haunting issues in clustering is to determine the optimal number of clusters. Selecting an inappropriately small number of clusters would cause the missing detection of some meaningful molecular trends and clusters, whereas an improperly large number of clusters may result in redundancy of cluster detection. There are several ways to assist the selection of the optimal number of clusters. One of the classic methods is called the elbow method, which calculates the within-cluster sum of squared error (wss). This method is widely applied; however, it gets tricky to determine the "elbow" point. Another way to survey this issue is to calculate minimal centroid distance, which is similar to the elbow method, aiming to find the "elbow" point to gain the minimal centroid distance. Another more efficient method is to calculate the correlations between cluster centroids, and decide on the optimal number of clusters once high positive correlations (e.g., 0.85) are detected.

Another method of clustering, supervised clustering, involves placing a priori information into a model before using the clustering algorithm. For example, if you have cases and controls, these may be entered into the data beforehand, or if you have longitudinal data with doses, the baseline controls may be contrasted with the doses. Categorical variables are required for this type of clustering, but they are an excellent method of assuring one will find analytes with similar expressions in the data [46].

4.7. Feature Selection for Covarying Analytes

A powerful tool in the arsenal of multi-omics researchers is feature selection. In some data, the analytes strongly covary. For example, in the metagenome, if one organism increases it will have an effect on every other organism in the system. In such circumstances it may be difficult to know which of these analytes to prioritize putting in a model. Least absolute shrinkage and selection operator (LASSO) and ridge regression tackle these problems. These functions will weight or eliminate the variables with the most and least explanatory power in your model. This way, future analyses may be performed on more manageable and more meaningful data, which may also increase statistical power. There are numerous feature selection methods, and descriptions and comparisons of all of them are obviously beyond the scope of our review. We mainly highlight two of them (LASSO and ridge) because they are widely applied penalized algorithms that reduce model complexity and prevent over-fitting which may result from simple linear regression. The main principle of these two regularization methods is to restrict or shrink the coefficients towards zero for the non-impactful features, in order to reach the goal of feature selection.

4.8. Machine Learning

Machine learning is an important subset of artificial intelligence, and nowadays has drawn attention in various fields. In omics studies, machine learning is widely applied on classification and prediction problems by using omics profiling data. Different suites of machine learning algorithms are suitable for classification and prediction scientific problems. Classification and prediction, as two main branches of machine learning, depend on the types of tasks or problems that are intended to be solved by machine learning and are either categorical (classification) or continuous (prediction). There are three main types of machine learning algorithms: unsupervised, supervised, and reinforcement learning.

Classification and regression are the two main prediction domains in the machine learning field. Classification is the problem of predicting a discrete class output, while regression is to predict a continuous quantity output. Due to the pronounced differences in principles for these two domains, the modeling algorithms applied on these two problems are different. Some algorithms can be used for both with minor modifications, e.g., decision trees and artificial neural networks, whereas some algorithms are only suitable for either classification or regression problems—e.g., logistic regression can only be used for classification, and linear regression is only for regression predictive modeling. More importantly, the matrices that are used to evaluate models varies for classification, e.g., accuracy, are usually used for assessing classification models but not regression algorithms, whereas root mean squared error (RMSE) is only for regression predictive models but not classification models.

One of the useful supervised machine learning algorithms in multi-omics is the random forest. A random forest is not a black box telling you which parameters are the most predictive of biology. Conversely, neural networks and deep learning are typically not appropriate for multi-omics datasets because of the structure of the multi-omics data, normally with more variables than sample size. Neural networks provide more accurate predictions when there are many samples, and relatively few measurements per sample. Multi-omic studies are typically the opposite, with relatively few samples but many, many measurements per sample. There is nothing in principle preventing neural networks from working on multi-omic datasets, but rather the practical considerations of how these studies are designed. Further, neural networks and deep learning are "black boxes" where the decisions of the algorithm are unknown to the researcher. For these reasons, random forests may provide better predictions in multiomis data, as measured by recall, area under receiver operating receiver curve, and the Mathews correlation coefficient [57]. Though more sophisticated than other analysis methods, these machine learning techniques are phenomenal for first, exploratory, unbiased passes on the data. They will determine which features are most predictive of data outcome, and what to look for as grounding during other analyses.

5. Conclusions

There are several limitations in multi-omic integration, including potential statistical overfitting, varying distributions between analytes, and limitations in throughput for some techniques [43,58,59]. Nonetheless, multi-omics are a suite of tools that allow researchers to answer questions with unparalleled depth. These measurements are not perfect in themselves, and consistency between omic measurements will ensure the discoveries are true to the underlying biological reality. Furthermore, there are no perfect methods for analyzing these data. A researcher should be confident in their findings when their discovery comes up in multiple omes but also when discovered through multiple analysis and statistical methods. What we have discussed here is not exhaustive of the excellent analysis methods that exist, but we are confident that any researcher employing these techniques will find the trends present in their multi-omic dataset successfully.

Author Contributions: Conceptualization, S.M.L., A.S., S.W., and M.P.S.; writing—original draft preparation, S.M.L.; writing—review and editing, S.M.L., A.S., S.W., and M.P.S.; visualization, S.M.L., and A.S.; supervision, M.P.S. All authors have read and agreed to the published version of the manuscript.

Funding: This research received no external funding.

Acknowledgments: We would like to acknowledge Brittany Ann Lee and Jeinffer V. Quijada for the consultations they provided during writing.

Conflicts of Interest: The authors declare no conflict of interest.

References

1. Venter, J.C.; Adams, M.D.; Myers, E.W.; Li, P.W.; Mural, R.J.; Sutton, G.G.; Smith, H.O.; Yandell, M.; Evans, C.A.; Holt, R.A.; et al. The sequence of the human genome. *Science* **2001**, *291*, 1304–1351. [CrossRef]
2. Hatfull, G.F. Bacteriophage genomics. *Curr. Opin. Microbiol.* **2008**, *11*, 447–453. [CrossRef]
3. Anders, S.; Huber, W. Differential expression analysis for sequence count data. *Genome Biol.* **2010**, *11*, R106. [CrossRef]
4. Segata, N.; Waldron, L.; Ballarini, A.; Narasimhan, V.; Jousson, O.; Huttenhower, C. Metagenomic microbial community profiling using unique clade-specific marker genes. *Nat. Methods* **2012**, *9*, 811–814. [CrossRef]
5. Ewing, B.; Hillier, L.; Wendl, M.C.; Green, P. Base-calling of automated sequencer traces using phred. I. Accuracy assessment. *Genome Res* **1998**, *8*, 175–185. [CrossRef]
6. Khan, A.R.; Pervez, M.T.; Babar, M.E.; Naveed, N.; Shoaib, M. A Comprehensive Study of De Novo Genome Assemblers: Current Challenges and Future Prospective. *Evol. Bioinform Online* **2018**, *14*, 1176934318758650. [CrossRef]
7. Li, H.; Durbin, R. Fast and accurate short read alignment with Burrows-Wheeler transform. *Bioinformatics* **2009**, *25*, 1754–1760. [CrossRef]
8. Langmead, B.; Salzberg, S.L. Fast gapped-read alignment with Bowtie 2. *Nat. Methods* **2012**, *9*, 357–359. [CrossRef]
9. Li, H.; Handsaker, B.; Wysoker, A.; Fennell, T.; Ruan, J.; Homer, N.; Marth, G.; Abecasis, G.; Durbin, R.; Genome Project Data Processing Subgroup. The Sequence Alignment/Map format and SAMtools. *Bioinformatics* **2009**, *25*, 2078–2079. [CrossRef] [PubMed]
10. Danecek, P.; Auton, A.; Abecasis, G.; Albers, C.A.; Banks, E.; DePristo, M.A.; Handsaker, R.E.; Lunter, G.; Marth, G.T.; Sherry, S.T.; et al. The variant call format and VCFtools. *Bioinformatics* **2011**, *27*, 2156–2158. [CrossRef]
11. Roadmap Epigenomics Consortium; Kundaje, A.; Meuleman, W.; Ernst, J.; Bilenky, M.; Yen, A.; Heravi-Moussavi, A.; Kheradpour, P.; Zhang, Z.; Wang, J.; et al. Integrative analysis of 111 reference human epigenomes. *Nature* **2015**, *518*, 317–330. [CrossRef]

12. Mouse, E.C.; Stamatoyannopoulos, J.A.; Snyder, M.; Hardison, R.; Ren, B.; Gingeras, T.; Gilbert, D.M.; Groudine, M.; Bender, M.; Kaul, R.; et al. An encyclopedia of mouse DNA elements (Mouse ENCODE). *Genome Biol.* **2012**, *13*, 418. [CrossRef] [PubMed]
13. Buenrostro, J.D.; Giresi, P.G.; Zaba, L.C.; Chang, H.Y.; Greenleaf, W.J. Transposition of native chromatin for fast and sensitive epigenomic profiling of open chromatin, DNA-binding proteins and nucleosome position. *Nat. Methods* **2013**, *10*, 1213–1218. [CrossRef] [PubMed]
14. Corces, M.R.; Trevino, A.E.; Hamilton, E.G.; Greenside, P.G.; Sinnott-Armstrong, N.A.; Vesuna, S.; Satpathy, A.T.; Rubin, A.J.; Montine, K.S.; Wu, B.; et al. An improved ATAC-seq protocol reduces background and enables interrogation of frozen tissues. *Nat. Methods* **2017**, *14*, 959–962. [CrossRef] [PubMed]
15. Feng, J.; Liu, T.; Qin, B.; Zhang, Y.; Liu, X.S. Identifying ChIP-seq enrichment using MACS. *Nat. Protoc.* **2012**, *7*, 1728–1740. [CrossRef] [PubMed]
16. Yan, F.; Powell, D.R.; Curtis, D.J.; Wong, N.C. From reads to insight: A hitchhiker's guide to ATAC-seq data analysis. *Genome Biol.* **2020**, *21*, 22. [CrossRef] [PubMed]
17. Dobin, A.; Davis, C.A.; Schlesinger, F.; Drenkow, J.; Zaleski, C.; Jha, S.; Batut, P.; Chaisson, M.; Gingeras, T.R. STAR: Ultrafast universal RNA-seq aligner. *Bioinformatics* **2013**, *29*, 15–21. [CrossRef]
18. Consortium, E.P. An integrated encyclopedia of DNA elements in the human genome. *Nature* **2012**, *489*, 57–74. [CrossRef]
19. Bray, N.L.; Pimentel, H.; Melsted, P.; Pachter, L. Near-optimal probabilistic RNA-seq quantification. *Nat. Biotechnol.* **2016**, *34*, 525–527. [CrossRef]
20. Patro, R.; Duggal, G.; Love, M.I.; Irizarry, R.A.; Kingsford, C. Salmon provides fast and bias-aware quantification of transcript expression. *Nat. Methods* **2017**, *14*, 417–419. [CrossRef]
21. Robinson, M.D.; McCarthy, D.J.; Smyth, G.K. edgeR: A Bioconductor package for differential expression analysis of digital gene expression data. *Bioinformatics* **2010**, *26*, 139–140. [CrossRef]
22. Ritchie, M.E.; Phipson, B.; Wu, D.; Hu, Y.; Law, C.W.; Shi, W.; Smyth, G.K. Limma powers differential expression analyses for RNA-sequencing and microarray studies. *Nucleic Acids Res.* **2015**, *43*, e47. [CrossRef]
23. Soneson, C.; Love, M.I.; Robinson, M.D. Differential analyses for RNA-seq: Transcript-level estimates improve gene-level inferences. *F1000Research* **2015**, *4*, 1521. [CrossRef] [PubMed]
24. Johnson, W.E.; Li, C.; Rabinovic, A. Adjusting batch effects in microarray expression data using empirical Bayes methods. *Biostatistics* **2007**, *8*, 118–127. [CrossRef] [PubMed]
25. Robinson, M.D.; Oshlack, A. A scaling normalization method for differential expression analysis of RNA-seq data. *Genome Biol.* **2010**, *11*, R25. [CrossRef] [PubMed]
26. da Huang, W.; Sherman, B.T.; Lempicki, R.A. Systematic and integrative analysis of large gene lists using DAVID bioinformatics resources. *Nat. Protoc.* **2009**, *4*, 44–57. [CrossRef] [PubMed]
27. Eden, E.; Navon, R.; Steinfeld, I.; Lipson, D.; Yakhini, Z. GOrilla: A tool for discovery and visualization of enriched GO terms in ranked gene lists. *BMC Bioinform.* **2009**, *10*, 48. [CrossRef] [PubMed]
28. Nguyen, T.M.; Shafi, A.; Nguyen, T.; Draghici, S. Identifying significantly impacted pathways: A comprehensive review and assessment. *Genome Biol.* **2019**, *20*, 203. [CrossRef] [PubMed]
29. Vallania, F.; Tam, A.; Lofgren, S.; Schaffert, S.; Azad, T.D.; Bongen, E.; Haynes, W.; Alsup, M.; Alonso, M.; Davis, M.; et al. Leveraging heterogeneity across multiple datasets increases cell-mixture deconvolution accuracy and reduces biological and technical biases. *Nat. Commun.* **2018**, *9*, 4735. [CrossRef]
30. Chen, B.; Khodadoust, M.S.; Liu, C.L.; Newman, A.M.; Alizadeh, A.A. Profiling Tumor Infiltrating Immune Cells with CIBERSORT. *Methods Mol. Biol.* **2018**, *1711*, 243–259. [CrossRef]
31. Franzosa, E.A.; McIver, L.J.; Rahnavard, G.; Thompson, L.R.; Schirmer, M.; Weingart, G.; Lipson, K.S.; Knight, R.; Caporaso, J.G.; Segata, N.; et al. Species-level functional profiling of metagenomes and metatranscriptomes. *Nat. Methods* **2018**, *15*, 962–968. [CrossRef] [PubMed]
32. Wood, D.E.; Salzberg, S.L. Kraken: Ultrafast metagenomic sequence classification using exact alignments. *Genome Biol.* **2014**, *15*, R46. [CrossRef] [PubMed]
33. Clarridge, J.E., 3rd. Impact of 16S rRNA gene sequence analysis for identification of bacteria on clinical microbiology and infectious diseases. *Clin. Microbiol. Rev.* **2004**, *17*, 840–862, table of contents. [CrossRef] [PubMed]
34. Woese, C.R.; Fox, G.E. Phylogenetic structure of the prokaryotic domain: The primary kingdoms. *Proc. Natl. Acad. Sci. USA* **1977**, *74*, 5088–5090. [CrossRef] [PubMed]

35. MacLean, B.; Tomazela, D.M.; Shulman, N.; Chambers, M.; Finney, G.L.; Frewen, B.; Kern, R.; Tabb, D.L.; Liebler, D.C.; MacCoss, M.J. Skyline: An open source document editor for creating and analyzing targeted proteomics experiments. *Bioinformatics* **2010**, *26*, 966–968. [CrossRef]
36. Tyanova, S.; Temu, T.; Sinitcyn, P.; Carlson, A.; Hein, M.Y.; Geiger, T.; Mann, M.; Cox, J. The Perseus computational platform for comprehensive analysis of (prote)omics data. *Nat. Methods* **2016**, *13*, 731–740. [CrossRef]
37. Rost, H.L.; Rosenberger, G.; Navarro, P.; Gillet, L.; Miladinovic, S.M.; Schubert, O.T.; Wolski, W.; Collins, B.C.; Malmstrom, J.; Malmstrom, L.; et al. OpenSWATH enables automated, targeted analysis of data-independent acquisition MS data. *Nat. Biotechnol.* **2014**, *32*, 219–223. [CrossRef]
38. Liu, Y.; Buil, A.; Collins, B.C.; Gillet, L.C.; Blum, L.C.; Cheng, L.Y.; Vitek, O.; Mouritsen, J.; Lachance, G.; Spector, T.D.; et al. Quantitative variability of 342 plasma proteins in a human twin population. *Mol. Syst. Biol.* **2015**, *11*, 786. [CrossRef]
39. Saigusa, D.; Okamura, Y.; Motoike, I.N.; Katoh, Y.; Kurosawa, Y.; Saijyo, R.; Koshiba, S.; Yasuda, J.; Motohashi, H.; Sugawara, J.; et al. Establishment of Protocols for Global Metabolomics by LC-MS for Biomarker Discovery. *PLoS ONE* **2016**, *11*, e0160555. [CrossRef]
40. Fan, S.; Kind, T.; Cajka, T.; Hazen, S.L.; Tang, W.H.W.; Kaddurah-Daouk, R.; Irvin, M.R.; Arnett, D.K.; Barupal, D.K.; Fiehn, O. Systematic Error Removal Using Random Forest for Normalizing Large-Scale Untargeted Lipidomics Data. *Anal. Chem.* **2019**, *91*, 3590–3596. [CrossRef]
41. Contrepois, K.; Mahmoudi, S.; Ubhi, B.K.; Papsdorf, K.; Hornburg, D.; Brunet, A.; Snyder, M. Cross-Platform Comparison of Untargeted and Targeted Lipidomics Approaches on Aging Mouse Plasma. *Sci. Rep.* **2018**, *8*, 17747. [CrossRef] [PubMed]
42. Xia, J.; Psychogios, N.; Young, N.; Wishart, D.S. MetaboAnalyst: A web server for metabolomic data analysis and interpretation. *Nucleic Acids Res.* **2009**, *37*, W652–W660. [CrossRef]
43. Subramanian, I.; Verma, S.; Kumar, S.; Jere, A.; Anamika, K. Multi-omics Data Integration, Interpretation, and Its Application. *Bioinform. Biol. Insights* **2020**, *14*, 1177932219899051. [CrossRef] [PubMed]
44. Misra, B.B.; Langefeld, C.D.; Olivier, M.; Cox, L.A. Integrated Omics: Tools, Advances, and Future Approaches. *J. Mol. Endocrinol.* **2018**. [CrossRef] [PubMed]
45. Gibbons, R.D.; Hedeker, D.; DuToit, S. Advances in analysis of longitudinal data. *Annu. Rev. Clin. Psychol.* **2010**, *6*, 79–107. [CrossRef]
46. Caruana, E.J.; Roman, M.; Hernandez-Sanchez, J.; Solli, P. Longitudinal studies. *J. Thorac. Dis.* **2015**, *7*, E537–E540. [CrossRef]
47. Huang, S.; Chaudhary, K.; Garmire, L.X. More Is Better: Recent Progress in Multi-Omics Data Integration Methods. *Front. Genet.* **2017**, *8*, 84. [CrossRef]
48. McInnes, L.; Healy, J.; Melville, J. UMAP: Uniform Manifold Approximation and Projection for Dimension Reduction. *arXiv* **2018**, arXiv:1802.03426.
49. Zhou, W.; Sailani, M.R.; Contrepois, K.; Zhou, Y.; Ahadi, S.; Leopold, S.R.; Zhang, M.J.; Rao, V.; Avina, M.; Mishra, T.; et al. Longitudinal multi-omics of host-microbe dynamics in prediabetes. *Nature* **2019**, *569*, 663–671. [CrossRef]
50. Contrepois, K.; Wu, S.; Moneghetti, K.J.; Hornburg, D.; Ahadi, S.; Tsai, M.S.; Metwally, A.A.; Wei, E.; Lee-McMullen, B.; Quijada, J.V.; et al. Molecular Choreography of Acute Exercise. *Cell* **2020**, *181*, 1112–1130.e16. [CrossRef]
51. Chen, R.; Mias, G.I.; Li-Pook-Than, J.; Jiang, L.; Lam, H.Y.; Chen, R.; Miriami, E.; Karczewski, K.J.; Hariharan, M.; Dewey, F.E.; et al. Personal omics profiling reveals dynamic molecular and medical phenotypes. *Cell* **2012**, *148*, 1293–1307. [CrossRef] [PubMed]
52. Csardi, G.; Nepusz, T. The Igraph Software Package for Complex Network Research. *InterJ. Complex Syst.* **2006**, *1695*, 1–9.
53. Handcock, M.S.; Hunter, D.R.; Butts, C.T.; Goodreau, S.M.; Morris, M. Statnet: Software Tools for the Representation, Visualization, Analysis and Simulation of Network Data. *J. Stat. Softw.* **2008**, *24*, 1548–7660. [CrossRef] [PubMed]
54. Kumar, L.; Matthias, E. Mfuzz: A software package for soft clustering of microarray data. *Bioinformation* **2007**, *2*, 5–7. [CrossRef] [PubMed]

55. Piening, B.D.; Zhou, W.; Contrepois, K.; Rost, H.; Gu Urban, G.J.; Mishra, T.; Hanson, B.M.; Bautista, E.J.; Leopold, S.; Yeh, C.Y.; et al. Integrative Personal Omics Profiles during Periods of Weight Gain and Loss. *Cell Syst.* **2018**, *6*, 157–170.e8. [CrossRef] [PubMed]
56. Stanberry, L.; Mias, G.I.; Haynes, W.; Higdon, R.; Snyder, M.; Kolker, E. Integrative analysis of longitudinal metabolomics data from a personal multi-omics profile. *Metabolites* **2013**, *3*, 741–760. [CrossRef]
57. Matthews, B.W. Comparison of the predicted and observed secondary structure of T4 phage lysozyme. *Biochim. Biophys. Acta* **1975**, *405*, 442–451. [CrossRef]
58. Canzler, S.; Schor, J.; Busch, W.; Schubert, K.; Rolle-Kampczyk, U.E.; Seitz, H.; Kamp, H.; von Bergen, M.; Buesen, R.; Hackermuller, J. Prospects and challenges of multi-omics data integration in toxicology. *Arch. Toxicol.* **2020**, *94*, 371–388. [CrossRef]
59. Pinu, F.R.; Beale, D.J.; Paten, A.M.; Kouremenos, K.; Swarup, S.; Schirra, H.J.; Wishart, D. Systems Biology and Multi-Omics Integration: Viewpoints from the Metabolomics Research Community. *Metabolites* **2019**, *9*, 76. [CrossRef]

MDPI

Article

From Prevention to Disease Perturbations: A Multi-Omic Assessment of Exercise and Myocardial Infarctions

Melanie T. Odenkirk [1], Kelly G. Stratton [2], Lisa M. Bramer [2], Bobbie-Jo M. Webb-Robertson [3,4], Kent J. Bloodsworth [3], Matthew E. Monroe [3], Kristin E. Burnum-Johnson [5] and Erin S. Baker [1,6,*]

1 Department of Chemistry, North Carolina State University, Raleigh, NC 27606, USA; mtodenki@ncsu.edu
2 National Security Division, Pacific Northwest National Laboratory, Richland, WA 99354, USA; kelly.stratton@pnnl.gov (K.G.S.); lisa.bramer@pnnl.gov (L.M.B.)
3 Biological Sciences Division, Pacific Northwest National Laboratory, Richland, WA 99354, USA; bobbie-jo.webb-robertson@pnnl.gov (B.-J.M.W.-R.); kent.bloodsworth@pnnl.gov (K.J.B.); matthew.monroe@pnnl.gov (M.E.M.)
4 Department of Biostatistics and Informatics, University of Colorado, Aurora, CO 80045, USA
5 Environmental Molecular Sciences Laboratory, Pacific Northwest National Laboratory, Richland, WA 99354, USA; kristin.burnum-johnson@pnnl.gov
6 Comparative Medicine Institute, North Carolina State University, Raleigh, NC 27695, USA
* Correspondence: ebaker@ncsu.edu

Abstract: While a molecular assessment of the perturbations and injury arising from diseases is essential in their diagnosis and treatment, understanding changes due to preventative strategies is also imperative. Currently, complex diseases such as cardiovascular disease (CVD), the leading cause of death worldwide, suffer from a limited understanding of how the molecular mechanisms taking place following preventive measures (e.g., exercise) differ from changes occurring due to the injuries caused from the disease (e.g., myocardial infarction (MI)). Therefore, this manuscript assesses lipidomic changes before and one hour after exercise treadmill testing (ETT) and before and one hour after a planned myocardial infarction (PMI) in two separate patient cohorts. Strikingly, unique lipidomic perturbations were observed for these events, as could be expected from their vastly different stresses on the body. The lipidomic results were then combined with previously published metabolomic characterizations of the same patients. This integration provides complementary insights into the exercise and PMI events, thereby giving a more holistic understanding of the molecular changes associated with each.

Keywords: lipidomics; metabolomics; multi-omics; planned myocardial infarction (PMI); myocardial infarction (MI); exercise; heart; cheminformatics

Citation: Odenkirk, M.T.; Stratton, K.G.; Bramer, L.M.; Webb-Robertson, B.M.; Bloodsworth, K.J.; Monroe, M.E.; Burnum-Johnson, K.E.; Baker, E.S. From Prevention to Disease Perturbations: A Multi-Omic Assessment of Exercise and Myocardial Infarctions. *Biomolecules* **2021**, *11*, 40. https://doi.org/10.3390/biom11010040

Received: 24 November 2020
Accepted: 24 December 2020
Published: 30 December 2020

1. Introduction

For decades, physical activity and diet have been considered the primary preventative strategies for numerous diseases, including cardiovascular disease (CVD). As the leading cause of death worldwide, rigorous characterization of CVD and the subsequent incidences of myocardial infarction (MI) are crucial for reducing its occurrence [1]. Despite the prevalence of CVD and resulting MI events worldwide, the complex pathophysiology underlying CVD origins has yet to be fully defined [2]. Even with advancements such as diagnosis with CK-MB and cTn assays and methods for CVD prediction from traditional risk factors alone or in tandem with molecular predictors, CVD-related events continue to be the leading cause of death worldwide [1,3–5]. Thus, improving our understanding of these disease mechanisms could serve to reduce the current morbidity rate of CVD by providing more effective prevention, intervention and treatment strategies.

In CVD and other diseases, such as type 2 diabetes, osteoporosis and some forms of cancer, there is a well-recognized, negative correlation with the intensity, duration and continuation of exercise events [6–8]. Since exercise subjects the heart to hemodynamic stress and overloading of pressure and volume [8], morphological adaptation of the heart occurs following recurrent exposure to exercise, effectively diminishing the risk of heart disease by reducing cholesterol and suppressing hypertension and atherogenesis [9,10]. However, over-exertion of the heart muscle from exercise can result in calcification that limits the capacity of the heart to pump blood, thereby increasing the risk of cardiovascular events [11]. On the other hand, a sedentary lifestyle along with high blood pressure, abnormal blood lipid profiles, smoking and obesity are all major risk factors for CVD, typically triggering the development of an intermediate phenotype prior to a MI [12–14]. Therefore, elucidating a balance between the beneficial and detrimental mechanisms of exercise is crucial for optimizing heart performance and reducing CVD risk and mortality rates [13].

While exercise and diet can be preventative, certain people are genetically predisposed to CVD and other heart diseases. Therefore, leveraging models of stroke and MI events responsible for 80% of CVD end-stage phenotypes provides additional molecular information about treatments and the induced injuries. Hypertrophic cardiomyopathy (HCM) is the most prevalent heritable cardiac disease, estimated to be present in 1 out of every 500 individuals [15,16]. Obstructive HCM (HOCM) is a subtype mechanistically defined by the barricaded outflow of the left ventricular heart cavity at rest (1/3 of cases) or at provocation (1/3 of cases) [15,16]. The reduction of left ventricular outflow in HOCM cases culminates in increased left ventricular pressure, high wall stress, impaired left ventricular filling, myocardial ischemia and a reduced cardiac output [16,17]. Currently, aspirin, β-blockers and pacemakers are all common remediation strategies to mitigate these symptoms [16,17]. Failure of these therapeutic approaches to alleviate left ventricular blockage, however, requires removal of obstructing tissues through either surgical excision or alcohol septal ablation (ASA), where an injection of alcohol triggers a planned myocardial infarction (PMI) and reduces the left ventricle blockage caused by systolic anterior motion of the mitral valve [16]. While both procedures have had similar patient outcomes and survival rates, ASA treatment and the resulting PMI have proven to be a less invasive approach preferable for surgically at-risk patients [16,17]. Evaluating the molecular changes occurring from a PMI also grants researchers tremendous insight into the pathophysiology of spontaneous MI events that plague one American every 40 s with a global mortality rate of CVD-related events accounting for 31% of the deaths in 2015 [18,19].

Mass spectrometry (MS) has become a popular analytical tool to characterize molecules changing in biological systems through omic studies. While the annotation of a singular "ome" (i.e., proteome, lipidome, metabolome) elucidates significant aspects of disease pathophysiology, comprehensively characterizing a disease through one class of biomolecules does not provide the holistic information often needed. Thus, multi-omic measurements, wherein multiple classes of biomolecules are analyzed and integrated, provide a greater understanding of molecular interplay and pathophysiology [20]. For example, since metabolites and lipids both reflect immediate changes occurring in a system, together they allow for an investigation into early-stage perturbations [21,22]. Furthermore, lipids have routinely been linked to exercise and MI mechanisms [2,23–25], so their combination with metabolites provides a complementary way to assess system dysregulation. In this study, lipidomic assessments were performed on plasma taken from two cohorts; the first cohort's samples were taken before and one hour after exercise performed with a specific treadmill testing procedure, and the second cohort's samples were taken before and one hour after a PMI. The lipidomic results were then compared to a targeted polar metabolite study of the same patient cohorts [26,27], and together, the multi-omic comparison provided a more comprehensive characterization of the various biomolecule classes altered upon different stressors of the body and heart. This comparison therefore allowed for the exploration of molecular differences between CVD-related events and preventative strategies within humans.

2. Materials and Methods

2.1. Sample Extraction and Data Collection

2.1.1. Human Sample Collection and Extraction

Both an exercise and a PMI cohort were evaluated in this manuscript, and informed consent was obtained from all human participants in the studies. In the exercise cohort, plasma samples were collected from the periphery veins of 25 patients before and one hour following exercise treadmill testing (ETT) [26]. In the PMI cohort, plasma samples were also collected from the periphery veins of an additional 20 patients before and one hour following a PMI [27]. The paired before and after sampling of the same patient for both studies was performed to yield a high statistical power despite the limited number of samples analyzed, since the before sample could be used as the control for each patient [28]. An overview of the patient demographics for both the ETT and PMI cohorts is given in Figure 1 and Supplemental Table S1. Additional cohort information is also expanded upon in the original publications [26,27]. Notably, the male demographic of the exercise cohort was large compared to the PMI study, wherein women were in the majority [26]. Additionally, in the exercise study, enrolled patients had to meet a normal exercise tolerance criteria, which included having an estimated peak VO_2 capacity over 70%, a heart response rate exceeding 85% predicted baseline and a pre-exercise fasting period of 4 h [26]. PMI patients were also monitored with CK-MB and troponin T assays, with peak levels observed at standard spontaneous MI times with CK-MB at 4.5 h and troponin T at 8 h following a PMI event [27,29]. The PMI derivation cohort were all primary HOCM cases with septal thickness $\geq$16 mm; resting outflow tract gradient $\geq$30 mmHg; inducible outflow tract gradient $\geq$50 mmHg; failed medical intervention; and appropriate coronary anatomy [27]. Targeted analysis of 210 metabolites was completed in the original publications for each cohort with a triple quadrupole mass spectrometer (AB4000Q; Applied Biosystem/Sciex, Farmingham, MA, USA), and detailed protocols on those methods can be found for the ETT study in Lewis 2010 [26] and PMI study in Lewis 2008 [27].

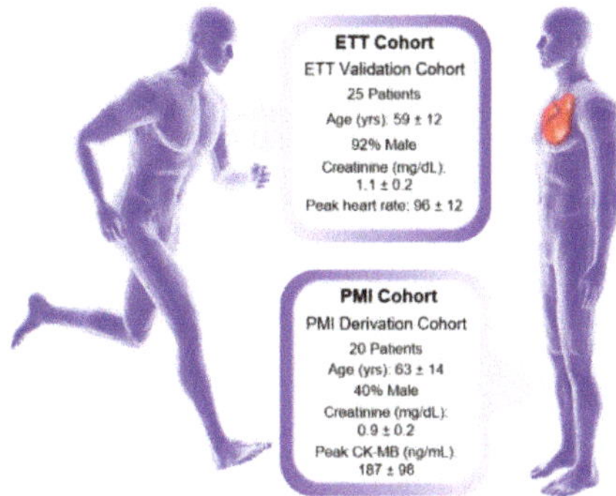

Figure 1. Demographics for the PMI cohort (**left**) and ETT cohort (**right**). Continuous variables are given as mean ± standard deviation and categorical variables are shown as percentages.

2.1.2. Lipid Extraction

For the lipidomic study, lipids were extracted in 2 mL Sorenson tubes from 25 µL aliquots of plasma following a modified Folch protocol [30,31]. Briefly, 600 µL of a 2:1 mixture of −20 °C chloroform/methanol was introduced to the plasma samples which was then vortexed for 30 s. A phase separation was induced by adding 150 µL aliquots of HPLC grade water and then vortexed again for an additional 30 s. The samples then rested for 5 min at room temperature prior to centrifugation at 12,000 rpm for 10 min at 4 °C. Samples were then placed on ice where 350 µL aliquots of the bottom organic layer were removed, dried in a speedvac and then re-suspended in 250 µL of 2:1 chloroform/methanol for storage at −20 °C. Immediately before instrumental analysis, the total lipid extracts were dried down and reconstituted in 5 μL chloroform and 100 μL methanol. Pooled case and control samples for the exercise and PMI studies were generated by combining 5 μL aliquots of each before plasma sample separately.

2.1.3. Lipidomic Instrumental Analysis

Lipidomic instrumental analysis of the 45 before and 45 after extracted human plasma samples was completed with an Agilent 6560 IM-QTOF MS platform (Santa Clara, CA, USA) outfitted with the commercial gas kit (Alternate Gas Kit, Agilent, Santa Clara, CA, USA) and a precision flow controller (640B, MKS Instruments, Andover, MA, USA). The LC–IMS–CID–MS data were collected in both positive and negative ESI from 50–1700 m/z with a 1 sec/spectra cycle time. Reverse phase liquid chromatography (RPLC) separation was completed with a 10 μL sample injection onto a Waters CSH column (3.0 mm × 150 mm × 1.7 μm particle size) on a Waters Acquity UPLC H class system (Waters Corporation, Milford, MA, USA). Separation of lipid species was achieved with a 34-min LC gradient (mobile phase A: acetonitrile/water (40:60) containing 10 mM ammonium acetate; mobile phase B: acetonitrile/isopropyl alcohol (10:90) containing 10 mM ammonium acetate) at a flow rate of 250 μL/min as described in Table 1. A 4-min column wash and 4-min equilibration were also used as described in Table 2.

Table 1. Lipid elution gradient.

Time	% MPA	% MPB	Flow Rate (mL/min)
0	60	40	0.25
2	50	50	0.25
3	40	60	0.25
12	30	70	0.25
15	25	75	0.25
17	22	78	0.25
19	15	85	0.25
22	8	92	0.25
25	1	99	0.25
34	1	99	0.25

Table 2. Lipid column wash.

Time	% MPA	% MPB	Flow Rate (mL/min)
34.5	60	40	0.3
35	1	99	0.3
35.5	1	99	0.3
36	60	40	0.35
37	60	40	0.3
38	60	40	0.25

2.2. Data Processing

2.2.1. Lipid Identification

Accurate mass tag (AMT) matching within LIQUID software was used to assign all lipid identifications [32]. The LC–IMS–CID–MS platform typically allows for the assignment of head group and fatty acyl (FA) structural moieties of each uniquely identified lipid species using the criterion of mass accuracy below 5 ppm, precursor and fragment peak alignment across dimensions, and CCS values < 2% different from the reference value. While head group annotation is largely unambiguous, FA assignment is more complex due to the propensity of isomers. From the collision induced dissociation (CID) measurements, the number of carbons and double bonds is generally achieved; however, additional specifics, such as *sn*-backbone position, double bond position or double bond orientation, are often indistinguishable in these studies [33]. Therefore, the most confident lipid speciation achieved through this analysis included the head group and individual fatty acyl groups with unknown *sn*-positions, as denoted by "_" (i.e., PC (16:0_18:2)) [34]. For lipids where individual FA constituents could not be identified, the summed carbon and double bond counts are noted, e.g., PC (34:2). Any features matching more than one lipid identification are separated by a ";" to denote both as potential matches. Furthermore, isomeric experimental observations were assigned "_A"; "_B"; etc., to denote the observed chromatographic and/or IMS separation of these species. The peak areas of the 352 lipids identified in the exercise study (262 from positive mode, 85 in negative mode and 5 in both modes) and the 299 lipids identified in the PMI study (225 in positive, 72 in negative and 2 in both modes) were exported as a ".csv" format for processing and statistical assessment regarding each before/after paired comparison (Supplemental Tables S2 and S3).

2.2.2. Data Processing and Statistics

Statistical analysis of the targeted polar metabolites was carried out as detailed previously [26,27]. Briefly, in the targeted annotation of 210 metabolites in the ETT and PMI studies, 20 were found to be statistically significant one hour following exercise (16 upregulated and 4 downregulated) and 13 were statistically significant one hour following a PMI (7 upregulated and 6 downregulated) at a Benjamini–Hochberg corrected $p \leq 0.005$ cut-off. Processing and statistics of the lipidomics data also followed the same procedures, where statistical significance was determined from $\log_2$ transformed abundances using MetaboAnalyst (version 4.0, Edmonton, AB, CA) [35]. The ETT statistical analysis was completed using a paired t-test, and the PMI comparisons were completed with a Wilcoxon signed-rank paired t-test due to their unequal variance. A Benjamini–Hochberg multiple comparison correction was also applied for both analyses with a significance cut-off of $p \leq 0.005$ to match the previously published metabolite statistics [36]. Interestingly, no statistically significant lipids were observed one hour following ETT, whereas the PMI study yielded 207 statistically significant lipids: 66 upregulated and 141 downregulated (Supplemental Tables S2 and S3). Comparison of sex in the PMI cohort and ischemia in the ETT cohort was completed to account for additional differentiation following the above protocols for each cohort. No significant species were detected from either comparison.

2.3. *Data Interpretation*

2.3.1. Lipidomics Data Interpretation

Lipidomic relationships were investigated using cheminformatics to interrogate structure-function associations across head groups and fatty acyl (FA) moieties [37–39]. Head group clustering was completed with the SCOPE toolbox [39]. Here, SMILES [40] obtained from LipidMaps [34] for each lipid identification were clustered by structural similarity using an ECFP_6 fingerprint [41], Tanimoto distance and complete linkage using the *fingerprint* and *ggtree* packages in R (Version 3.6.2, Vienna, Austria) [42,43]. Lipids with multiple LipidMaps matches were cataloged by a representative SMILES for hierarchical clustering. To facilitate the visualization of head-group trends, pigmentation of dendrogram nodes was used to denote lipid classes. FA tail presence was further assessed by selectively parsing out lipids by FA composition. For our analyses, most *sn*-1 and *sn*-2 fatty acyl positions were unknown, so all possible positions were considered to account for potential *sn*-positional effects. Lipids with multiple identities were partitioned into all possible identifications to visualize each potential FA contribution to significance. Summary statistics (adjusted *p*-value, $\log_2$ fold change) of lipids were subsequently overlaid with the *pheatmap* package in R [42,44]. Color gradients of red (upregulated) and blue (downregulated) were applied to visualize significance with darker colors indicating a larger fold change ($\log_2$FC) or smaller *p*-value (adjusted *p*-value), while grey values represented identified but not statistically significant lipids.

2.3.2. Multi-omics Data Interpretation

Hierarchical clustering was again utilized to assess the multi-omic association of statistically significant metabolites and lipids. Dendrograms provided visualization of the structurally similar and statistically significant species (BH adjusted p-value ≤ 0.005), both individually and in tandem. Metabolite clustering was accomplished with MAACS keys fingerprint, Tanimoto distance and complete linkage using *fingerprint* and *ggtree* packages in R (Version 3.6.2, Vienna, Austria) from each SMILES representation [42,43]. The resulting metabolite dendrograms allowed for a summary of the significant species following exercise and PMI events where adjusted p-values followed the same gradient as described above. Node color in the metabolite dendrogram was used to annotate the biological roles attributed to each metabolite. Conversely, in the multi-omics dendrogram built using ECFP_6 fingerprint, Tanimoto distance and complete linkage, all metabolites were grouped together in a single node color because of the relatively small number of statistically significant metabolites relative to lipids.

3. Results

The previous targeted metabolomic study for both the ETT and PMI cohorts provided great insight into statistically significant polar metabolites [26,27,45], but overlooked important nonpolar molecules changing due to each event. The recent annotation of lipids in both CVD and exercise has elucidated the critical roles these molecules serve in each event [24,46–56]. Therefore, the inclusion of lipidomic and multi-omic assessments in this manuscript provides a more in-depth profile of ETT and PMI molecular mechanisms.

3.1. *Lipid Identifications and Statistical Significance*

To perform both the ETT and PMI lipidomic analyses, multi-dimensional assessments were carried out by leveraging a LC–IMS–CID–MS instrumental platform [32,38]. The LC–IMS–CID–MS analyses yielded a total of 352 unique lipid identifications for the ETT cohort and 299 for the PMI cohort across the same five lipid categories: glycerolipids, sphingolipids, phospholipids, fatty acids and sterols [57]. The 352 ETT lipids were composed of 216 phospholipids, 88 glycerolipids, 39 sphingolipids, 5 sterols and 4 fatty acids (Figure 2a, left); the PMI cohort had 185 phospholipids, 71 glycerolipids, 31 sphingolipids, 7 sterols and 5 fatty acids (Figure 2a, right). The breakdown of lipid category designation into classes showed both studies having: three phospholipids (phosphatidylinositols (PIs), phosphatidylcholines (PCs) and phosphatidylethanolamines (PEs)), three sphingolipids (sphingomyelins (SM), ceramides (Cer) and hexose ceramides (HexCer)), two glycerolipids (triacylglycerolipids (TGs) and diacylglycerolipids (DGs)), one sterol (cholesteryl ester (CE)) and one FA (carnitine) (Figure 2b). Additional diversity within the phospholipids was observed in the FA linkages (including alkenyl ether (plasmalogen; P-) and alkyl ether (O-)) and FA numbers (e.g., lyso and diacyl species). Only a few lipid species were specific to each cohort including a ganglioside (GM3) belonging to the sphingolipid category in the ETT cohort and a monoacylglycerol (MG) from the glycerolipid category observed in the PMI cohort.

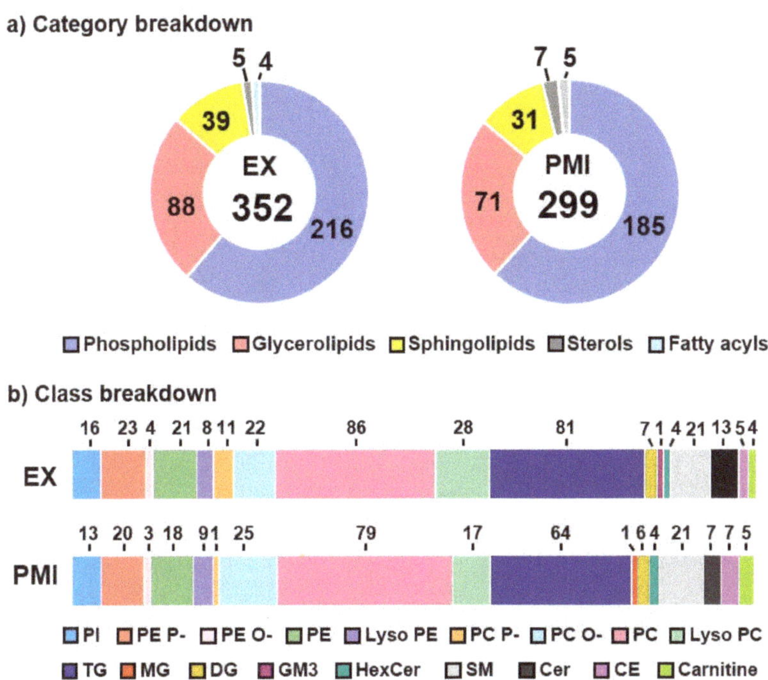

Figure 2. Identified lipid category and class coverage. (**a**) Five lipid categories were observed for plasma from patients in both the ETT (**left**) and PMI (**right**) cohorts. (**b**) In the class breakdown, the majority of the lipids fall within the sphingolipid, glycerolipid and phospholipid categories.

Of the identified lipids, a drastic difference was observed in statistical significance for the exercise and PMI cohorts. One hour after ETT, no lipids were found to be statistically significant, even after further assessment of metadata, including gender, age and BMI. However, we do note our statistical criteria were very stringent to compare them with the previous metabolomics studies, so directly above our significance cutoff we observed lipids of interest within the lyso PC, GM3, PE P-, DG and carnitine classes. Specifically, we noted the largest fold changes for PC (20:5_0:0), carnitine (10:1) and carnitine (14:1), which had values of −1.28, −1.29 and −1.18 FC. The lipidome changes in PMI, however, told a completely different story. An hour after a PMI, 207 lipids (69% identified) were statistically significant, even with the stringent criteria, with 141 downregulated and 66 upregulated (Figure 3a). To further evaluate the PMI lipids, we utilized head group and FA composition to visualize structure–function relationships of the statistically significant species. Head group associations of all identified lipids were clustered by their structural similarity [34,40,42]. The resulting circular dendrogram is shown in Figure 3a with the adjusted p-value in the inner ring and $\log_2$FC on the outer ring. The most consistent observation relating to head groups was the upregulation of PC O-, PC P- and PE P-. The upregulation of SM lipids, another component of lipid bilayers abundantly present in lipid rafts and integral in cholesterol homeostasis, was also observed in the PMI study [58]. Conversely, PC lipids which have overlapping roles as charged species enriched within the outer lipid membrane layer were downregulated following a PMI [59]. Additionally, a general downregulation of glycerolipids was also detected following a PMI event, a contradictory finding to the positive correlation of TGs and MI incidence [60,61] This finding may instead reflect FAs serving as the primary energy substrates of the heart where non-esterified FAs, products of glycerolipids degradation, are rapidly complexed with CoA [62]. Notably, ceramides which have previously been positively correlated with cardiac disease risk were not observed to be statistically significant in our PMI cohort [63]. Exceptions to the head group trends, however, were noted for almost every class discussed herein. For example, we observed split dysregulation across SMs, CEs and other classes, illustrating effects beyond just head group influence.

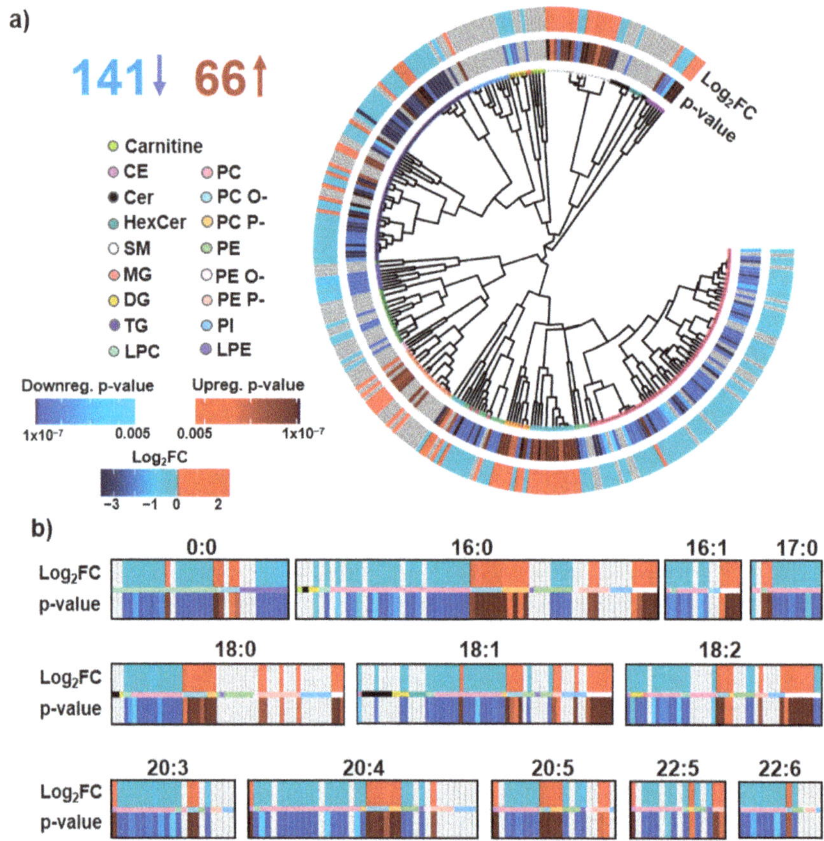

Figure 3. Lipids detected and statistically significant in the PMI comparison. (**a**) Of the 299 uniquely identified lipids, 141 were statistically downregulated and 66 were upregulated following a PMI with a p-value cut-off of 0.005. The lipid head group associations are visualized in a circular dendrogram with p-values (inner ring) and Log2FC (outer ring) statistics overlaid simultaneously for each lipid identification. (**b**) FA lipid composition was also investigated by plotting all unique FA components. Statistically significant lipids are shown in red and blue for up- and downregulation, and identified lipids lacking statistical significance are shown in grey. The magnitude of variation for Log_2FC and adjusted p-values are visualized through a color gradient, with darker colors indicating a more significant p-value or larger fold change.

Discrepancies between lipid head group composition and biological dysregulation suggest additional selectivity likely attributable to the FA components of lipid structures. Within FAs, important differences include chain length, and double bond number, position and orientation [64]. Previous efforts have elucidated FA chain length to directly influence cardiac pathology, but plasma studies have been less successful in capturing this effect [24]. To explore these associations, we further interrogated FA dysregulation in the identified lipids, as shown in Figure 3b. While the findings in these plots mainly correlated with the head group analyses, a few FA-specific observations could be extracted. First, a FA dependence of CE differential expression was observed—with 18 carbon-containing CEs being downregulated and CEs with 20 and 22 carbon PUFAs being upregulated.

Long chain polyunsaturated fatty acids (LC-PUFAs) are a class of FAs characterized as having 18 or more carbons and at least two double bonds, often serving as precursors to lipid mediators [65]. PUFA dysregulation was also recognized among PE, PC and PE P- lipids; PE and PC lipids containing PUFA tails were downregulated, while the majority of significant PE P- lipids were upregulated. In an additional assessment of the summed FA double bond number, it was observed that while the majority of glycerolipids were statistically downregulated, the upregulated TG species had a greater number of unsaturation sites compared to the downregulated species. This is in agreement with models for predicting CVD onset, which have included unsaturated TG species [4,24]. However individual FA information was not attained for the majority of the TG species due to difficulties in assigning their MS/MS spectra.

3.2. Multi-Omics Results

To assess how polar and nonpolar molecules change in both the ETT and PMI cohorts, the lipidomic results were integrated with the previously performed targeted analysis of 210 polar metabolites [26,27]. Results from these analyses elucidated both unique and shared statistically significant metabolites and biological processes across both events (Figure 4). For example, glycolysis and TCA cycle metabolites (red, pink and peach nodes) were upregulated following ETT, a finding agreeable with the known mechanisms of burning energy through high-intensity exercise [66]. Additionally, niacinamide, a component of NADH that is also associated with energy through its interaction with insulin, was found to be statistically upregulated with exercise. In PMI, the dietary metabolites of PC lipids previously shown to predict CVD risk, choline and trimethylamine N-oxide (TMAO), were downregulated and clustered next to each other to affirm their structural relationship [18]. Amino acid dysregulation was also observed following both ETT and PMI, as alanine was statistically significant through its upregulation immediately following exercise but downregulation following a PMI. Both ETT and PMI also shared an upregulation of xanthine and hypoxanthine, metabolites involved in purine metabolism and ATP degradation, which are notably upregulated following cellular damage. These xanthine metabolites can also interact with xanthine oxidase to produce reactive oxygen species, a mechanism well characterized in heart failure [67].

From our analysis comparing the lipidomic changes in ETT and PMI, we note unique profiles where plasma metabolite signals best characterized mechanistic changes following high-intensity exercise training. Conversely, we demonstrated an overwhelming dysregulation of lipids following a PMI in the end-stage phenotype of CVD, in addition to the metabolomic findings that were previously published. While the metabolomic analyses elucidated changes for both the ETT and PMI cohorts with slight overlap between each characterization, the lipidomics results were quite different. While no statistically significant lipids were noted in the ETT study, the sheer number of statistically significant lipid associations in the PMI cohort (207, 69% of identified lipids) provide striking evidence for the integral role of the lipidome immediately following a PMI event (Figure 5). The findings from the ETT cohort were, however, in accordance with other exercise studies which have observed lipidome disruption being proportional to the duration and intensity of exercise [68]. Previous characterizations of lipid variation in exercise have centered on the decrease in free carnitine and increase in short-chain acylcarnitine through its crucial capability of shuttling FAs into mitochondria within muscle tissue for energy usage [25,69]. Dysregulation of carnitines has faced some disagreement in literature, likely from the lack of correlation between muscle and plasma sampling [69]. Further, the energy sources of exercise differ substantially as low intensity training relies on fat as a primary fuel source, while high intensity training uses carbohydrates as an immediate energy supply [23]. From the observation of metabolite intermediates of glycolysis and the TCA cycle such as lactate being upregulated, we feel we can confidently state that known mechanisms of high intensity training were taking place in our cohort [26]. From the dysregulation of energy processes in the metabolomics data and the variation in carnitine species observed just above the significance cutoff, it is possible that these species were in fact perturbed in our system as has been noted by others (Supplemental Table S2) [52,69]. A variety of factors may preclude this annotation, including age-based impairment of the acyl carnitine pathway that diminishes FA oxidation and study-to-study variation from different exercise training regimes [54,70]. A lack of differential expression of the lipidome following exercise may also reflect that lipid variation is not always immediate [24]. The singular treadmill training event for this analysis, therefore, may be too short to assess any additional lipidomic changes [52]. From the known pathophysiology of over-exercise triggering the calcification of the heart muscle, the activation of lipid enzymes by Ca^{2+} may suggest more drastic lipidome dysregulation would be observed with repeated exercise training [24].

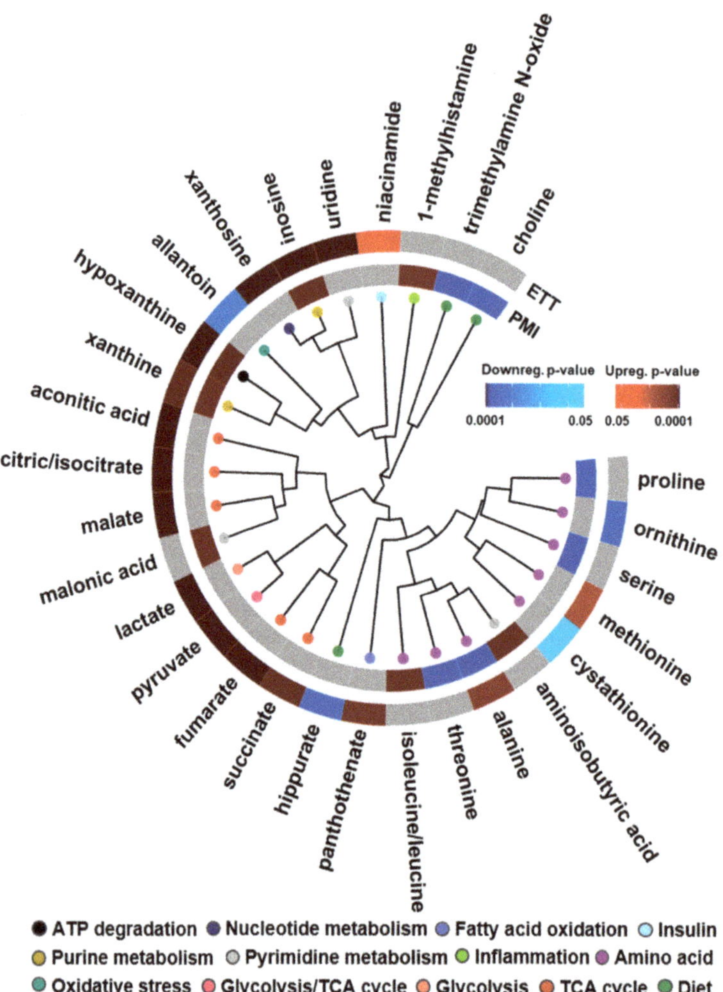

Figure 4. Statistically significant metabolites in the ETT and PMI studies. A circular dendrogram is utilized to showcase the differentially expressed metabolites in the PMI (inner ring) and ETT (outer ring) cohorts. Red and blue are used for up- and downregulation with a color gradient visualizing the magnitude of the adjusted *p*-value observed. Grey metabolites were detected but not statistically significant.

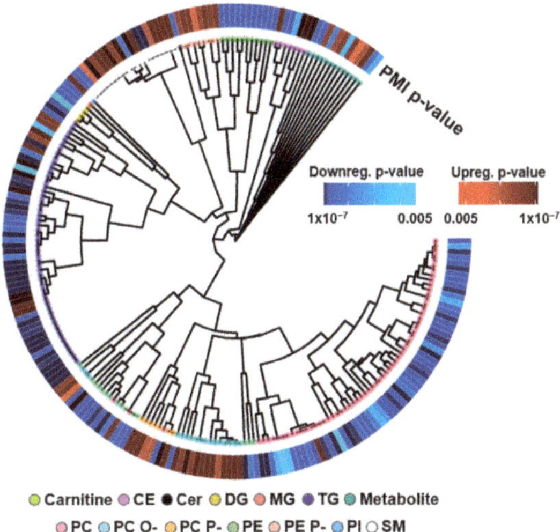

Figure 5. The multi-omic assessment of statistically significant lipids and metabolites from a PMI event. Adjusted *p*-values for each molecule are shown around the dendrogram. Red and blue are used for up- and downregulation with a color gradient to visualize magnitude.

From our analysis of 20 patients before and after a PMI event, we observed several instances of lipid dysregulation with substantial biological implications. Ether lipids (PC O-, PC P-, PE P-) have been shown to be disproportionately abundant in brain and heart tissues as components of the lipid bilayer with unknown biological implications [71]. While the biological significance of ether lipids overexpressed in heart tissue is not fully understood, the upregulation of these species in plasma following a PMI is likely indicative of tissue degradation following ASA treatment. However, only a subset of membrane lipids were upregulated, suggesting these lipid classes carry additional significance for ASA-induced PMI. This finding could potentially be explained by the preferential oxidation of O- and P- linkage sites that serve to protect the *sn*-2 FA group from oxidation [72]. In the heart, dysregulation of PC lipids in tandem with increased activity of phospholipase enzymes has been observed in CVD, where lysophospholipids contribute to atherosclerosis and vascular damage through their role in inflammation, as was observed here in their downregulation [73,74]. Altogether, the obstruction of heart tissue following a PMI could serve dual purposes, reflecting both a breakdown of ablated cellular tissue and dysregulation of essential biological processes, such as energy production and inflammation. PUFAs with a double bond on the third carbon (n-3) have previously been shown to serve preventative roles in CVD through their antiatherogenic effects and may explain the dysregulation of PUFA-containing lipids [75]. Metabolomic analysis of the PMI samples showed the most dysregulation among amino acids, where branched amino acids are precursors for glutamine and alanine synthesis in muscles [76]. Conversely, amino acids associated with cardiac remodeling (proline) were downregulated [77].

3.3. Study Comparison

Combining the two complementary stories of metabolite and lipid dysregulation before and after exercise and a PMI provides an important assessment of their biological changes (Figures 3 and 4). This comparison is incredibly insightful for understanding CVD pathophysiology, as shown by comparing our results with the general consensus of molecular dysregulation from several exercise, CVD onset and MI studies (Figure 6) [4,24,51,52,61,69,78]. Results from the CVD onset studies have illustrated upregulation of sphingolipids and carnitines, and shown downregulation of lyso PC and DGs. CVD onset metabolomic studies have also elucidated distinct molecular changes to include mechanisms of oxidative stress and PC degradation products promoting atherogenesis [55]. In the comparison of exercise, CVD onset and a PMI, there is quite substantial overlap in the perturbed processes, but the molecules being dysregulated are often unique. For example, different energy processes were dysregulated in both exercise and PMI, as glycerolipids were largely downregulated in PMI and TCA/glycolysis metabolites were upregulated with exercise. Differential expression of both 1-methyl histamine and lysophospholipids was also observed, and since both have been linked to roles in inflammation, this suggests a possible response to ASA treatment [73,74]. Uniquely, ether lipids, which were upregulated in PMI, are also recognized regulators of ion channels [71]. Relative to CVD onset, we noted a number of lipid and metabolite species dysregulated in both the ETT and PMI studies (Figure 6). For example, lysoPCs (LPC) were downregulated across PMI and are largely corroborated by literature [4,56]. The further annotation of choline and TMAO degradation products of PC lipids suggests an even greater significance in PC lipids for the development of end-stage perturbations, however the direction of change between CVD and the PMI model differed [55]. We also noted opposite trends when comparing CVD onset results from the literature and our ETT cohort; carnitines have been reported to be downregulated in exercise but upregulated in CVD onset, further reflecting the importance of the shift in energy processes between PMI and exercise [69,78]. These findings are significant for further elucidating the mechanisms of CVD, which we and others have shown reflect drastic changes in the lipidome but are missed from polar metabolomics experiments [4,48,51,53]. We would, however, like to note that the limited size of our patient study fails to capture sex-based differentiation of CVD onset established previously [79–82]. We also note limitations in our ETT analysis from a singular bout of exercise and disparities among patients from variables such as cardiovascular health history and ischemia that may hinder the elucidation of exercise-based lipid dysregulation.

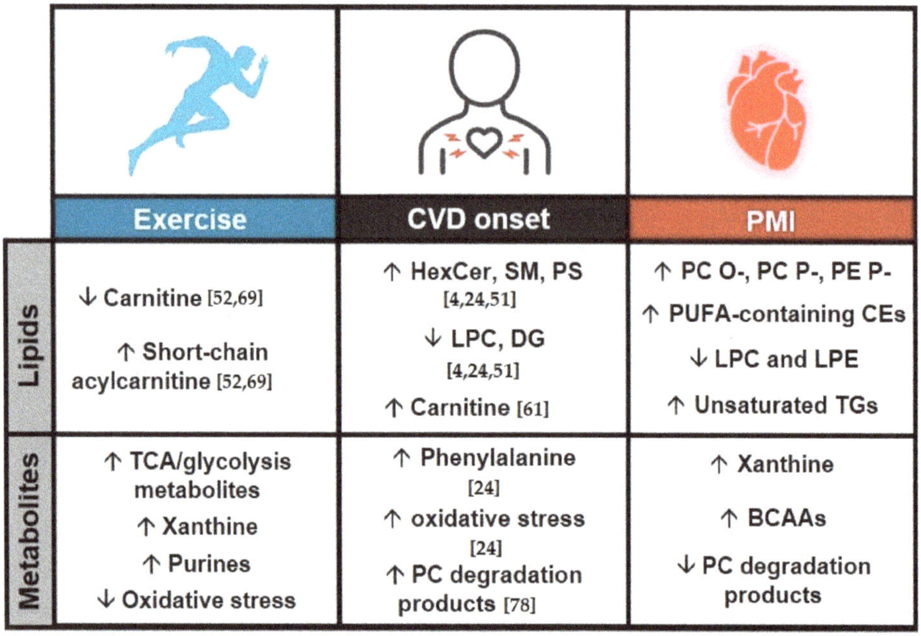

	Exercise	CVD onset	PMI
Lipids	↓ Carnitine [52,69] ↑ Short-chain acylcarnitine [52,69]	↑ HexCer, SM, PS [4,24,51] ↓ LPC, DG [4,24,51] ↑ Carnitine [61]	↑ PC O-, PC P-, PE P- ↑ PUFA-containing CEs ↓ LPC and LPE ↑ Unsaturated TGs
Metabolites	↑ TCA/glycolysis metabolites ↑ Xanthine ↑ Purines ↓ Oxidative stress	↑ Phenylalanine [24] ↑ oxidative stress [24] ↑ PC degradation products [78]	↑ Xanthine ↑ BCAAs ↓ PC degradation products

Figure 6. Comparison of lipidomic and metabolomic trends for exercise (**left**), cardiovascular disease (CVD) onset (**middle**) and planned myocardial infarction (PMI) model (**right**). Results include a summary of observed results from this and the referenced previous studies noted by citation number in the figure [4,24,51,52,61,69,78].

4. Conclusions

The metabolomic and lipidomic findings observed for the exercise and PMI cohorts showcased their unique pathophysiology. Of the statistically significant metabolites observed for both events, little overlap was found, implicating unique molecular processes for each [26,27]. Since the insights from a singular class of biomolecules are inherently limited, we expanded the metabolomic analyses to include lipids. Novel instrumentation platforms and cheminformatics tools were applied to provide confident lipid identifications and investigate lipid variation [38]. The lipidomic analyses illustrated how the exercise cohort had no statistically significant lipids after treadmill testing, while 69% of identified lipids were dysregulated one hour after a PMI. This finding was in itself very interesting and distinguished the molecular mechanisms for the two events. As such, the polar metabolites were more informative for the exercise study, while the lipidomic results provided a better assessment of the PMI cohort. Specifically, one hour following a PMI, lipid species with head groups including PC O-, PC P- and PE P- were all upregulated, while SMs were mainly upregulated and PCs were mostly downregulated. PUFAs were also selectively dysregulated across lipid head groups following a PMI. However, even with the lipid structural insight achieved, discrepancies in class trends were still observed, since LC–IMS–CID–MS analyses allow for the confident assignment of lipids, but analytical improvements are necessary to probe the roles of double bond position and orientation in these discrepancies. Interestingly, integrating the multi-omic exercise and PMI studies showed perturbation of energy processes across both events. The multi-omic analyses also corroborated findings from singular omic analyses where inflammation and atherogenic processes are heavily implicated in PMI. Furthermore, their comparison with CVD onset studies showed strong agreement between the lipid and metabolite dysregulation observed in the PMI cohort, and less agreement with the ETT cohort results, as expected. Ultimately, the integration of the lipid and metabolite data elucidated unique biological roles within molecular classes, providing complementary profiles for how preventative strategies and MI events greatly differ in their molecular mechanisms.

Supplementary Materials: The following are available online at https://www.mdpi.com/2218-273X/11/1/40/s1, Table S1: ETT and PMI cohort information. Table S2: Exercise results. Table S3: PMI results.

Author Contributions: M.T.O. and E.S.B. wrote the manuscript. K.G.S., L.M.B., B.-J.M.W.-R. and M.T.O. performed statistical analyses; and K.J.B., M.E.M. and E.S.B. performed the experiments. E.S.B. and K.E.B.-J. designed the experiments and supervised different aspects of the project. All authors have read and agreed to the published version of the manuscript.

Funding: Portions of this research were supported by grants from the NIH National Institute of Environmental Health Sciences (P30ES025128, P42 ES027704 and P42ES031009) and startup funds from North Carolina State University.

Institutional Review Board Statement: All blood sampling was performed as part of the human studies protocols approved by the Massachusetts General Hospital Institutional Review Board.

Informed Consent Statement: Written informed consent was obtained from all subjects. All samples in this manuscript were also de-identified prior to the study.

Data Availability Statement: Raw data is available through MassIVE (https://massive.ucsd.edu/MSV000086620).R Code for recreating data visualization within this manuscript is available at https://github.com/BakerLabNCSU/PMI_Exercise_Multiomics.

Acknowledgments: The data were collected in the W. R. Wiley Environmental Molecular Sciences Laboratory (EMSL) (grid.436923.9), a DOE Office of Science User Facility sponsored by the Office of Biological and Environmental Research and located at Pacific Northwest National Laboratory (PNNL). PNNL is a multiprogram national laboratory operated by Battelle for the Department of Energy (DOE) under Contract DE-AC05-76RLO 1830.

Conflicts of Interest: The manuscript authors declare no conflict of interest.

References

1. Nowbar, A.N.; Gitto, M.; Howard, J.P.; Francis, D.P.; Al-Lamee, R. Mortality from Ischemic Heart Disease. *Circ. Cardiovasc. Qual. Outcomes* **2019**, *12*, e005375. [CrossRef] [PubMed]
2. Dokken, B.B. The Pathophysiology of Cardiovascular Disease and Diabetes: Beyond Blood Pressure and Lipids. *Diabetes Spectr.* **2008**, *21*, 160–165. [CrossRef]
3. Chapman, A.R.; Adamson, P.D.; Shah, A.S.; Anand, A.; Strachan, F.E.; Ferry, A.V.; Lee, K.K.; Berry, C.; Findlay, I.; Cruikshank, A.; et al. High-Sensitivity Cardiac Troponin and the Universal Definition of Myocardial Infarction. *Circulation* **2020**, *141*, 161–171. [CrossRef]
4. Meikle, P.J.; Wong, G.; Tsorotes, D.; Barlow, C.K.; Weir, J.M.; Christopher, M.J.; MacIntosh, G.L.; Goudey, B.; Stern, L.; Kowalczyk, A.; et al. Plasma Lipidomic Analysis of Stable and Unstable Coronary Artery Disease. *Arter. Thromb. Vasc. Biol.* **2011**, *31*, 2723–2732. [CrossRef]
5. Xu, R.-Y.; Zhu, X.-F.; Yang, Y.; Ye, P. High-sensitive cardiac troponin T. *J. Geriatr. Cardiol.* **2013**, *10*, 102–109.
6. Booth, F.W.; Roberts, C.K.; Laye, M.J. Lack of Exercise Is a Major Cause of Chronic Diseases. *Compr. Physiol.* **2012**, *2*, 1143–1211. [CrossRef]
7. Klassen, G.A.; Wolf, H.K. Exercise and cardiovascular health. *Can. J. Cardiol.* **1992**, *8*, 22.
8. Nystoriak, M.A.; Bhatnagar, A. Cardiovascular Effects and Benefits of Exercise. *Front. Cardiovasc. Med.* **2018**, *5*, 135. [CrossRef]
9. Blair, S.N.; Jackson, A.S. Physical fitness and activity as separate heart disease risk factors: A meta-analysis. *Med. Sci. Sports Exerc.* **2001**, *33*, 762–764. [CrossRef]
10. Thompson, P.D.; Buchner, D.; Pina, I.L.; Balady, G.J.; Williams, M.A.; Marcus, B.H.; Berra, K.; Blair, S.N.; Costa, F.; Franklin, B.; et al. American Heart Association Council on Clinical Cardiology Subcommittee on Exercise, Rehabilitation, and Prevention; American Heart Association Council on Nutrition, Physical Activity, and Metabolism Subcommittee on Physical Activity. Exercise and physical activity in the prevention and treatment of atherosclerotic cardiovascular disease: A statement from the Council on Clinical Cardiology (Subcommittee on Exercise, Rehabilitation, and Prevention) and the Council on Nutrition, Physical Activity, and Metabolism (Subcommittee on Physical Activity). *Circulation* **2003**, *107*, 3109–3116.
11. Laddu, D.R.; Rana, J.S.; Murillo, R.; Sorel, M.E.; Quesenberry, C.P., Jr.; Allen, N.B.; Gabriel, K.P.; Carnethon, M.R.; Liu, K.; Reis, J.P.; et al. 25-Year Physical Activity Trajectories and Development of Subclinical Coronary Artery Disease as Measured by Coronary Artery Calcium: The Coronary Artery Risk Development in Young Adults (CARDIA) Study. *Mayo Clin. Proc.* **2017**, *92*, 1660–1670. [CrossRef] [PubMed]
12. Smyth, A.; O'Donnell, M.J.; Lamelas, P.; Teo, K.; Rangarajan, S.; Yusuf, S.; Investigators, I. Physical Activity and Anger or Emotional Upset as Triggers of Acute Myocardial Infarction: The INTERHEART Study. *Circulation* **2016**, *134*, 1059–1067. [CrossRef] [PubMed]
13. Yusuf, S.; Hawken, S.; Ôunpuu, S.; Dans, T.; Avezum, A.; Lanas, F.; McQueen, M.; Budaj, A.; Pais, P.; Varigos, J.; et al. Effect of potentially modifiable risk factors associated with myocardial infarction in 52 countries (the INTERHEART study): Case-control study. *Lancet* **2004**, *364*, 937–952. [CrossRef]
14. Yusuf, S.; Reddy, S.; Ounpuu, S.; Anand, S. Global burden of cardiovascular diseases: Part I: General considerations, the epidemiologic transition, risk factors, and impact of urbanization. *Circulation* **2001**, *104*, 2746–2753. [CrossRef] [PubMed]
15. Nishimura, R.A.; Seggewiss, H.; Schaff, H.V. Hypertrophic Obstructive Cardiomyopathy: Surgical Myectomy and Septal Ablation. *Circ. Res.* **2017**, *121*, 771–783. [CrossRef]
16. Liebregts, M.; Vriesendorp, P.A.; Ten Berg, J.M. Alcohol Septal Ablation for Obstructive Hypertrophic Cardiomyopathy: A Word of Endorsement. *J. Am. Coll. Cardiol.* **2017**, *70*, 481–488. [CrossRef]
17. Gersh, B.J.; Maron, B.J.; Bonow, R.O.; Dearani, J.A.; Fifer, M.A.; Link, M.S.; Naidu, S.S.; Nishimura, R.A.; Ommen, S.R.; Rakowski, H.; et al. 2011 ACCF/AHA guideline for the diagnosis and treatment of hypertrophic cardiomyopathy: A report of the American College of Cardiology Foundation/American Heart Association Task Force on Practice Guidelines. *Circulation* **2011**, *124*, e783–e831. [CrossRef]
18. Benjamin, E.J.; Muntner, P.; Alonso, A.; Bittencourt, M.S.; Callaway, C.W.; Carson, A.P.; Chamberlain, A.M.; Chang, A.R.; Cheng, S.; Das, S.R.; et al. Heart disease and stroke statistics—2019 update: A report from the American heart association. *Circulation* **2019**, *139*, e56–e528. [CrossRef]
19. Mishra, R.; Monica, J. Determinants of cardiovascular disease and sequential decision-making for treatment among women: A Heckman's approach. *SSM Popul. Heal.* **2019**, *7*, 100365. [CrossRef]
20. Rotroff, D.M.; Motsinger-Reif, A.A. Embracing Integrative Multiomics Approaches. *Int. J. Genom.* **2016**, *2016*, 1715985. [CrossRef]
21. Pinu, F.R.; Beale, D.J.; Paten, A.M.; Kouremenos, K.; Swarup, S.; Schirra, H.J.; Wishart, D.S. Systems Biology and Multi-Omics Integration: Viewpoints from the Metabolomics Research Community. *Metabolites* **2019**, *9*, 76. [CrossRef] [PubMed]
22. Lagarde, M.; Géloën, A.; Record, M.; Vance, D.; Spener, F. Lipidomics is emerging. *Biochim. Biophys. Acta (BBA) Mol. Cell Biol. Lipids* **2003**, *1634*, 61. [CrossRef] [PubMed]
23. Kannan, U.; Vasudevan, K.; Balasubramaniam, K.; Yerrabelli, D.; Shanmugavel, K.; John, N.A. Effect of Exercise Intensity on Lipid Profile in Sedentary Obese Adults. *J. Clin. Diagn. Res.* **2014**, *8*, BC08–BC10. [CrossRef] [PubMed]
24. Tham, Y.K.; Bernardo, B.C.; Huynh, K.; Ooi, J.Y.; Gao, X.M.; Kiriazis, H.; Giles, C.; Meikle, P.J.; McMullen, J.R. Lipidomic Profiles of the Heart and Circulation in Response to Exercise versus Cardiac Pathology: A Resource of Potential Biomarkers and Drug Targets. *Cell Rep.* **2018**, *24*, 2757–2772. [CrossRef]

25. Wang, Y.; Xu, D. Effects of aerobic exercise on lipids and lipoproteins. *Lipids Heal. Dis.* **2017**, *16*, 1–8. [CrossRef]
26. Lewis, G.D.; Farrell, L.; Wood, M.J.; Martinovic, M.; Arany, Z.; Rowe, G.C.; Souza, A.; Cheng, S.; McCabe, E.L.; Yang, E.; et al. Metabolic Signatures of Exercise in Human Plasma. *Sci. Transl. Med.* **2010**, *2*, 33–37. [CrossRef]
27. Lewis, G.D.; Wei, R.; Liu, E.; Yang, E.; Shi, X.; Martinovic, M.; Farrell, L.; Asnani, A.; Cyrille, M.; Ramanathan, A.; et al. Metabolite profiling of blood from individuals undergoing planned myocardial infarction reveals early markers of myocardial injury. *J. Clin. Investig.* **2008**, *118*, 3503–3512. [CrossRef]
28. Stevens, J.R.; Herrick, J.S.; Wolff, R.K.; Slattery, M.L. Power in pairs: Assessing the statistical value of paired samples in tests for differential expression. *BMC Genom.* **2018**, *19*, 953. [CrossRef]
29. Zimmerman, J.; Fromm, R.; Meyer, D.; Boudreaux, A.; Wun, C.-C.C.; Smalling, R.; Davis, B.; Habib, G.; Roberts, R. Diagnostic Marker Cooperative Study for the Diagnosis of Myocardial Infarction. *Circulation* **1999**, *99*, 1671–1677. [CrossRef]
30. Folch, J.; Lees, M.; Stanley, G.H.S. A simple method for the isolation and purification of total lipides from animal tissues. *J. Biol. Chem.* **1957**, *226*, 497–509.
31. Nakayasu, E.S.; Nicora, C.D.; Sims, A.C.; Burnum-Johnson, K.E.; Kim, Y.-M.; Kyle, J.E.; Matzke, M.M.; Shukla, A.K.; Chu, R.K.; Schepmoes, A.A.; et al. MPLEx: A Robust and Universal Protocol for Single-Sample Integrative Proteomic, Metabolomic, and Lipidomic Analyses. *mSystems* **2016**, *1*, e00043-16. [CrossRef] [PubMed]
32. Kyle, J.; Crowell, K.L.; Casey, C.P.; Fujimoto, G.M.; Kim, S.; Dautel, S.E.; Smith, R.D.; Payne, S.H.; Metz, T.O. LIQUID: An-open source software for identifying lipids in LC-MS/MS-based lipidomics data. *Bioinformatics* **2017**, *33*, 1744–1746. [CrossRef] [PubMed]
33. Koelmel, J.P.; Ulmer, C.Z.; Jones, C.M.; Yost, R.A.; Bowden, J.A. Common cases of improper lipid annotation using high resolution tandem mass spectrometry data and corresponding limitations in biological interpretation. *Biochim. Biophys. Acta Mol. Cell Biol. Lipids* **2017**, *1862*, 766–770. [CrossRef] [PubMed]
34. Fahy, E.; Subramaniam, S.; Murphy, R.C.; Nishijima, M.; Raetz, C.R.H.; Shimizu, T.; Spener, F.; Van Meer, G.; Wakelam, M.J.O.; Dennis, E.A. Update of the LIPID MAPS comprehensive classification system for lipids. *J. Lipid Res.* **2009**, *50*, S9–S14. [CrossRef] [PubMed]
35. Chong, J.; Soufan, O.; Li, C.; Caraus, I.; Li, S.; Bourque, G.; Wishart, D.S.; Xia, J. MetaboAnalyst 4.0: Towards more transparent and integrative metabolomics analysis. *Nucleic Acids Res.* **2018**, *46*, W486–W494. [CrossRef]
36. Benjamini, Y.; Hochberg, Y. Controlling the False Discovery Rate—A Practical and Powerful Approach to Multiple Testing. *J. R. Stat. Soc. Ser. B-Methodol.* **1995**, *57*, 289–300. [CrossRef]
37. Ash, J.; Kuenemann, M.A.; Rotroff, D.M.; Motsinger-Reif, A.A.; Fourches, A.P.D. Cheminformatics approach to exploring and modeling trait-associated metabolite profiles. *J. Cheminform.* **2019**, *11*, 43. [CrossRef]
38. Odenkirk, M.T.; Stratton, K.G.; Gritsenko, M.A.; Bramer, L.M.; Webb-Robertson, B.-J.M.; Bloodsworth, K.J.; Weitz, K.K.; Lipton, A.K.; Monroe, M.E.; Ash, J.R.; et al. Unveiling molecular signatures of preeclampsia and gestational diabetes mellitus with multi-omics and innovative cheminformatics visualization tools. *Mol. Omics* **2020**, *16*, 521–532. [CrossRef]
39. Odenkirk, M.T.; Zin, P.P.K.; Ash, J.R.; Reif, D.; Fourches, D.; Baker, E.S. Structural-based connectivity and omic phenotype evaluations (SCOPE): A cheminformatics toolbox for investigating lipidomic changes in complex systems. *Analyst* **2020**, *145*, 7197–7209. [CrossRef]
40. Weininger, D. SMILES, a chemical language and information system. 1. Introduction to methodology and encoding rules. *J. Chem. Inf. Comput. Sci.* **1988**, *28*, 31–36. [CrossRef]
41. Rogers, D.; Hahn, M. Extended-Connectivity Fingerprints. *J. Chem. Inf. Model.* **2010**, *50*, 742–754. [CrossRef] [PubMed]
42. Team RC. *R: A Language and Environment for Statistical Computing*; Team RC: Vienna, Austria, 2017.
43. Yu, G.; Smith, D.K.; Zhu, H.; Guan, Y.; Lam, T.T. ggtree: An r package for visualization and annotation of phylogenetic trees with their covariates and other associated data. *Methods Ecol. Evol.* **2017**, *8*, 28–36. [CrossRef]
44. Kolde, R. Pheatmap: Pretty Heatmaps, R Package v. 16 (R Foundation for Statistical Computing, 2012). Available online: https://rdrr.io/cran/pheatmap/ (accessed on 14 August 2020).
45. Saito, K.; Yagi, H.; Maekawa, K.; Nishigori, M.; Ishikawa, M.; Muto, S.; Osaki, T.; Iba, Y.; Minatoya, K.; Ikeda, Y.; et al. Lipidomic signatures of aortic media from patients with atherosclerotic and nonatherosclerotic aneurysms. *Sci. Rep.* **2019**, *9*, 15472. [CrossRef] [PubMed]
46. Kelly, R.S.; Kelly, M.P.; Kelly, P. Metabolomics, Physical Activity, Exercise and Health: A Review of the Current Evidence. *Biochim. Biophys. Acta (BBA) Mol. Basis Dis.* **2020**, 165936. [CrossRef]
47. Sakaguchi, C.A.; Nieman, D.C.; Signini, E.F.; Abreu, R.M.; Catai, A.M. Metabolomics-Based Studies Assessing Exercise-Induced Alterations of the Human Metabolome: A Systematic Review. *Metabolites* **2019**, *9*, 164. [CrossRef]
48. Kohno, S.; Keenan, A.L.; Ntambi, J.M.; Miyazaki, M. Lipidomic insight into cardiovascular diseases. *Biochem. Biophys. Res. Commun.* **2018**, *504*, 590–595. [CrossRef]
49. Monnerie, S.; Comte, B.; Ziegler, D.; Morais, J.A.; Pujos-Guillot, E.; Gaudreau, P. Metabolomic and Lipidomic Signatures of Metabolic Syndrome and its Physiological Components in Adults: A Systematic Review. *Sci. Rep.* **2020**, *10*, 669. [CrossRef]
50. Goulart, V.A.M.; Santos, A.K.; Sandrim, V.C.; Batista, J.M.; Pinto, M.C.X.; Cameron, L.C.; Resende, R.R. Metabolic Disturbances Identified in Plasma Samples from ST-Segment Elevation Myocardial Infarction Patients. *Dis. Markers* **2019**, *2019*, 7676189. [CrossRef]
51. Burrello, J.; Biemmi, V.; Cas, M.D.; Amongero, M.; Bolis, S.; Lazzarini, E.; Bollini, S.; Vassalli, G.; Paroni, R.; Barile, L. Sphingolipid composition of circulating extracellular vesicles after myocardial ischemia. *Sci. Rep.* **2020**, *10*, 16182. [CrossRef]

52. Hoene, M.; Li, J.; Li, Y.; Runge, H.; Zhao, X.; Häring, H.-U.; Lehmann, R.; Xu, G.; Weigert, C. Muscle and liver-specific alterations in lipid and acylcarnitine metabolism after a single bout of exercise in mice. *Sci. Rep.* **2016**, *6*, 22218. [CrossRef]
53. Park, J.Y.; Lee, S.-H.; Shin, M.-J.; Hwang, G.-S. Alteration in Metabolic Signature and Lipid Metabolism in Patients with Angina Pectoris and Myocardial Infarction. *PLoS ONE* **2015**, *10*, e0135228. [CrossRef] [PubMed]
54. Huynh, K.; Barlow, C.K.; Jayawardana, K.S.; Weir, J.M.; Mellett, N.A.; Cinel, M.; Magliano, D.J.; Shaw, J.E.; Drew, B.G.; Meikle, P.J. High-Throughput Plasma Lipidomics: Detailed Mapping of the Associations with Cardiometabolic Risk Factors. *Cell Chem. Biol.* **2019**, *26*, 71–84.e4. [CrossRef] [PubMed]
55. Wang, Z.; Klipfell, E.; Bennett, B.J.; Koeth, R.; Levison, B.S.; Dugar, B.; Feldstein, A.E.; Britt, E.B.; Fu, X.; Chung, Y.-M.; et al. Gut flora metabolism of phosphatidylcholine promotes cardiovascular disease. *Nat. Cell Biol.* **2011**, *472*, 57–63. [CrossRef] [PubMed]
56. Tomczyk, M.M.; Dolinsky, V.W. The Cardiac Lipidome in Models of Cardiovascular Disease. *Metabolites* **2020**, *10*, 254. [CrossRef]
57. Quehenberger, O.; Dennis, E.A. The Human Plasma Lipidome. *N. Engl. J. Med.* **2011**, *365*, 1812–1823. [CrossRef]
58. Kikas, P.; Chalikias, G. Cardiovascular Implications of Sphingomyelin Presence in Biological Membranes. *Eur. Cardiol. Rev.* **2018**, *13*, 42–45. [CrossRef]
59. van Meer, G.; Voelker, D.R.; Feigenson, G.W. Membrane lipids: Where they are and how they behave. *Nat. Rev. Mol. Cell Biol.* **2008**, *9*, 112–124. [CrossRef]
60. Talayero, B.G.; Sacks, F.M. The Role of Triglycerides in Atherosclerosis. *Curr. Cardiol. Rep.* **2011**, *13*, 544–552. [CrossRef]
61. Jiao, Z.-Y.; Li, X.-T.; Li, Y.-B.; Zheng, M.-L.; Cai, J.; Chen, S.-H.; Wu, S.-L.; Yang, X.-C. Correlation of triglycerides with myocardial infarction and analysis of risk factors for myocardial infarction in patients with elevated triglyceride. *J. Thorac. Dis.* **2018**, *10*, 2551–2557. [CrossRef]
62. Schulze, P.C.; Drosatos, K.; Goldberg, I.J. Lipid Use and Misuse by the Heart. *Circ. Res.* **2016**, *118*, 1736–1751. [CrossRef]
63. Havulinna, A.S.; Sysi-Aho, M.; Hilvo, M.; Kauhanen, D.; Hurme, R.; Ekroos, K.; Salomaa, V.; Laaksonen, R. Circulating Ceramides Predict Cardiovascular Outcomes in the Population-Based FINRISK 2002 Cohort. *Arter. Thromb. Vasc. Biol.* **2016**, *36*, 2424–2430. [CrossRef] [PubMed]
64. Goonewardena, S.N.; Prevette, L.E.; Desai, A.A. Metabolomics and Atherosclerosis. *Curr. Atheroscler. Rep.* **2010**, *12*, 267–272. [CrossRef] [PubMed]
65. Gill, I.; Valivety, R. Polyunsaturated fatty acids, part 1: Occurrence, biological activities and applications. *Trends Biotechnol.* **1997**, *15*, 401–409. [CrossRef]
66. Hargreaves, M.; Spriet, L.L. Skeletal muscle energy metabolism during exercise. *Nat. Metab.* **2020**, *2*, 817–828. [CrossRef] [PubMed]
67. Tamariz, L.; Hare, J.M. Xanthine oxidase inhibitors in heart failure: Where do we go from here? *Circulation* **2015**, *131*, 1741–1744. [CrossRef] [PubMed]
68. Pronk, N.P. Short Term Effects of Exercise on Plasma Lipids and Lipoproteins in Humans. *Sports Med.* **1993**, *16*, 431–448. [CrossRef]
69. Hiatt, W.R.; Regensteiner, J.G.; Wolfel, E.E.; Ruff, L.; Brass, E.P. Carnitine and acylcarnitine metabolism during exercise in humans. Dependence on skeletal muscle metabolic state. *J. Clin. Investig.* **1989**, *84*, 1167–1173. [CrossRef]
70. Horowitz, J.F.; Klein, S. Lipid metabolism during endurance exercise. *Am. J. Clin. Nutr.* **2000**, *72*, 558S–563S. [CrossRef]
71. Dean, J.M.; Lodhi, I.J. Structural and functional roles of ether lipids. *Protein Cell* **2018**, *9*, 196–206. [CrossRef]
72. André, A.; Juanéda, P.; Sébédio, J.; Chardigny, J. Plasmalogen metabolism-related enzymes in rat brain during aging: Influence of n-3 fatty acid intake. *Biochimie* **2006**, *88*, 103–111. [CrossRef]
73. Paapstel, K.; Kals, J.; Eha, J.; Tootsi, K.; Ottas, A.; Piir, A.; Jakobson, M.; Lieberg, J.; Zilmer, M. Inverse relations of serum phosphatidylcholines and lysophosphatidylcholines with vascular damage and heart rate in patients with atherosclerosis. *Nutr. Metab. Cardiovasc. Dis.* **2018**, *28*, 44–52. [CrossRef] [PubMed]
74. Djekic, D.; Pinto, R.; Repsilber, D.; Hyotylainen, T.; Henein, M. Serum untargeted lipidomic profiling reveals dysfunction of phospholipid metabolism in subclinical coronary artery disease. *Vasc. Health Risk Manag.* **2019**, *15*, 123–135. [CrossRef] [PubMed]
75. Ander, B.P.; Dupasquier, C.M.; Prociuk, M.A.; Pierce, G.N. Polyunsaturated fatty acids and their effects on cardiovascular disease. *Exp. Clin. Cardiol.* **2003**, *8*, 164–172. [PubMed]
76. Szpetnar, M.; Pasternak, K.; Boguszewska, A. Branched chain amino acids (BCAAs) in heart diseases (ischaemic heart disease and myocardial infarction). *Ann. UMCS Med.* **2004**, *59*, 91–95.
77. Wang, J.; Xue, Z.; Lin, J.; Wang, Y.; Ying, H.; Lv, Q.; Hua, C.; Wang, M.; Chen, S.; Zhou, B. Proline improves cardiac remodeling following myocardial infarction and attenuates cardiomyocyte apoptosis via redox regulation. *Biochem. Pharmacol.* **2020**, *178*, 114065. [CrossRef] [PubMed]
78. Shah, S.H.; Bain, J.R.; Muehlbauer, M.J.; Stevens, R.D.; Crosslin, D.R.; Haynes, C.; Dungan, J.; Newby, L.K.; Hauser, E.R.; Ginsburg, G.S.; et al. Association of a Peripheral Blood Metabolic Profile With Coronary Artery Disease and Risk of Subsequent Cardiovascular Events. *Circ. Cardiovasc. Genet.* **2010**, *3*, 207–214. [CrossRef]
79. Lichtman, J.H.; Leifheit, E.C.; Safdar, B.; Bao, H.; Krumholz, H.M.; Lorenze, N.P.; Daneshvar, M.; Spertus, J.A.; D'Onofrio, G. Sex Differences in the Presentation and Perception of Symptoms Among Young Patients With Myocardial Infarction: Evidence from the VIRGO Study (Variation in Recovery: Role of Gender on Outcomes of Young AMI Patients). *Circulation* **2018**, *137*, 781–790. [CrossRef]
80. Maas, A.; Appelman, Y. Gender differences in coronary heart disease. *Neth. Heart J.* **2010**, *18*, 598–603. [CrossRef]
81. Mehta, P.A.; Cowie, M.R. Gender and heart failure: A population perspective. *Heart* **2006**, *92*, iii14–iii18. [CrossRef]
82. Woodward, M. Cardiovascular Disease and the Female Disadvantage. *Int. J. Environ. Res. Public Health* **2019**, *16*, 1165. [CrossRef]

Article

Chronic High-Fat Diet Induces Early Barrett's Esophagus in Mice through Lipidome Remodeling

Jeffrey Molendijk [1,2], Thi-My-Tam Nguyen [2], Ian Brown [3,4], Ahmed Mohamed [1], Yenkai Lim [2], Johanna Barclay [5], Mark P. Hodson [2,6,7], Thomas P. Hennessy [2,8], Lutz Krause [2], Mark Morrison [2] and Michelle M. Hill [1,2,*]

1 QIMR Berghofer Medical Research Institute, Herston, Brisbane, QLD 4006, Australia; jeff.molendijk@unimelb.edu.au (J.M.); Ahmed.Mohamed@qimrberghofer.edu.au (A.M.)
2 UQ Diamantina Institute, Faculty of Medicine, The University of Queensland, Woolloongabba, Brisbane, QLD 4072, Australia; t.nguyen6@uq.edu.au (T.-M.-T.N.); y.lim2@uq.edu.au (Y.L.); M.Hodson@victorchang.edu.au (M.P.H.); thomas.hennessy@agilent.com (T.P.H.); lutz.krause@microba.com (L.K.); m.morrison1@uq.edu.au (M.M.)
3 Envoi Specialist Pathologists, Herston, Brisbane, QLD 4059, Australia; IanBrown@envoi.com.au
4 Faculty of Medicine, The University of Queensland, Brisbane, QLD 4072, Australia
5 Mater Research Institute, The University of Queensland, Woolloongabba, Brisbane, QLD 4059, Australia; johanna.barclay@unsw.edu.au
6 Metabolomics Australia, Australian Institute for Bioengineering and Nanotechnology, The University of Queensland, St Lucia, Brisbane, QLD 4072, Australia
7 School of Pharmacy, The University of Queensland, Woolloongabba, Brisbane, QLD 4102, Australia
8 Agilent Technologies, Mulgrave, VIC 3170, Australia
* Correspondence: michelle.hill@qimrberghofer.edu.au

Received: 1 April 2020; Accepted: 12 May 2020; Published: 16 May 2020

Abstract: Esophageal adenocarcinoma (EAC) incidence has been rapidly increasing, potentially associated with the prevalence of the risk factors gastroesophageal reflux disease (GERD), obesity, high-fat diet (HFD), and the precursor condition Barrett's esophagus (BE). EAC development occurs over several years, with stepwise changes of the squamous esophageal epithelium, through cardiac metaplasia, to BE, and then EAC. To establish the roles of GERD and HFD in initiating BE, we developed a dietary intervention model in C57/BL6 mice using experimental HFD and GERD (0.2% deoxycholic acid, DCA, in drinking water), and then analyzed the gastroesophageal junction tissue lipidome and microbiome to reveal potential mechanisms. Chronic (9 months) HFD alone induced esophageal inflammation and metaplasia, the first steps in BE/EAC pathogenesis. While 0.2% deoxycholic acid (DCA) alone had no effect on esophageal morphology, it synergized with HFD to increase inflammation severity and metaplasia length, potentially via increased microbiome diversity. Furthermore, we identify a tissue lipid signature for inflammation and metaplasia, which is characterized by elevated very-long-chain ceramides and reduced lysophospholipids. In summary, we report a non-transgenic mouse model, and a tissue lipid signature for early BE. Validation of the lipid signature in human patient cohorts could pave the way for specific dietary strategies to reduce the risk of BE in high-risk individuals.

Keywords: lipid; lipidomics; cardiac metaplasia; Barrett's esophagus; esophageal adenocarcinoma; microbiota

---

1. Introduction

There are two main forms of esophageal cancer: esophageal squamous cell carcinoma and esophageal adenocarcinoma (EAC) [1]. Over a period of three decades, the incidence of EAC has risen sixfold, while esophageal squamous cell carcinoma has remained relatively stable [2,3]. In the United

States, incidence of EAC was estimated to increase from 0.40 to 2.58 cases per 100,000 between 1975 and 2009 [4]. From less than 5% of all esophageal cancer cases before the mid-1970s [5], EAC now represents almost half of all cases [2,3], making it one of the most rapidly increasing cancers in Western populations. Despite recent advances in surveillance and treatment protocols, the prognosis for patients with advanced EAC is poor, with a 5 year survival rate of less than 16%, and a median survival of less than 1 year [6,7].

EAC is widely accepted to develop via a stepwise sequence, as a consequence of gastroesophageal reflux disease (GERD). GERD leads to chronic inflammation in the esophagus and reflux esophagitis [8]. In ~10%–15% of GERD patients, the damaged squamous epithelium of the distal esophagus is replaced by cardiac mucosa with intestinal metaplasia, a condition termed Barrett's esophagus (BE) [9,10]. Although BE itself has limited adverse health effects, patients with BE have a 30–60-fold increased risk of developing EAC [11], with estimated annual progression rate of ~0.1%–0.5% per year [12,13].

In addition to GERD and BE, epidemiology studies have identified male gender, tobacco smoking and obesity as risk factors for EAC [14]. To investigate the causality and to delineate the molecular mechanisms of GERD, surgical rodent models have been reported [15], but with high mortality rates due to the challenging surgeries. An alternative approach using dietary intervention was reported by Quante et al. [16], using 0.2% deoxycholic acid (DCA) in drinking water as a mimic of GERD to induce Barrett's-like metaplasia in interleukin 1β transgenic mice. A follow-up study showed that high-fat diet (HFD) accelerated tumor development in the interleukin 1β transgenic mouse model [17]. While the authors report an increased inflammatory tumor microenvironment and altered intestinal microbiome as potential mechanisms, HFD may also promote EAC through lipid dyshomeostasis and esophageal dysbiosis. Circumstantial evidence suggests roles for both lipids and the esophageal microbiome in BE/EAC pathogenesis. Patients receiving cholesterol-lowering statin therapy exhibit reduced incidence of BE [18,19] and EAC [20–23]. Alterations to the esophageal microbiome have been reported in human esophageal tissues during BE/EAC disease progression [24,25], while gastric *Helicobacter pylori* infection, or altered gastric microbiota, may influence EAC development by modulating refluxate composition or frequency [26,27].

To evaluate the impact of obesity and/or GERD on esophageal tissue morphology, and to address the hypotheses that the pathogenic mechanisms of HFD or GERD involve esophageal microbiome and/or tissue lipids, we employed HFD dietary intervention and 0.2% DCA exposure in non-transgenic mice, to mimic obesity and GERD, respectively. The mouse model mimicking early BE was adapted from a previous report using BE transgenic mouse [16,17]. We found that a 9 month HFD increased esophageal tissue inflammation and cardiac metaplasia. DCA in drinking water increased the severity of HFD-induced esophageal inflammation and metaplasia segment length, potentially via increased esophageal microbiome diversity. Tissue lipidomics analyses revealed a phospholipid and sphingolipid signature associated with esophageal inflammation and cardia development.

2. Materials and Methods

2.1. Animal Experiments

The study was approved by The University of Queensland Animal Ethics Committee.

2.1.1. Materials

Chow diet (Irradiated Rat and Mouse Diet) and HFD (SF04-001) were obtained from Specialty Feeds (Western Australia). Both diets were produced as cylindrical pellets with a diameter of 12 mm and comparable fiber contents of 5.2% and 5.4% respectively. The standard chow provides 12% of digestible energy from fat, 23% from protein and 65% from carbohydrates, and contained 0.78% saturated fats, 2.06% monounsaturated fats and 1.88% polyunsaturated fats by weight. The HFD provides 43% of calories from fat, 21% from protein and 36% from carbohydrates, and contained 10.03% saturated fats, 8.24% monounsaturated fats and 5.11% polyunsaturated fats by weight. Both diets were wheat- and soy-based, but differed in the primary source of fat; namely, fish meal, mixed vegetable oils and canola oil for the standard chow, or lard and soybean oil for the HFD. Deoxycholate was obtained from Sigma (Missouri, USA).

2.1.2. Dietary Treatments

Eight-week-old male C57BL/6 mice were randomly assigned to one of four treatment groups for 9 months (n = 12).

A. ad libitum standard chow diet and drinking water
B. ad libitum standard chow diet and 0.2% deoxycholic acid (unconjugated bile acid, pH 7) in drinking water [16]
C. ad libitum HFD and drinking water
D. ad libitum HFD and 0.2% deoxycholic acid in drinking water.

Mice were housed in groups in autoclaved standard shoe-box cages in a ventilated rack system. Drinking water with or without deoxycholate was prepared and replaced fresh weekly. All interventions were performed during the light period of a 12 h/12 h light/dark cycle.

2.1.3. Tissue and Serum Collection

Tissue was collected within the same 3 h window to avoid discrepancies due to circadian variations. Blood was collected via cardiac puncture under isoflurane anesthesia followed by cervical dislocation. Blood was centrifuged at 5000× g for 10 min at 4 °C, and serum removed and stored at −80 °C. Distal esophagus and gastroesophageal junction tissues were collected from each mouse. The entire gastroesophageal junction was fixed for histology, while distal esophageal tissues were cut in half lengthwise. One half was fixed in formalin for histology, and one half snap frozen in liquid nitrogen for 16S ribosomal DNA (rDNA) sequencing for microbiome analysis.

2.1.4. Histology

Tissues were fixed in 10% formalin for 24 h and embedded in paraffin. Embedded tissue blocks were cut into 4 μm sections and used for hematoxylin and eosin (H&E) staining. Histological evaluation and grading was performed by a specialist gastrointestinal pathologist (IB). For grading, inflammation was graded on a scale of 0 to 3 (0 = nil inflammation; 1 = mild; 2 = moderate; and 3 = severe). The presence and length of cardiac-type mucosa was recorded.

2.2. Lipidomics Experiments

2.2.1. Materials

SPLASH LipidoMix Mass Spec Standard mixture (#330707), containing deuterated lipids of 14 species at various concentrations, and the Ceramide/Sphingoid Internal Standard Mixture II (#LM-6005), were purchased from Avanti Polar Lipids, Inc. (Alabaster, U.S.A). ESI-L low concentration tuning mix (#G1969-85000) was purchased from Agilent Technologies (Mulgrave, VIC, Australia).

2.2.2. Lipid Extraction

All steps except for sonication and sample blowdown were performed on ice. Serum and tissue samples were homogenized differently but lipids were extracted using the same methyl-tert-butyl ether (MTBE)/methanol extraction method [28].

Mouse serum (30 μL) was added to 215 μL of ice-cold methanol containing 50 μg/mL butylated hydroxytoluene (BHT). Samples were homogenized by three rounds of vortex mixing for 30 s, freezing in liquid nitrogen for 1 min, thawing for 2 min and sonicating for 10 min at 15 °C, power 100% in a Grant XUB18 bath sonicator.

Tissue wet weight was determined using a Mettler-Toledo XS105 balance (Mettler-Toledo, Melbourne, Australia). Biopsies were transferred to Eppendorf tubes containing 500 μL ice-cold methanol, 50 μg/mL BHT and one steel bead and homogenized in a TissueLyzer LT (Qiagen, Melbourne, Australia) for six minutes at 50 Hz. Homogenate was transferred to new tubes and the original tube was washed with 400 μL methanol and transferred. Samples were dried down under nitrogen flow and resuspended in 20 μL water and 200 μL methanol (50 μg/mL BHT). Samples were homogenized by three rounds of vortex mixing for 30 s, freezing in liquid nitrogen for 1 min, thawing for 2 min and sonicating for 10 min at 15 °C, power 100% in a Grant XUB18 bath sonicator.

SPLASH LipidoMix Mass Spec Standard (10 μL) and Cer/Sph mixture II (10 μL) internal standards mixes from Avanti Polar Lipids were then added to each sample. After overnight incubation at −30 °C, 750 μL MTBE was added and each tube was vortex mixed for 10 s and shaken for 10 min on a tube rotator (4 °C). MilliQ water (188 μL) was then added, and the tube was vortex mixed for 30 s to form a biphasic separation. After centrifuging for 15 min at 15,000× *g*, 700 μL of the clear upper phase containing lipids in MTBE was transferred to another tube and dried down using a gentle stream of nitrogen. After drying down of lipids, extracts were resuspended in 50 μL methanol (containing 50 μg/mL BHT)/toluene (90%/10%, *v/v*). Dry weight of the remaining pellets from tissue samples was determined in triplicate using a Mettler-Toledo XS105 balance. Dry weights were used to normalize lipid injection volumes of tissue samples prior to mass spectrometry analysis. For serum samples equal volumes were injected.

2.2.3. Untargeted Lipidomics

An Agilent Technologies 1290 Infinity II UHPLC system with an Agilent ZORBAX Eclipse plus C18 1.8-micron column (#959757-902) and guard column (#821725-901), coupled online to an Agilent 6550A iFunnel QTOF mass spectrometry system, was used for untargeted lipidomics. The mass spectrometer was tuned in the low mass range (1700 *m/z*), high sensitivity slicer mode and the instrument mode was set to Extended Dynamic Range (2 GHz). The quadrupole and time-of-flight (TOF) sections of the mass spectrometer were both tuned prior to each experiment. The quadrupole was tuned to reference masses 118.09, 622.03 and 1221.99 in positive ionization mode. Experiments were performed if the quadrupole component passed the check tune for each reference mass in wide, medium and narrow modes. The TOF component was tuned using reference masses 118.09, 322.05, 622.03, 922.00, 1221.99 and 1521.97 in positive ionization mode. TOF mass calibration indicated that at around 110–120 *m/z* the resolution was ~12,000–13,000 and increased to 20,000–21,000 around 600–620 *m/z* range. The ion source used was Dual Agilent Jet Stream electrospray ionization, which allows for the simultaneous introduction of sample and reference masses into the mass spectrometer. Source capillary voltages were set to 4000 V for positive ionization mode whilst the nozzle voltage was set to 0 V, fragmentor was set to 365 and octopoleRFPeak to 750. Nitrogen gas temperature was set to 250 °C at a flow of 15 L/minute and a sheath gas temperature of 400 °C at a flow of 12 L/min. During the experiment reference masses were enabled (121.05 and 922.01 Da) to enable auto-recalibration of compounds with known masses. MS1 data was acquired between 100–1700 *m/z* at a scan rate of 2.5 spectra per second.

The sample dilution and injection volume used for experiments was determined by testing a representative sample prior to analyzing the cohort. Reversed phase buffers A and B contained 25 millimolar (mM) ammonium formate and 0.1% formic acid in 60%/40% (*v/v*) acetonitrile/water or 90%/10% (*v/v*) isopropanol/water respectively. The separation gradient was run at a flow rate of 0.5 mL/min to separate the lipids during a 16 min gradient. The method started at 15% B and increased to 30% B at 2:00, 48% B at 2:30, 82% B at 11:00, 99% B at 11:30. The gradient was retained at 99% B until 13:00 and retained at the starting condition of 15% B between 13:06 and 16:00. The column compartment was maintained at 60 °C for the duration of the experiment.

2.2.4. Targeted Lipidomics

Targeted lipidomics were performed on an Agilent Technologies 1290 Infinity UHPLC system with an Agilent HILIC Plus RRHD 2.1 × 100 mm 1.8 micron column, coupled online to an Agilent 6490A Triple Quadrupole mass spectrometer with iFunnel and Agilent Jet Stream electrospray ionization source, operated in dynamic MRM mode. The source nitrogen gas temperature was set to 250 °C at a flow rate of 15 L/min, and the sheath gas temperature set to 400 °C at a flow rate of 12 L/min. The capillary voltage was set to 4000 V for positive mode and 5000 V for negative mode and the nebulizer operated at 30 psi. Ion funnel low and high pressure in positive mode were 150 and 60, and in negative mode 150 and 120, respectively. Check tunes were performed in wide, unit and enhanced modes prior to each experiment to confirm the performance of the mass spectrometer. The quadrupole was tuned to reference masses 118.09, 322.05, 622.03, 922.01 and 1221.99 in positive ionization mode, and 112.99, 302.00, 601.98, 1033.99 and 1333.97 in negative ionization mode.

Each sample was analyzed in 3 separate dynamic MRM runs using two different HILIC buffer systems, both using 50%/50% (*v/v*) acetonitrile/water as Buffer A and 95% acetonitrile/water (*v/v*) as buffer B. The buffers were supplemented with 25 mM ammonium formate, pH 4.6 and 0.1% formic acid (denoted methods F1, F2) or 10 mM ammonium acetate, pH 7.6 (denoted method A). As detailed in Table S1, the methods had 155 (F1), 156 (F2) and 126 (A) transitions, including internal standards. The minimum dwell times were 4.2 milliseconds (ms), 4.1 ms and 3.1 ms respectively for methods F1, F2 and A. The method started at 0.1% A and increased to 40% A at 8:00, 90% A at 9:30 until 10:30. The gradient decreased to 0.1% A between 10:30 and 11:30 and was retained at the starting conditions of 0.1% A until 14:00. The column compartment was maintained at 30 °C for the duration of the experiment. A pooled quality control (QC) sample was injected multiple times to condition the HPLC column prior to analyzing samples, and also queued after every 6–7 biological samples to monitor mass spectrometry performance for the duration of the experiment [29,30].

2.2.5. Data Treatment and Analysis

Feature integration of untargeted lipidomics data was performed using the XCMS Centwave method and retention time alignment was performed using the Obiwarp method [31]. Features were grouped and peak filling was performed using the fill ChromPeaks method. Finally, feature information and abundances per samples were exported as a .csv file format. Lipid identification was performed using MS-DIAL version 3.90 (RIKEN Center for Sustainable Resource Science, Kanagawa, Japan) and the included FiehnRT (v47) lipid database [32]. Identifications were made based on accurate mass, retention time and database matching, and then manually confirmed. The MS1 tolerance was set to 0.01 Da and the tolerance for MS2 peaks was set to 0.05 Da. Database retention times were not used for scoring in the lipid identification. An identification score cut-off of 70 was set to remove most inaccurate identifications. The possible adduct ions were set to $[M + H]^+$, $[M + NH_4]^+$ and $[M\text{-}H]^-$. Manual confirmation included the visual inspection of all database matches, assessing the dot and reverse dot product similarity scores. Ambiguous identifications of features with multiple likely identifications were excluded from the analysis. Lipid identifications, accurate masses and retention times were exported from MS-DIAL and integrated into the data exported from XCMS.

Acquired targeted lipidomics data was imported into Skyline (MacCoss Lab, Department of Genome Sciences, University of Washington) [33], peak integration was automated but manually confirmed and corrected if required. Internal standard retention time was used to confirm correct peak integration of lipids belonging to the same class. Peak areas were exported from Skyline for further analysis in R (R Foundation for Statistical Computing, Vienna, Austria) [34].

The datasets were filtered to remove any lipids with a coefficient of variation greater than 20% among the quality control samples. Missing values were imputed using the MinDet method from the imputeLCMD R package using the default q-value of 0.01. All datasets were log2 transformed and normalized using the probabilistic quotient normalization method as described by Dieterle et al. [35]. Lipid information such as lipid class, number of unsaturated bonds and fatty acid chain lengths were parsed from the original lipid names using the lipidr R package [36]. Further analyses and visualizations, including principal component analysis (PCA) and lipid class boxplots were produced using lipidr [36]. The enrichment of lipid classes was determined using the LSEA (lipid set enrichment analysis method) [36]. Pearson correlation was used to determine the correlation between total lipid fatty acid chain lengths and the development of disease conditions.

2.3. *Microbiome Profiling*

2.3.1. DNA Extraction

Unless otherwise stated, solvents were purchased from Sigma (Missouri, USA). Mouse tissues were preincubated with lysis buffer (20 nanomolar (nM) Tris/HCl; 2 mM EDTA; 1% Triton X-100; pH 8; supplemented with 20 mg/mL lysozyme) for 60 min at 37 °C, then with 25 μL Proteinase K (20 mg/mL; Ambion, CA, USA) at 56 °C until completely lysed. DNA was extracted using the ISOLATE II Genomic DNA Kit (Bioline, London, UK) following manufacturer's standard protocol. The DNA samples were eluted in two lots of 50 μL Elution Buffer G from the kit.

2.3.2. Library Preparation and Sequencing

Library preparation was performed in batch. Polymerase chain reaction (PCR) preparation was conducted in a designated DNA template-free room. Sequencing library preparation of the samples and control (no DNA template) was based on the 16S Metagenomic Sequencing Library Preparation guidelines provided by Illumina. Q5 Hot Start High-Fidelity 2× Mastermix polymerase (NEB, Ipswich, MA, USA) was used for the Amplicon PCR step. Primers used for the amplification of the V6–V8 region of the 16S ribosomal RNA gene were primers 927-Forward (AAACTYAAAKGAATTGRCGG; universal) and 1392-Reverse (ACGGGCGGTG WGTRC; universal) with Illumina adapter sequences. Samples were barcoded using the Illumina dual-index system (Nextera XT v2 Index Kit Set A) for the Index PCR step. PCR products were purified using AMPure XP beads (Beckman Coulter, Brea, CA, USA). The DNA concentration for each barcoded amplicon mixture was quantified following manufacturer's instructions (Quantus, Promega, Madison, WI, USA) and all samples were pooled to provide 4 nanomol of each amplicon. The pooled libraries were sequenced using the Illumina MiSeq platform (Illumina, San Diego, CA, USA) and the MiSeq Reagent Kit v3 (2 × 300 bp) by the Australian Centre for Ecogenomics, located at the University of Queensland.

2.3.3. Bioinformatics and Statistical Analysis

Raw sequencing reads were processed and analyzed using Quantitative Insights Into Microbial Ecology 2 (QIIME 2, version 2019.7) according to the developer's recommendations [37]. Sequence quality control was carried out using the DADA2 algorithm, a QIIME 2 plugin-software to filter low-quality sequences as well as to identify and remove chimeric sequences. Amplicon sequence variants (ASVs) were generated from the filtered sequences and the SILVA_132 99% reference database was used to train the feature classifiers and provide taxonomic assignment accordingly. An ASV table was generated and normalized using total sum normalization (TSS) for all further analyses using Calypso (version 8.84) [38].

3. Results

3.1. *High-Fat Diet and Bile Acid Exposure as a Mouse Model for the Development of Esophageal Inflammation and Cardiac Metaplasia*

Chronic treatment with the unconjugated bile acid, deoxycholic acid (DCA, 0.2%), in drinking water was previously reported to accelerate Barrett's-like metaplasia development in an interleukin-1β transgenic mouse model [16]. We hypothesized that obesity induced by chronic HFD will replicate the chronic inflammation due to interleukin-1β overexpression, and leads to Barrett's-like epithelium development in wild-type mice. To test this hypothesis, male C57BL/6 mice were fed with standard chow diet or HFD with and without 0.2% DCA, for 9 months prior to sacrifice (n = 11 per group). Chow and HFD diets had comparable fiber (5.2% vs 5.4%) and protein (23% vs 21%) content, but the digestible energy from fat increased from 12% in chow to 43% in HFD, while carbohydrate reduced from 65% to 36%.

Body weight was monitored weekly, and HFD +/− DCA mice had significantly higher body weight than Chow +/− DCA ($q < 0.0001$), but no difference in body weight was observed between mice +/− DCA in either diet group (Figure 1a). Interestingly, weight gain in mice in the HFD + DCA group was delayed compared to the HFD + water group, potentially indicative of DCA-induced esophageal damage reducing food intake and subsequent recovery (Figure 1a).

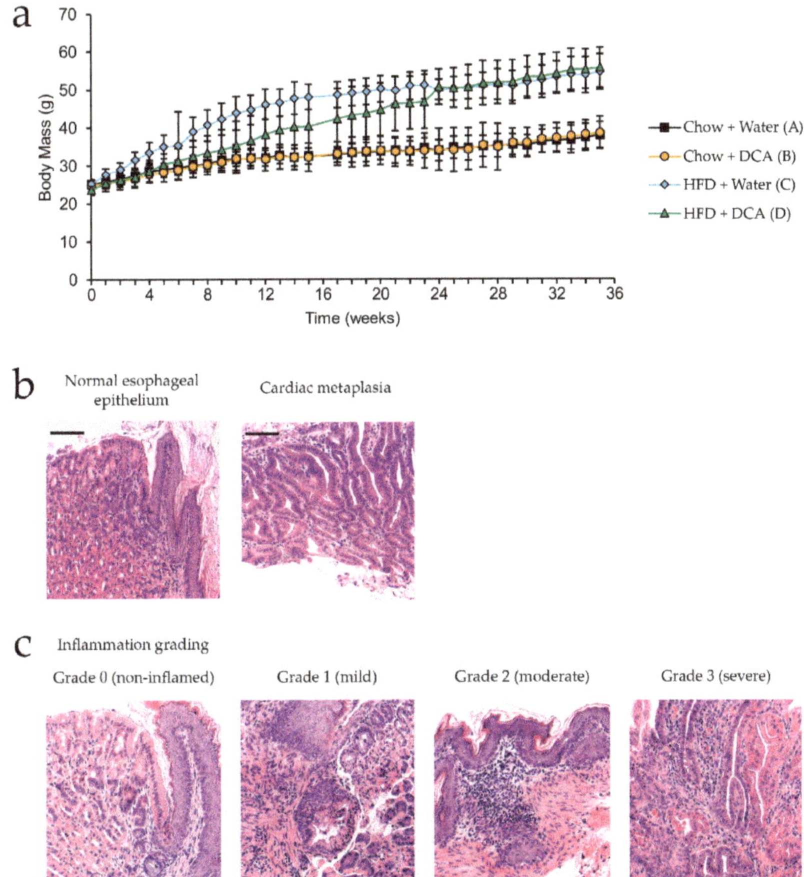

Figure 1. Chronic high-fat and/or bile acid dietary intervention in wild-type mice induces chronic inflammation and cardiac metaplasia development at the gastroesophageal junction. C57BL/6 mice (n = 11 per group) were given +/− high-fat diet (HFD) and +/− 0.2% deoxycholic acid (DCA) over a 9 month period, and gastroesophageal junction tissue morphology evaluated in hematoxylin & eosin (H&E) stained sections for inflammation and epithelial changes. (**a**) Body weight over time for each of the four groups. Values are mean ± SD; (**b**) Example esophageal epithelium morphology for normal and cardiac metaplasia. (**c**) Example inflammation grading. (200×; scale bar 100 μm).

Next, we examined the impact of the dietary treatments on tissue morphology of the gastroesophageal junction, where BE arises. H&E stained tissues were evaluated, and graded for inflammation severity and metaplasia length by an expert gastrointestinal pathologist in a blinded manner. Figure 1b shows the morphology of the normal squamous epithelium of the gastroesophageal junction, which was observed in most samples. In contrast, cardiac metaplasia with neutral mucin-producing glands was observed immediately adjacent to the squamous epithelium, observed in all four groups with varying frequency. Furthermore, varying grades of inflamed esophageal tissue were observed (Figure 1c). Inflammation grade 0 lacks inflammatory cells in the lamina propria, whereas mild inflammation with small numbers of lymphocytes and eosinophils are observed in inflammation grade 1. Inflammation grade 2 is marked by moderate inflammation, with a prominent infiltration of the lamina propria by lymphocytes and small numbers of eosinophils. Additionally, lymphocytes infiltrate the squamous epithelium. In severe inflammation, grade 3, a prominent infiltration of the lamina propria by lymphocytes, plasma cells, eosinophils and neutrophils is observed. Neutrophils and eosinophils are present within the epithelium.

Quantitative analysis revealed a basal level of mild inflammation in ~20% of the control and DCA treatment groups (Figure 2a). The combined HFD + DCA increased the overall incidence of inflammation to 67%, and was the only group with a grade of severe inflammation (Figure 2a). HFD alone slightly increased inflammation incidence to 27%, but induced a similarly high level of metaplasia (64%–67%) as the combined HFD + DCA (Figure 2b). However, all of the instances of metaplasia for the HFD + DCA group were long segment, while metaplasia induced by HFD alone comprised short, medium and long segments (Table S2).

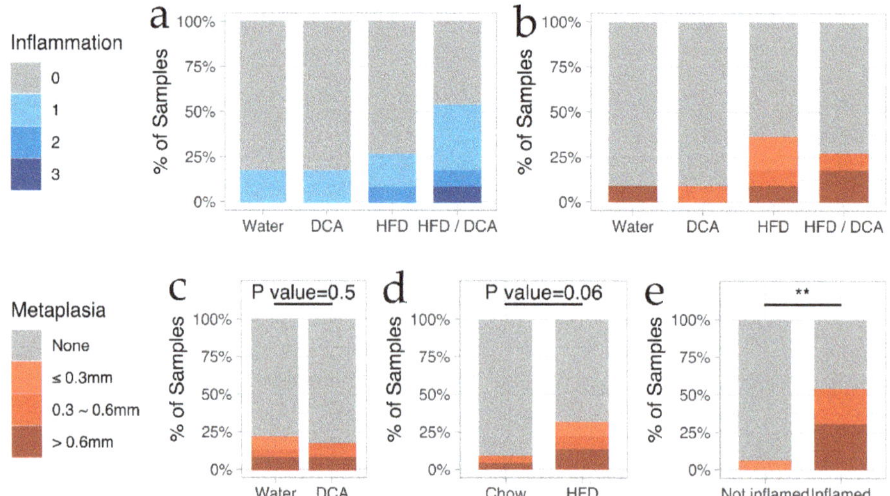

Figure 2. Synergistic action of chronic HFD and DCA promotes inflammation and cardiac metaplasia at the gastroesophageal junction. H&E stained tissue sections graded for the degree of inflammation (mild, moderate or severe), and the length of cardiac metaplasia (short, medium or long) were analyzed for (**a**) the occurrence and degree of inflammation, and (**b**) length of cardiac metaplasia in the four treatment groups. Correlation between presence of cardiac metaplasia was further compared for: (**c**) all mice treated with DCA compared to water control; (**d**) all mice on HFD diet compared to chow diet; and (**e**) any level of inflammation. The significance for plots c–e was calculated using the Fisher's exact test. ** p-value < 0.05.

The above results demonstrate that chronic HFD with DCA (mimicking GERD) induces the hallmarks of early BE, namely, tissue inflammation and metaplasia. To further evaluate the correlation between each dietary treatment, we next asked if the inflammation or metaplasia incidence correlate with HFD or DCA treatment. When all samples from DCA treatment groups were compared against all groups treated with water, no significant difference was detected for incidence of cardiac metaplasia

(Figure 2c). Similarly, HFD, with or without DCA, did not significantly increase the development of cardiac metaplasia (Figure 2d). Finally, we asked whether the incidence of inflammation and metaplasia was correlated, and found a significant relationship, with 6% of mice without esophageal inflammation and 54% of mice with inflammation developing cardiac metaplasia (Figure 2e). Furthermore, among the mice that developed cardiac metaplasia, the mice with inflammation developed a longer metaplastic tissue (Figure 2e).

3.2. Esophageal Tissue Microbiome Diversity Increases with DCA

After confirming the induction of gastroesophageal junction inflammation and cardiac metaplasia by chronic HFD + DCA treatment, we went on to profile the esophageal microbiota of 43 samples from the four study groups, using 16S ribosomal RNA gene sequencing. One sample gave no sequences and was removed from subsequent analysis. In total, 21,708 high quality sequences were obtained, with an average of 504.84 sequences per sample. From these sequences, four major phyla (*Actinobacteria*, *Bacteroidetes*, *Firmicutes*, and *Proteobacteria*) were identified, and a total of 23 ASVs were detected at 99% sequence identity threshold via SILVA_132 database.

We first compared microbial diversity (Shannon index) and richness between treatment groups using rank test in Calypso. No significant differences in microbial richness was observed, but a higher microbial diversity was observed in DCA alone, and HFD + DCA groups (Figure 3a). To further test the relationship between DCA and microbial diversity, we then re-grouped the data into HFD-treated and DCA-treated groups, as previously done (Figure 2). While no significant differences in microbial diversity or richness were detected for HFD treatment (Figure 3b), a significant increase of microbial diversity in DCA-treated groups was detected, with a similar but non-significant increase in richness (Figure 3c).

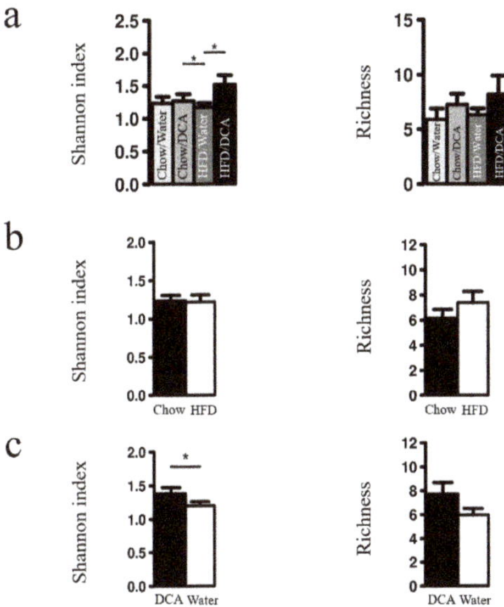

Figure 3. Esophageal microbiome diversity is increased by HFD + DCA treatment. Shannon index and microbial richness of esophageal microbiome data was measured using rank test for (**a**) each of the four treatment groups, (**b**) combining HFD/Water and HFD/DCA groups into the HFD group, and Chow/Water + Chow/DCA into the Chow group, or (**c**) combining Chow/DCA and HFD/DCA groups into the DCA group, and Chow/Water + Chow/DCA into the Water group. * p-value < 0.01.

3.3. Lipidomic Changes Associated with Dietary Interventions

In parallel to the esophageal microbiome analysis, we conducted lipidomics analyses on the collected serum and gastroesophageal junction samples, to determine associations between the respective lipidomes and dietary treatments (HFD or DCA), inflammation or cardiac metaplasia. A combined approach of untargeted and targeted lipidomics was conducted, to quantitate 339 and 197 mammalian lipid species in the serum and gastroesophageal junction samples, respectively.

While we observed no separation of gastroesophageal junction lipidome as a result of dietary treatments by PCA in the first two principal components (Figure 4b), the serum lipidome showed clear separation and clustering according to dietary intervention groups (Figure 4a).

Differential expression analysis was conducted on the lipidomics data of both datasets. Lipid class enrichment was conducted to determine if specific lipid classes were selectively altered. The boxplots in Figure 4c,d summarize the $\log_2$ fold change for each lipid class for each group, for serum and gastroesophageal junction tissue lipids, respectively. Statistically significant changes are colored in blue. Gastroesophageal junction tissue lipid class analysis (Figure 4d) revealed overlapping impacts of HFD and DCA treatments. All three treatment groups showed elevated lysophosphatidylcholine (LPC), as well as decreased phosphatidylcholine (PC) and phosphatidylethanolamine (PE) (Figure 4d). While triacylglycerol (TAG) was elevated only in group B (DCA alone), phosphatidylglycerol (PG) was elevated in HFD-treated groups (Figure 4d). For serum lipids, both HFD-treated groups (C and D) show similar changes, with elevated ceramide (Cer), PG and sphingomyelin (SM), and reduced lysophosphatidylethanolamine (LPE), PE and phosphatidylinositol (PI) (Figure 4c). In contrast, DCA treatment alone (Group B) showed a large decrease in ether-PC, with modest changes in PI and SM (Figure 4c). Interestingly, the reduction in ether-PC was not observed in the combined HFD + DCA treatment (Group D), suggesting HFD rather than DCA is the main driver of the serum lipidome.

3.4. Lipidomic Changes Associated with Early Tissue Pathology

Since esophageal inflammation or metaplasia occurred in ~10% to 70% of mice in each group, we next investigated the association between serum and gastroesophageal junction tissue lipidome with early esophageal pathology. To this end, lipid class enrichment analysis was conducted on metaplasia vs normal samples, and inflamed vs normal samples. Apart from elevated serum ether-PC, the serum lipidome returned minor changes of $< 25\%$ magnitude (Figure 5a). In contrast, the tissue lipidome showed similar changes for metaplasia and inflammation, characterized by reduced lysolipids and elevated ceramides (Figure 5b). This result revealed major differences between the lipidome associated with dietary intervention (Figure 4) and that associated with esophageal pathology (Figure 5). Specifically, differences were observed for ceramides and the lysolipids LPC and LPE. Elevated tissue ceramide was associated with metaplasia and inflamed tissue, but not with any dietary treatment, even in the HFD + DCA treatment group, where 66.7% of cases were inflamed (Figure 5). Reductions in the lysolipids LPC and LPE were associated with metaplasia and inflammation (Figure 5b), but elevated tissue LPC was associated with HFD and DCA treatment (Figure 4d). These results strongly implicate roles for elevated ceramides and reduced lysolipids in metaplasia development due to chronic inflammation.

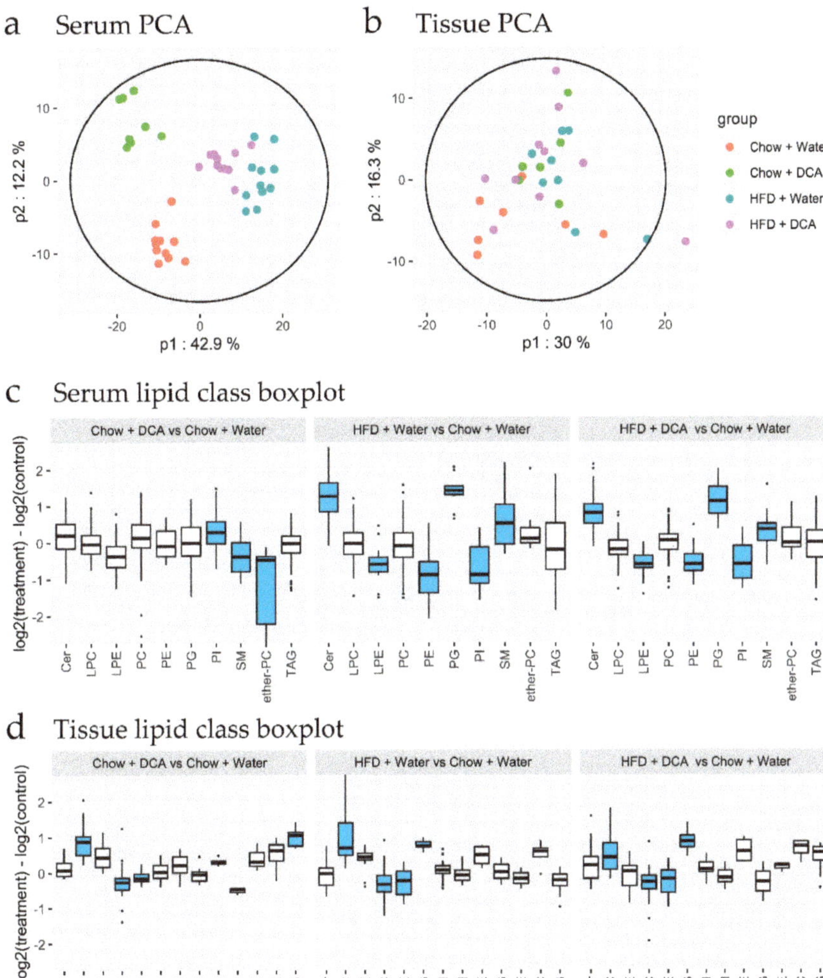

Figure 4. The impact of dietary interventions on tissue and serum lipidome. After 9 months high fat diet (HFD) +/- 0.2% deoxycholate (DCA), mouse gastroesophageal junction tissue and serum samples were subjected to lipidomics analyses. (**a**) Principal component analysis score plot of mean-centered unit variance-scaled untargeted serum lipidome data ($n = 38$). (**b**) Principal component analysis score plot of mean-centered unit variance-scaled untargeted gastroesophageal junction tissue lipidome data ($n = 29$). Plot ellipses represents the 95% Hotelling's T2 confidence intervals for the multivariate data. (**c**,**d**) Lipid class boxplots for serum and gastroesophageal junction tissue lipids, showing the distribution of log2 differences between the treatment group and control. Positive values represent lipids that are more abundant in the treatment group than in the control group. Blue color indicates significant enrichment using the fast gene set enrichment analysis (fgsea) method. Cer—Ceramide, LPC—lysophosphatidylcholine, LPE—lysophosphatidylethanolamine, PC—phosphatidylcholine, PE—phosphatidylethanolamine, PG—phosphatidylglycerol, PI—phosphatidylinositol, SM—sphingomyelin, DAG—diacylglycerol, TAG—triacylglycerol.

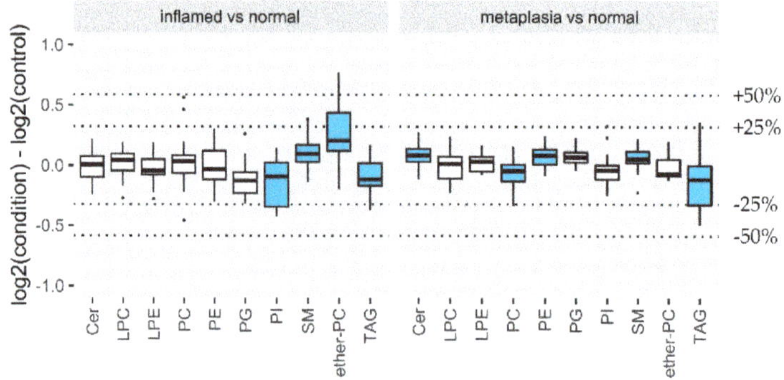

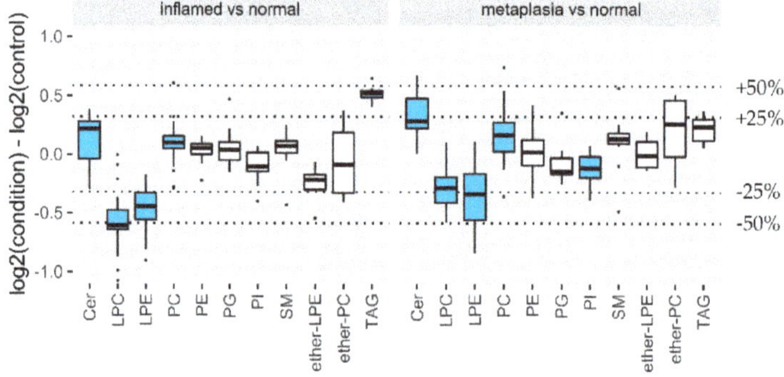

Figure 5. Lipid classes associated with gastroesophageal junction tissue pathology. (**a**,**b**) lipid class boxplots for serum and tissue lipids, showing the distribution of log2 differences between the disease condition and control. The disease conditions cardia and inflammation were visualized after applying the removeBatchEffect function from the limma R package. Positive values represent lipids that are more abundant in the disease condition group than in the control group. Blue color indicates significant enrichment using the fast gene set enrichment analysis (fgsea) method. Cer—Ceramide, LPC—lysophosphatidylcholine, LPE—lysophosphatidylethanolamine, PC—phosphatidylcholine, PE—phosphatidylethanolamine, PG—phosphatidylglycerol, PI—phosphatidylinositol, SM—sphingomyelin, DAG—diacylglycerol, TAG—triacylglycerol.

As differing fatty acid chain lengths on a lipid can greatly impact biological function in cancer development [39], we next determined whether fatty acid chain lengths were associated with inflammation or metaplasia for the Cer, LPC and LPE classes. Ceramides comprise a single fatty acid chain with a sphingoid backbone (commonly 18:1, as illustrated in Figure 6a). Figure 6a plots the log2 fold change for different total fatty acid chain lengths of each measured ceramide species. As evident in Figure 6a, a significant correlation was found between very long chain ceramides and the disease conditions inflammation and metaplasia. On the other hand, specificity in fatty acid chain lengths were not observed for LPC in either metaplasia or inflamed tissues (Figure 6b). Increased LPE chain lengths were significantly correlated with metaplasia, but not with inflammation (Figure 6c).

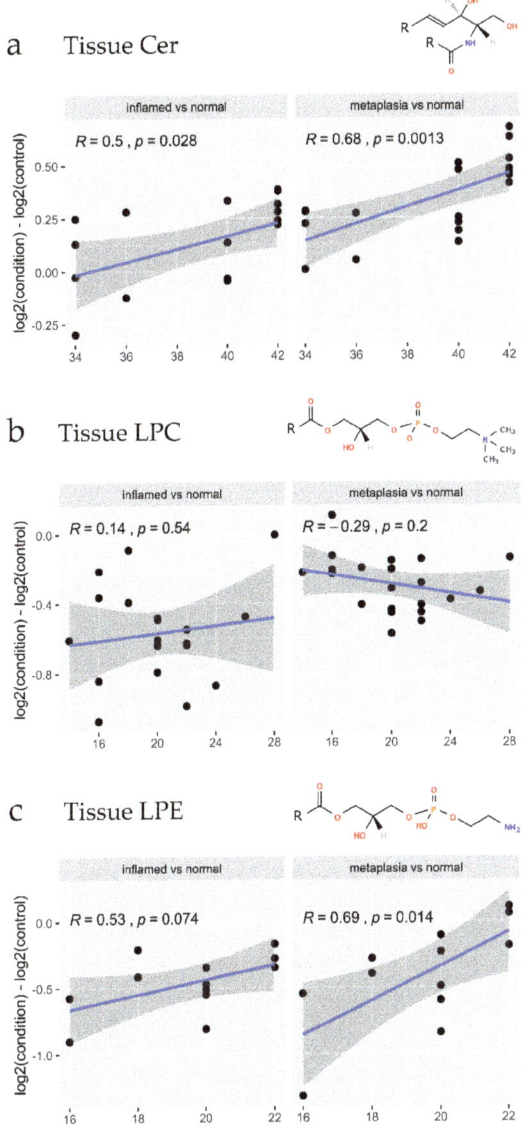

Figure 6. Association of tissue ceramide, LPC and LPE fatty acid chain lengths with gastroesophageal junction tissue pathology. (**a**–**c**) Total chain length plots for tissue lipids, showing the alterations in log2 abundances between the disease condition and control. The x-axis labels refer to the total fatty acid chain length of the measured lipid. The disease conditions metaplasia and inflammation were visualized after applying the removeBatchEffect function from the limma R package. Positive values represent lipids that are more abundant in the disease condition group than in the control group. The metrics shown in the plots refer to the Pearson correlation coefficient (R) and *p*-value. The smoothed line and 95% confidence interval were drawn using geom_smooth, by fitting a linear model. The structures above each plot represent lipid species of the Ceramide, LPC and LPE classes, where the R groups refer to the hydrocarbon chains of varying lengths.

4. Discussion

This is the first study to demonstrate that chronic HFD in non-transgenic mice is sufficient to induce esophageal inflammation and cardiac metaplasia, the first steps in BE/EAC pathogenesis. While DCA in drinking water had no effect on esophageal morphology on its own, it increased the severity of inflammation and length of metaplasia when combined with HFD. HFD clearly induced obesity and serum lipid derangements, but only a proportion of HFD-treated mice developed esophageal inflammation and cardiac metaplasia. Intriguingly, the esophageal tissue lipidome showed a similar signature for inflammation and metaplasia, which was not associated with HFD. These results suggest that homeostatic mechanisms can buffer HFD/obesity-induced lipidome derangement to an extent, beyond which inflammation and metaplasia ensue.

Obesity increases the risk of several cancer types, and the mechanisms of specific lipids on carcinogenesis are beginning to be revealed [39]. In this study, we identified an esophageal tissue lipid signature for inflammation and metaplasia, which is characterized by elevated very long chain ceramides and reduced lysolipids, LPC and LPE. Very long chain ceramides have been reported to increase cancer proliferation, and evade growth suppressor and apoptotic signals [39]. A link between HFD and tissue ceramide levels was recently reported by Zalewska et al. [40] for submandibular gland ceramide following HFD treatment in mice. The authors suggested that elevated ceramide increased mitochondrial reactive oxygen species (ROS) production and respiratory chain, leading to inflammation [40].

Phospholipid remodeling has recently emerged as playing an important role in disease pathogenesis, through the characterization of the lysophosphatidylcholine acyltransferase (LPCAT) family [41]. Lysolipids LPC and LPE contain a single fatty acyl chain, while the more abundant PC and PE contain two fatty acyl chains. Due to the differing biophysical properties, altered lysolipid:phospholipid ratio can lead to altered membrane curvature and fluidity, which could translate to organelle remodeling and altered signal transduction in pathology [41].

Warnecke-Eberz et al. [42] identified the LPCAT1 gene to be elevated in late- and early-stage esophageal adenocarcinoma tissue, compared to adjacent normal tissue. Elevated LPCAT1 could explain the decreased LPC and increased PC that we identified for inflamed and cardia gastroesophageal junction tissue (Figure 5). LPCAT1 enzyme and LPC are elevated in several other cancers, including colorectal cancer [43], hepatocellular carcinoma [44], gastric cancer [45] and clear cell renal carcinoma [46]. Interestingly, body fatness is a risk factor for all these cancers [47]. In a recent study of western diet-associated non-alcoholic steatohepatitis, LPCAT1 and LPCAT2 are in the top 10 liver genes/transcripts most significantly elevated in mice fed western style diets compared to standard diets [48]. Together, these data suggest a mechanistic link between high-fat diet, activation of LPCAT transcripts, altered LPC:PC ratio, and induction of esophageal inflammation and metaplasia development.

As GERD is a well-established risk factor for BE, the lack of esophageal pathology from the mice treated with DCA alone was somewhat surprising. This result may suggest that 0.2% DCA in drinking water does not fully mimic GERD, or that GERD is less damaging to mice esophagus compared to human. Nevertheless, as expected for the additive effect of risk factors, DCA treatment in addition to HFD increased the severity of inflammation and length of metaplasia, compared to HFD treatment alone. DCA treatment increased the esophageal microbiome diversity, which is consistent with previous reports describing the effect that levels of bile acids in the gut have on the major division/phyla level taxa of the gut microbiome [49]. These effects could potentially extend to the esophagus, given that the composition of the esophageal microbiome depends on the oral and gut microbiome [50]. Previous studies have reported a depletion of Gram-positive *Streptococcus*, and enrichment of Gram-negative taxa, including *Veillonella* and *Prevotella*, in BE [25,51]. Interestingly, dysplasia and esophageal adenocarcinoma were reported to have reduced esophageal microbiome diversity [24]. Further studies will be required to establish the cause–effect relationship and mechanisms of esophageal microbiota in BE/EAC pathogenesis.

5. Conclusions

In conclusion, we report the results of a dietary intervention model for early BE, and a lipidomic signature for inflamed and metaplastic esophageal tissue. In non-transgenic mice, chronic HFD was sufficient to induce inflammation and cardiac metaplasia at the gastroesophageal junction. As a GERD-mimic, bile acid in drinking water in addition to HFD increased the severity of inflammation and length of metaplasia. GERD, but not HFD, increased the esophageal microbiome diversity. The causality of microbiome in BE development remains to be established.

Supplementary Materials: The following are available online at http://www.mdpi.com/2218-273X/10/5/776/s1, Table S1: Targeted lipidomics methods. Table S2: Tissue pathology grades. Tissue and serum lipidomics data are available on Metabolomics Workbench with Study IDs ST001323 and ST001336.

Author Contributions: Conceptualization, J.B. and M.M.H.; methodology, J.M., L.K., M.P.H. and M.M.; formal analysis, J.M., A.M., Y.L., I.B.; investigation, J.M., T.-M.-T.N. and J.B.; resources, T.P.H., M.M.H.; writing—original draft preparation, J.M. and M.M.H.; writing—review and editing, T.-M.-T.N., A.M., L.K., Y.L., M.P.H., T.P.H.; supervision, M.M., M.P.H., T.P.H. and M.M.H.; funding acquisition, J.B., T.P.H. and M.M.H. All authors have read and agreed to the published version of the manuscript.

Funding: This research was funded by Translational Research Institute Spore Grant to M.M.H., J.B. and T.P.H. J.M. and T.-M.-T.N. were supported by Australian Postgraduate Research Awards. Metabolomics Australia is part of the Bioplatforms Australia network, funded through the Australian Government's National Collaborative Research Infrastructure Strategy (NCRIS). Translational Research Institute is supported by a grant from the Australian Government.

Conflicts of Interest: The authors declare no conflict of interest. The funders had no role in the design of the study; in the collection, analyses, or interpretation of data; in the writing of the manuscript, or in the decision to publish the results.

References

1. Daly, J.M.; Fry, W.A.; Little, A.G.; Winchester, D.P.; McKee, R.F.; Stewart, A.K.; Fremgen, A.M. Esophageal cancer: Results of an American College of Surgeons Patient Care Evaluation Study. *J. Am. Coll. Surg.* **2000**, *190*, 562–572. [CrossRef]
2. Pohl, H.; Welch, H.G. The role of overdiagnosis and reclassification in the marked increase of esophageal adenocarcinoma incidence. *J. Natl. Cancer Inst.* **2005**, *97*, 142–146. [CrossRef] [PubMed]
3. Powell, J.; McConkey, C.C.; Gillison, E.W.; Spychal, R.T. Continuing rising trend in oesophageal adenocarcinoma. *Int J. Cancer* **2002**, *102*, 422–427. [CrossRef] [PubMed]
4. Hur, C.; Miller, M.; Kong, C.Y.; Dowling, E.C.; Nattinger, K.J.; Dunn, M.; Feuer, E.J. Trends in esophageal adenocarcinoma incidence and mortality. *Cancer* **2013**, *119*, 1149–1158. [CrossRef]
5. Lagergren, J. Adenocarcinoma of oesophagus: What exactly is the size of the problem and who is at risk? *Gut* **2005**, *54* (Suppl. 1), i1–i5. [CrossRef]
6. Rubenstein, J.H.; Shaheen, N.J. Epidemiology, Diagnosis, and Management of Esophageal Adenocarcinoma. *Gastroenterology* **2015**, *149*, 302–317.e301. [CrossRef]
7. Enzinger, P.C.; Mayer, R.J. Esophageal cancer. *N. Engl. J. Med.* **2003**, *349*, 2241–2252. [CrossRef]
8. Vakil, N.; van Zanten, S.V.; Kahrilas, P.; Dent, J.; Jones, R.; Global Consensus, G. The Montreal definition and classification of gastroesophageal reflux disease: A global evidence-based consensus. *Am. J. Gastroenterol.* **2006**, *101*, 1900–1920; quiz 1943. [CrossRef]
9. Schlottmann, F.; Patti, M.G.; Shaheen, N.J. From Heartburn to Barrett's Esophagus, and Beyond. *World J. Surg* **2017**, *41*, 1698–1704. [CrossRef]
10. Chandrasoma, P. Controversies of the cardiac mucosa and Barrett's oesophagus. *Histopathology* **2005**, *46*, 361–373. [CrossRef]
11. Cossentino, M.J.; Wong, R.K. Barrett's esophagus and risk of esophageal adenocarcinoma. *Semin. Gastrointest. Dis.* **2003**, *14*, 128–135. [PubMed]
12. Kroep, S.; Lansdorp-Vogelaar, I.; Rubenstein, J.H.; de Koning, H.J.; Meester, R.; Inadomi, J.M.; van Ballegooijen, M. An Accurate Cancer Incidence in Barrett's Esophagus: A Best Estimate Using Published Data and Modeling. *Gastroenterology* **2015**, *149*, 577–585.e4. [CrossRef] [PubMed]

13. Hvid-Jensen, F.; Pedersen, L.; Drewes, A.M.; Sorensen, H.T.; Funch-Jensen, P. Incidence of adenocarcinoma among patients with Barrett's esophagus. *N. Engl. J. Med.* **2011**, *365*, 1375–1383. [CrossRef] [PubMed]
14. Coleman, H.G.; Xie, S.H.; Lagergren, J. The Epidemiology of Esophageal Adenocarcinoma. *Gastroenterology* **2018**, *154*, 390–405. [CrossRef] [PubMed]
15. Read, M.D.; Krishnadath, K.K.; Clemons, N.J.; Phillips, W.A. Preclinical models for the study of Barrett's carcinogenesis. *Ann. N. Y. Acad. Sci.* **2018**, *1434*, 139–148. [CrossRef]
16. Quante, M.; Bhagat, G.; Abrams, J.A.; Marache, F.; Good, P.; Lee, M.D.; Lee, Y.; Friedman, R.; Asfaha, S.; Dubeykovskaya, Z.; et al. Bile acid and inflammation activate gastric cardia stem cells in a mouse model of Barrett-like metaplasia. *Cancer Cell* **2012**, *21*, 36–51. [CrossRef]
17. Munch, N.S.; Fang, H.Y.; Ingermann, J.; Maurer, H.C.; Anand, A.; Kellner, V.; Sahm, V.; Wiethaler, M.; Baumeister, T.; Wein, F.; et al. High-Fat Diet Accelerates Carcinogenesis in a Mouse Model of Barrett's Esophagus via Interleukin 8 and Alterations to the Gut Microbiome. *Gastroenterology* **2019**, *157*, 492–506.e492. [CrossRef]
18. Nguyen, T.; Khalaf, N.; Ramsey, D.; El-Serag, H.B. Statin use is associated with a decreased risk of Barrett's esophagus. *Gastroenterology* **2014**, *147*, 314–323. [CrossRef] [PubMed]
19. Beales, I.L.; Dearman, L.; Vardi, I.; Loke, Y. Reduced Risk of Barrett's Esophagus in Statin Users: Case-Control Study and Meta-Analysis. *Dig. Dis. Sci.* **2016**, *61*, 238–246. [CrossRef] [PubMed]
20. Beales, I.L.; Vardi, I.; Dearman, L. Regular statin and aspirin use in patients with Barrett's oesophagus is associated with a reduced incidence of oesophageal adenocarcinoma. *Eur. J. Gastroenterol. Hepatol.* **2012**, *24*, 917–923. [CrossRef] [PubMed]
21. Kantor, E.D.; Onstad, L.; Blount, P.L.; Reid, B.J.; Vaughan, T.L. Use of statin medications and risk of esophageal adenocarcinoma in persons with Barrett's esophagus. *Cancer Epidemiol. Biomark. Prev. A Publ. Am. Assoc. Cancer Res. Cosponsored Am. Soc. Prev. Oncol.* **2012**, *21*, 456–461. [CrossRef] [PubMed]
22. Beales, I.L.; Vardi, I.; Dearman, L.; Broughton, T. Statin use is associated with a reduction in the incidence of esophageal adenocarcinoma: A case control study. *Dis. Esophagus* **2013**, *26*, 838–846. [CrossRef] [PubMed]
23. Nguyen, T.; Duan, Z.; Naik, A.D.; Kramer, J.R.; El-Serag, H.B. Statin Use Reduces Risk of Esophageal Adenocarcinoma in US Veterans With Barrett's Esophagus: A Nested Case-Control Study. *Gastroenterology* **2015**, *149*, 1392–1398. [CrossRef] [PubMed]
24. Elliott, D.R.F.; Walker, A.W.; O'Donovan, M.; Parkhill, J.; Fitzgerald, R.C. A non-endoscopic device to sample the oesophageal microbiota: A case-control study. *Lancet Gastroenterol. Hepatol.* **2017**, *2*, 32–42. [CrossRef]
25. Yang, L.; Lu, X.; Nossa, C.W.; Francois, F.; Peek, R.M.; Pei, Z. Inflammation and intestinal metaplasia of the distal esophagus are associated with alterations in the microbiome. *Gastroenterology* **2009**, *137*, 588–597. [CrossRef]
26. Kountouras, J.; Doulberis, M.; Papaefthymiou, A.; Polyzos, S.A.; Vardaka, E.; Tzivras, D.; Dardiotis, E.; Deretzi, G.; Giartza-Taxidou, E.; Grigoriadis, S.; et al. A perspective on risk factors for esophageal adenocarcinoma: Emphasis on Helicobacter pylori infection. *Ann. N. Y. Acad. Sci.* **2019**, *1452*, 12–17. [CrossRef]
27. Wang, Z.; Shaheen, N.J.; Whiteman, D.C.; Anderson, L.A.; Vaughan, T.L.; Corley, D.A.; El-Serag, H.B.; Rubenstein, J.H.; Thrift, A.P. Helicobacter pylori Infection Is Associated With Reduced Risk of Barrett's Esophagus: An Analysis of the Barrett's and Esophageal Adenocarcinoma Consortium. *Am. J. Gastroenterol.* **2018**, *113*, 1148–1155. [CrossRef]
28. Matyash, V.; Liebisch, G.; Kurzchalia, T.V.; Shevchenko, A.; Schwudke, D. Lipid extraction by methyl-tert-butyl ether for high-throughput lipidomics. *J. Lipid Res.* **2008**, *49*, 1137–1146. [CrossRef]
29. Sangster, T.; Major, H.; Plumb, R.; Wilson, A.J.; Wilson, I.D. A pragmatic and readily implemented quality control strategy for HPLC-MS and GC-MS-based metabonomic analysis. *Analyst* **2006**, *131*, 1075–1078. [CrossRef]
30. Broadhurst, D.; Goodacre, R.; Reinke, S.N.; Kuligowski, J.; Wilson, I.D.; Lewis, M.R.; Dunn, W.B. Guidelines and considerations for the use of system suitability and quality control samples in mass spectrometry assays applied in untargeted clinical metabolomic studies. *Metabolomics* **2018**, *14*, 72. [CrossRef]
31. Tautenhahn, R.; Bottcher, C.; Neumann, S. Highly sensitive feature detection for high resolution LC/MS. *Bmc Bioinform.* **2008**, *9*, 504. [CrossRef] [PubMed]
32. Tsugawa, H.; Cajka, T.; Kind, T.; Ma, Y.; Higgins, B.; Ikeda, K.; Kanazawa, M.; VanderGheynst, J.; Fiehn, O.; Arita, M. MS-DIAL: Data-independent MS/MS deconvolution for comprehensive metabolome analysis. *Nat. Methods* **2015**, *12*, 523–526. [CrossRef] [PubMed]
33. Peng, B.; Ahrends, R. Adaptation of Skyline for Targeted Lipidomics. *J. Proteome Res.* **2016**, *15*, 291–301. [CrossRef]

34. R Core Team. *R: A Language and Environment for Statistical Computing*; R Foundation for Statistical Computing: Vienna, Austria, 2017.
35. Dieterle, F.; Ross, A.; Schlotterbeck, G.; Senn, H. Probabilistic quotient normalization as robust method to account for dilution of complex biological mixtures. Application in 1H NMR metabonomics. *Anal. Chem.* **2006**, *78*, 4281–4290. [CrossRef]
36. Mohamed, A.; Molendijk, J.; Hill, M.M. lipidr: A Software Tool for Data Mining and Analysis of Lipidomics Datasets. *J. Proteome Res.* **2020**. [CrossRef]
37. Bolyen, E.; Rideout, J.R.; Dillon, M.R.; Bokulich, N.A.; Abnet, C.C.; Al-Ghalith, G.A.; Alexander, H.; Alm, E.J.; Arumugam, M.; Asnicar, F.; et al. Reproducible, interactive, scalable and extensible microbiome data science using QIIME 2. *Nat. Biotechnol.* **2019**, *37*, 852–857. [CrossRef] [PubMed]
38. Zakrzewski, M.; Proietti, C.; Ellis, J.J.; Hasan, S.; Brion, M.J.; Berger, B.; Krause, L. Calypso: A user-friendly web-server for mining and visualizing microbiome-environment interactions. *Bioinformatics* **2016**. [CrossRef] [PubMed]
39. Molendijk, J.; Robinson, H.; Djuric, Z.; Hill, M.M. Lipid mechanisms in hallmarks of cancer. *Mol. Omics* **2020**. [CrossRef] [PubMed]
40. Zalewska, A.; Maciejczyk, M.; Szulimowska, J.; Imierska, M.; Blachnio-Zabielska, A. High-Fat Diet Affects Ceramide Content, Disturbs Mitochondrial Redox Balance, and Induces Apoptosis in the Submandibular Glands of Mice. *Biomolecules* **2019**, *9*, 877. [CrossRef]
41. Wang, B.; Tontonoz, P. Phospholipid Remodeling in Physiology and Disease. *Annu. Rev. Physiol.* **2019**, *81*, 165–188. [CrossRef]
42. Warnecke-Eberz, U.; Metzger, R.; Holscher, A.H.; Drebber, U.; Bollschweiler, E. Diagnostic marker signature for esophageal cancer from transcriptome analysis. *Tumour Biol.* **2016**, *37*, 6349–6358. [CrossRef]
43. Mansilla, F.; da Costa, K.A.; Wang, S.; Kruhoffer, M.; Lewin, T.M.; Orntoft, T.F.; Coleman, R.A.; Birkenkamp-Demtroder, K. Lysophosphatidylcholine acyltransferase 1 (LPCAT1) overexpression in human colorectal cancer. *J. Mol. Med. (Berl)* **2009**, *87*, 85–97. [CrossRef]
44. Morita, Y.; Sakaguchi, T.; Ikegami, K.; Goto-Inoue, N.; Hayasaka, T.; Hang, V.T.; Tanaka, H.; Harada, T.; Shibasaki, Y.; Suzuki, A.; et al. Lysophosphatidylcholine acyltransferase 1 altered phospholipid composition and regulated hepatoma progression. *J. Hepatol.* **2013**, *59*, 292–299. [CrossRef] [PubMed]
45. Uehara, T.; Kikuchi, H.; Miyazaki, S.; Iino, I.; Setoguchi, T.; Hiramatsu, Y.; Ohta, M.; Kamiya, K.; Morita, Y.; Tanaka, H.; et al. Overexpression of Lysophosphatidylcholine Acyltransferase 1 and Concomitant Lipid Alterations in Gastric Cancer. *Ann. Surg Oncol* **2016**, *23* (Suppl. 2), S206–S213. [CrossRef] [PubMed]
46. Du, Y.; Wang, Q.; Zhang, X.; Wang, X.; Qin, C.; Sheng, Z.; Yin, H.; Jiang, C.; Li, J.; Xu, T. Lysophosphatidylcholine acyltransferase 1 upregulation and concomitant phospholipid alterations in clear cell renal cell carcinoma. *J. Exp. Clin. Cancer Res.* **2017**, *36*, 66. [CrossRef]
47. Lauby-Secretan, B.; Scoccianti, C.; Loomis, D.; Grosse, Y.; Bianchini, F.; Straif, K. International Agency for Research on Cancer Handbook Working, G. Body Fatness and Cancer–Viewpoint of the IARC Working Group. *N. Engl. J. Med.* **2016**, *375*, 794–798. [CrossRef] [PubMed]
48. Garcia-Jaramillo, M.; Spooner, M.H.; Lohr, C.V.; Wong, C.P.; Zhang, W.; Jump, D.B. Lipidomic and transcriptomic analysis of western diet-induced nonalcoholic steatohepatitis (NASH) in female Ldlr -/- mice. *PLoS ONE* **2019**, *14*, e0214387. [CrossRef]
49. Ridlon, J.M.; Kang, D.J.; Hylemon, P.B.; Bajaj, J.S. Bile acids and the gut microbiome. *Curr. Opin. Gastroenterol.* **2014**, *30*, 332–338. [CrossRef]
50. Gorkiewicz, G.; Moschen, A. Gut microbiome: A new player in gastrointestinal disease. *Virchows Arch.* **2018**, *472*, 159–172. [CrossRef]
51. Liu, N.; Ando, T.; Ishiguro, K.; Maeda, O.; Watanabe, O.; Funasaka, K.; Nakamura, M.; Miyahara, R.; Ohmiya, N.; Goto, H. Characterization of bacterial biota in the distal esophagus of Japanese patients with reflux esophagitis and Barrett's esophagus. *BMC Infect. Dis* **2013**, *13*, 130. [CrossRef]

MDPI

Article

Eicosanoid Content in Fetal Calf Serum Accounts for Reproducibility Challenges in Cell Culture

Laura Niederstaetter [1,†], Benjamin Neuditschko [1,2,†], Julia Brunmair [1], Lukas Janker [1], Andrea Bileck [1,3], Giorgia Del Favero [4,5] and Christopher Gerner [1,3,*]

1 Department of Analytical Chemistry, Faculty of Chemistry, University of Vienna, 1090 Vienna, Austria; laura.niederstaetter@univie.ac.at (L.N.); benjamin.neuditschko@univie.ac.at (B.N.); julia.brunmair@univie.ac.at (J.B.); lukas.janker@univie.ac.at (L.J.); andrea.bileck@univie.ac.at (A.B.)
2 Department of Inorganic Chemistry, Faculty of Chemistry, University of Vienna, 1090 Vienna, Austria
3 Joint Metabolome Facility, Faculty of Chemistry, University of Vienna, 1090 Vienna, Austria
4 Department of Food Chemistry and Toxicology, Faculty of Chemistry, University of Vienna, 1090 Vienna, Austria; giorgia.del.favero@univie.ac.at
5 Core Facility Multimodal Imaging, Faculty of Chemistry, University of Vienna, 1090 Vienna, Austria
* Correspondence: christopher.gerner@univie.ac.at
† Contributed equally to the work.

Abstract: Reproducibility issues regarding in vitro cell culture experiments are related to genetic fluctuations and batch-wise variations of biological materials such as fetal calf serum (FCS). Genome sequencing may control the former, while the latter may remain unrecognized. Using a U937 macrophage model for cell differentiation and inflammation, we investigated whether the formation of effector molecules was dependent on the FCS batch used for cultivation. High resolution mass spectrometry (HRMS) was used to identify FCS constituents and to explore their effects on cultured cells evaluating secreted cytokines, eicosanoids, and other inflammatory mediators. Remarkably, the FCS eicosanoid composition showed more batch-dependent variations than the protein composition. Efficient uptake of fatty acids from the medium by U937 macrophages and inflammation-induced release thereof was evidenced using C13-labelled arachidonic acid, highlighting rapid lipid metabolism. For functional testing, FCS batch-dependent nanomolar concentration differences of two selected eicosanoids, 5-HETE and 15-HETE, were balanced out by spiking. Culturing U937 cells at these defined conditions indeed resulted in significant proteome alterations indicating HETE-induced PPARγ activation, independently corroborated by HETE-induced formation of peroxisomes observed by high-resolution microscopy. In conclusion, the present data demonstrate that FCS-contained eicosanoids, subject to substantial batch-wise variation, may modulate cellular effector functions in cell culture experiments.

Keywords: batch variations; eicosanoids; fetal calf serum; mass spectrometry; peroxisomes; proteomics

Citation: Niederstaetter, L.; Neuditschko, B.; Brunmair, J.; Janker, L.; Bileck, A.; Del Favero, G.; Gerner, C. Eicosanoid Content in Fetal Calf Serum Accounts for Reproducibility Challenges in Cell Culture. *Biomolecules* **2021**, *11*, 113. https://doi.org/10.3390/biom11010113

Received: 9 December 2020
Accepted: 14 January 2021
Published: 15 January 2021

Publisher's Note: MDPI stays neutral with regard to jurisdictional claims in published maps and institutional affiliations.

1. Introduction

Problems with the inter-laboratory reproducibility of results obtained with in vitro cell culture models are increasingly being recognized [1,2]. The need to reduce the use of animal models for research purposes relies also on the use of accurate in vitro test models [3,4]. Important decisions such as the choice of drug candidates to be evaluated in clinical studies may be based on such experiments [5]. Thus, the identification of influencing factors potentially modulating such in vitro data is mandatory. Biological materials and reference materials have been recognized as the main contributors for irreproducibility, resulting in a current focus on the investigation of genetic heterogeneity and genetic instability of cell culture models [6,7]. Here we present Fetal Bovine Serum (FBS; also fetal calf serum, FCS) as another relevant contributor to reproducibility issues. FCS is commonly used as cell culture supplement sustaining the growth and duplication of mammalian cells in vitro.

Since its introduction in the 1950s its use has been established world-wide, irrespective of evident limitations regarding scientific as well as ethical points of view [8]. Fetal serum is basically a by-product of meat production collected from the still beating hearts of living fetuses. While efforts are made to reduce the use of FCS, they have shown rather limited successes [9].

As for other supplements of natural origin, the main variability source associated to FBS can be traced back to largely uncharacterized bioactive components. Due to low concentrations or lack of experimental standard measurements, they may remain poorly controlled, but may still influence the outcome of cell-based experiments. While some effort has been spent to define the composition of FCS, the main bioactive constituents subjected to meaningful variation are hardly known [8]. Batch-dependent variations have been described to affect biological outcomes but such considerations remain limited to rather specialized topics such as hormone regulation [10]. Chemically defined media (CDM) represent a general and consequent solution for these problems, but have only been established and available for a limited number of cell model systems [9].

The focus of the present study was to investigate whether it was possible to identify bioactive compounds in FCS accounting for relevant batch-specific effects and correlate this information with proven and biochemically evident readouts on cell functions. Variations of amino acid and metabolite composition of FCS may be considered as less likely as they should be subjected to the homoeostatic control of the organism and, further limited by the dilution of FCS in the accurately produced cell culture media also containing these molecules (typically 5–10% FCS is used). Thus proteins as well as eicosanoids and other polyunsaturated fatty acids (PUFAs) are profiled as the most relevant bioactive candidate molecules to account for inter-batch variation. Indeed, they are responsible for regulating biological processes associated with inflammation [11] and inflammation-associated pathomechanisms [12–14]. The monocyte cell line U937, a well-established cell model for macrophages [15–17], was chosen for these investigations. Overall, the data demonstrated significant effects of FCS-contained eicosanoids with batch-dependent variations on relevant cell functions, proving that bioactive lipid content in serum contributes to reproducibility issues in cell culture experiments.

2. Results

Formation of bioactive pro-inflammatory mediators by macrophages may be influenced by FCS batch effects.

In order to systematically investigate cell culture reproducibility issues resulting from FCS batch effects, a proteome profiling experiment using a U937 macrophage differentiation and activation model was performed. A single batch of U937 cells was seeded into 24 identical aliquots, forming four groups subsequently sub-cultured with four different FCS batches (Table 1, see Materials and Methods). All cells were differentiated using phorbol 12-myristate 13-acetate (PMA) to induce macrophage formation as verified by FACS (fluorescence activated cell sorting) analysis (Supplementary Figure S1), while three aliquots of each group (FCS batch) were subsequently treated with lipopolysaccharides (LPS) to induce inflammatory stimulation, the other three per group serving as untreated controls. The formation of inflammatory mediators was investigated by comparative secretome analysis resulting in the identification of 488 proteins (Supplementary Table S3) and 54 eicosanoids and fatty acid precursor molecules (Supplementary Table S4). Whereas most molecules such as the chemokine CCL3 and CXCL5 showed rather little variation between the groups, reproducibility issues of differentiated U937 macrophages were evidenced by FCS batch-dependent significant (FDR (false discovery rate) < 0.05) differences in the formation of the chemokine CCL5, the cell growth regulator IGFBP2, and the cell migration and fibrinolysis regulator SERPINE1 (PAI1) and MMP1 (Figure 1A). In line, the amount of bioactive eicosanoids comprising the hydroxyeicosatetraenoic acids 11-, 12-, and 15-HETE, hydroxydocosahexaenoic acid 17-HDoHE, the prostaglandin PGJ2 and others were found to differ significantly (FDR < 0.05) depending on the FCS batch used for cell culture (Figure 1B).

Table 1. Tested FCS batches stating Vender, Lot number, expiration date as well as letter used in this work.

Nomenclature	Vendor	Lot Number	Expiration Date	Origin	Processed
A	Sigma	BCBT4187	07.2021		-
B	Gibco	42Q5650K	06.2020	Brazil	-
C	Gibco	42G8378K	11.2022	Brazil	-
D	Gibco	08Q8082K	02.2023	Brazil	Heat inactivated

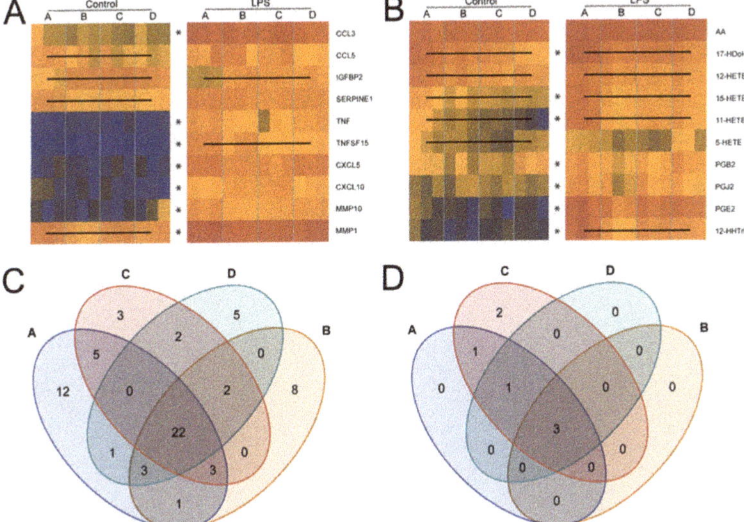

Figure 1. Heatmaps of selected proteins (**A**) and eicosanoids (**B**) determined in secretomes of control and LPS-treated U937 cells cultured with the indicated batch of FCS, A, B, C or D. Lines within a heatmap indicate a significant difference of the given molecule within at least two batches. Asterisks (*) indicate that LPS-treatment induced a significant increase. Venn diagrams of significantly up- and downregulated (**C**) proteins (S0 = 2, FDR = 0.01) and (**D**) eicosanoids comparing LPS activation with control samples for all four FCS batches (**A–D**).

Inflammatory stimulation with LPS induced the secretion of a total of 67 proteins (FDR < 0.05, Supplementary Table S3), including tumor necrosis factors TNF and TNFSF15, chemokines such as CCL3, CXCL5, CXCL10, metalloproteinases including MMP1 and MMP10 and other promoters and mediators of inflammation (Figure 1A and Supplementary Table S3). Only 22 of those 67 proteins were found uniformly regulated independent of the FCS batch (Figure 1C), whereas other bioactive molecules such as IGFBP2 and TNFSF15 again showed FCS batch-dependent expression patterns (Figure 1C). Similarly, LPS treatment induced the formation of lipid mediators of inflammation such as 15-HETE, PGE2, PGJ2 and others (Figure 1B, Supplementary Table S4). The formation of eicosanoids varied rather strongly dependent on the FCS batch used, four out of seven LPS-induced eicosanoids showing significant batch-dependent alterations (FDR < 0.05, Figure 1D).

2.1. The Eicosanoid Content of FCS Varies in a Batch-Dependent Fashion

The induction of inflammatory activities of cells may be subject to modulation by a delicate balance of pro- and anti-inflammatory molecules. Thus we investigated whether the above-described batch effects may be caused by differences of the protein and eicosanoid content of the FCS batches used for the cell culture experiments. Remarkably, the protein profile comprising 289 identified proteins (FDR < 0.01) of the four different FCS batches was rather consistent (Figure 2A, Supplementary Table S5). Several significant abundance differences between batches were observed (Supplementary Table S5), and a principle component analysis showed fairly good clustering of the FCS samples according to batches (Figure 2B). The analysis of FCS eicosanoid contents revealed even stronger batch-dependent differences, comprising mainly COX and LOX-products (Figure 2C, Supplementary Table S6). Here, an unbiased PCA clustered the FCS batches with clear distances between batch clusters (Figure 2D), and demonstrated that batch dependent differences of eicosanoid content exceeded the differences of protein content.

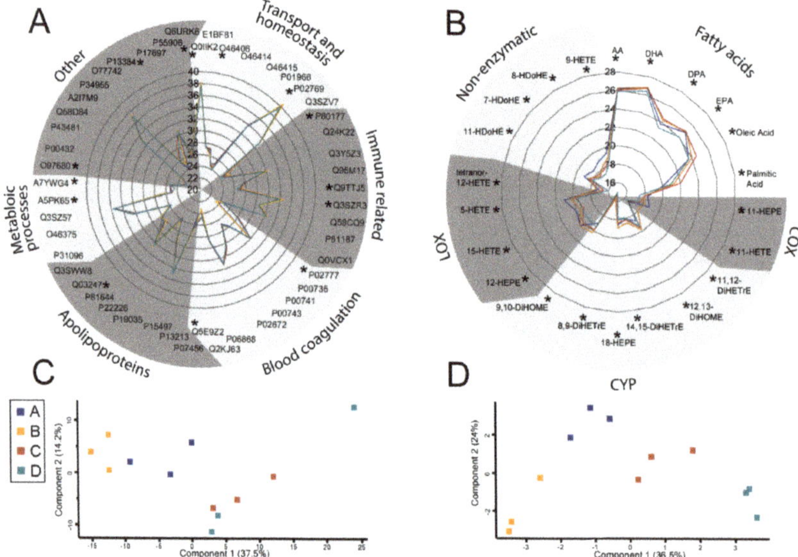

Figure 2. Radar plot for selected proteins (**A**) and fatty acids (**B**) identified in 4 different FCS batches without incubation with cells (baseline levels). Principal component analysis of protein (**C**) and eicosanoid (**D**) measurements of the same FCS batches, as indicated by different colors, demonstrates superior clustering in the case of eicosanoids. Asterisks mark significantly regulated molecules.

2.2. Cell Culture Subjects Fatty Acids to a High Turnover

Growing cells require medium supply of fatty acids and fetal calf serum is rich in polyunsaturated fatty acids. In order to mediate biological effects as assumed for the FCS-contained fatty acids described above (Figure 2B), cells are supposed to take up fatty acids from the medium. In order to estimate to what extent cultured U937 cells may be able to take up fatty acids from the medium and release fatty acids back into the medium upon stimulation, we investigated the intracellular to extracellular exchange dynamics of the eicosanoid precursor molecule arachidonic acid (AA). For this purpose, cultured U937 cells were spiked with stable isotope labelled AA at a concentration double that of the endogenous AA (1.6 µM). Stable isotope labelled AA can be clearly distinguished from endogenous AA by mass spectrometry. As demonstrated in Figure 3, upon differentiation to macrophages, U937 cells apparently picked up more than 99% of the labelled and spiked

AA within 72 h and less than 1% C13-AA remained detectable in the medium after that period. As expected, subsequent LPS treatment triggered a significant increase of the amount of C13-AA in the supernatant (Figure 3). This demonstrated that phospholipase activity was capable of releasing previously incorporated AA back into the medium. When supplementation with C13-AA was performed after PMA differentiation but before LPS treatment, the outcome was similar. Around 95% of the available AA was incorporated, but still a significant release of C13-AA was observed upon LPS treatment, clearly indicating a high turnover rate of AA. Concomitant measurement of endogenous C12-AA confirmed that AA was consumed substantially during cell culture and released back in the medium again in a smaller proportion upon LPS treatment.

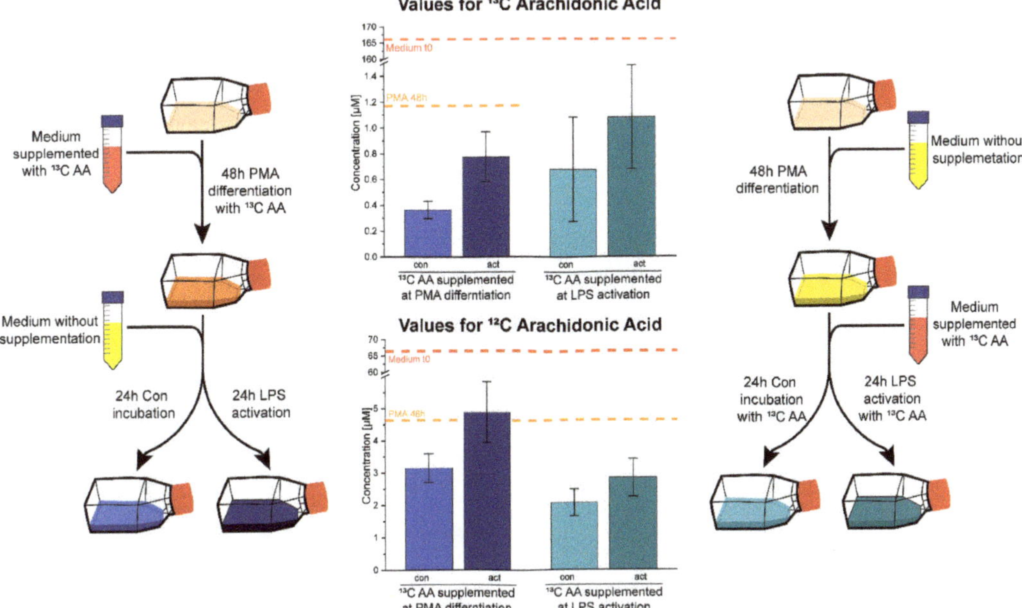

Figure 3. Experimental setup and results from AA spiking experiments. Medium was supplemented with 13C AA either before differentiation (left hand side) or before LPS treatment (right hand side). AA determination of cell supernatants by LC-MS/MS revealed AUC values as indicated. Medium levels at the beginning of cell culture are indicated by lines. Error bars are derived from three independent experiments. Con, untreated cells; act, LPS-treated cells. Note that AA concentration values strongly decrease upon cell cultivation but increase again upon LPS treatment.

2.3. Supplementation of 5-HETE and 15-HETE in the Nanomolar Concentration Range Induces the Formation of Peroxisomes in U937 Macrophages

In order to demonstrate that the detected differences in eicosanoid content of FCS could originate from the observed batch effects, we performed an additional proteome profiling experiment with U937 cells at conditions only differing with regard to two selected eicosanoids, 5-HETE and 15-HETE. To this aim, we supplemented the FCS batch found to have the lowest levels of these two eicosanoids, here designated FCS-B (containing 5 nM 5-HETE and 12 nM 15-HETE), with the pure chemicals to levels close to those observed in case of FCS batch A (FCS-A, containing 42 nM 5-HETE and 49 nM 15-HETE, Figure 4A). Thus, U937 cells were grown and differentiated as before, using either FCS-A, supplemented FCS-A, or FCS-B, and subsequently subjected to proteome profiling of cytoplasmic proteins (Figure 4B). Indeed, spiking in of the two HETEs was associated with distinct proteome alterations (Figure 4B) including down-regulation of PKM and up-regulation of PEX 16, a peroxisomal membrane biogenesis protein [18]. To independently verify with a complementary method that this was a relevant observation, peroxisome formation was analyzed using immunofluorescence with an anti-PMP70 antibody. Nuclei and mitochondria were additionally stained to demonstrate uniform appearance of these organelles serving as background control. Indeed, treatment of U937 cells with increasing concentration of HETEs induced the formation of peroxisomes in an apparently concentration dependent manner (Figure 4C).

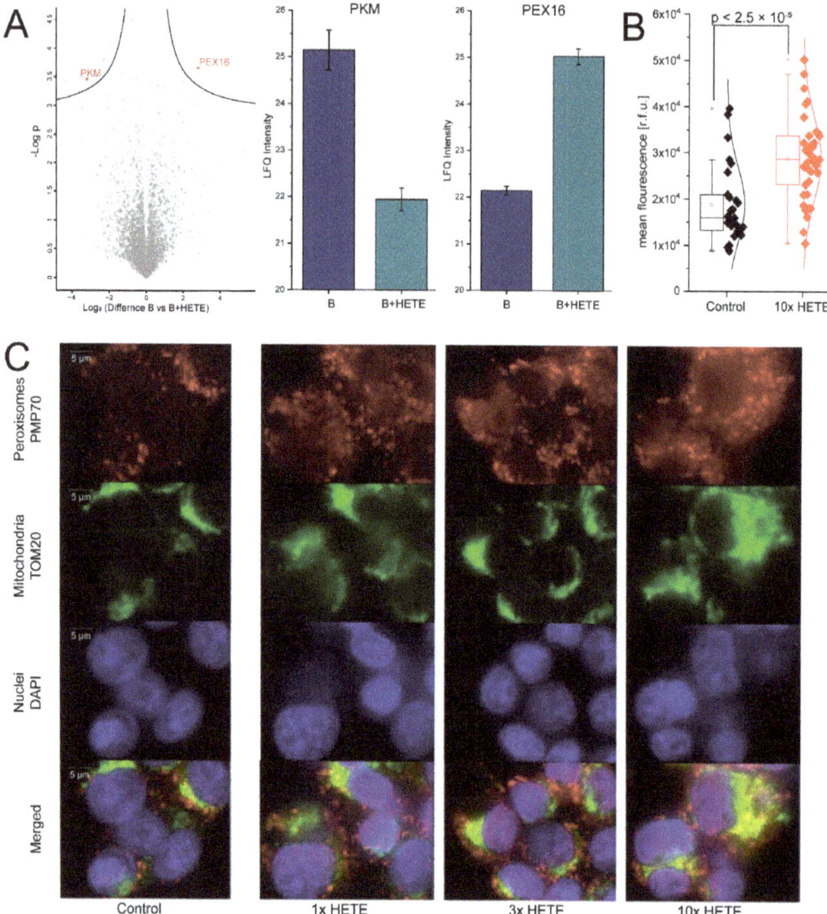

Figure 4. (**A**) Eicosanoid levels of 5-HETE and 15-HETE for FCS-B (before spiking), FCS-A and FCS-B after spiking with 5-HETE and 15-HETE. (**B**) Volcano plot for cytoplasmic proteins obtained from U937 cells after PMA differentiation when cultured in either FCS-B or FCS-B supplemented HETEs. Bar plots exemplify the significantly regulated proteins PKM and PEX16. (**C**) Immunofluorescence detection of peroxisomes (red, PMP70 antibody), mitochondria (green, TOM20 antibody) and nuclei (blue, DAPI) shown for control and increasing concentrations of supplemented HETEs (addition of 1, 3, or 10 times of the spiked HETE mix).

3. Discussion

The present data demonstrates that variations in the eicosanoid content of FCS may account for substantial batch effects with regard to functional readouts of a cell culture model reporting inflammatory mediators. This finding may be of great relevance for a large number of laboratories working with cell culture and using FCS, as FCS-contained eicosanoids have hardly been considered to have major implications for cell culture experiments and have thus, to the best of our knowledge, not yet been subjected to rigorous control. There are reasons, why relevant effects of eicosanoids contained in FCS were hardly expected. First, these molecules are generally considered to be short-lived and to act mainly in situ [19]. Second, eicosanoids were detected in FCS in the lower nanomolar concentration range, this is much less than the concentration range applied for functional assays in vitro, which is typically around 1 μM [20,21]. Furthermore, fatty acids including eicosanoids contained in serum are bound to albumin and only about 0.1% is actually free from associated molecules [22–24]. This free pool has a high turnover rate of about 2 min accounting for the redistribution of albumin-bound fatty acids in vivo to distant organs such as muscles or the liver.

When investigating FCS batch effects, we initially expected proteins to represent the most plausible candidates contributing to reproducibility issues. Proteins as well as metabolites are strictly regulated in vivo to ensure homeostasis and consequently stable viability of the organism. While proteins may be rather stable in biological environments, metabolites such as fatty acids are much more vulnerable to chemical reactions such as oxidation, which may occur also during processing of FCS and are hard to control. Thus it was somewhat unexpected to see that FCS eicosanoid profiles were stable and clustered the FCS samples according to batches (Figure 2). This finding, supported by older and current literature reporting remarkable biological effects of eicosanoids [25–27], motivated us to focus on this class of molecules. Functional analyses were based on spiking experiments with the U937 cells. As a first step, the efficient and fast uptake of albumin bound arachidonic acid (AA) was verified in the present cell model system using stable isotope labelled AA. The subsequent release of labelled AA upon LPS stimulation of the cells strongly indicated the previous uptake and incorporation of AA into more complex lipids, from where AA was apparently released by the action of LPS-induced phospholipase A2 [28]. In order to test potential biological effects of eicosanoids on U937 cells, a decision was made in favor of commercially available eicosanoids, 5-HETE and 15-HETE, which were found to show remarkable concentration differences among the FCS batches. Hydroxyeicosatetraenoic acids (HETEs) are formed with AA by the action of lipoxygenases ALOX5 and ALOX15, expressed typically by epithelial cells as well as phagocytes such as neutrophils and macrophages [29,30]. Beside their effects on cell proliferation and differentiation, they are known activators of PPARγ [31]. Actually, peroxisome proliferator-activated receptors are known to induce the uptake and metabolism of fatty acids and to strongly modulate immune functions [32]. As fatty acid metabolism takes place in peroxisomes [33], the 5-HETE/15-HETE induced up-regulation of PEX16 (Figure 4), a peroxisome biogenesis protein indicative for peroxisome proliferation [18], indicated that this treatment caused an increased demand for these organelles. The concomitant down-regulation of PKM (Figure 4), a key enzyme for glycolysis [34], may suggest that HETE-treatment of U937 cells induced a metabolic shift increasing beta oxidation and attenuating glycolysis. This interpretation was independently supported by the concentration-dependent HETE-induced formation of peroxisomes (Figure 4) observed by immunofluorescence staining using a PMP70 antibody [35].

4. Conclusions and Outlook

The present data demonstrate that batch-dependent differences of eicosanoids contained in FCS may have a profound effect on cellular functions as observed with the U937 in vitro cell culture model for differentiation and inflammatory stimulation. Eicosanoids affect many relevant cellular events far beyond that, suggesting that they may represent the main contributors for reproducibility issues in cell culture. The establishment of a strict quality control regime controlling eicosanoid content in FCS may alleviate this challenging problem.

5. Materials and Methods

5.1. Cell Culture

U937 cell line was cultured in RPMI medium (1X with L-Glutamine; Gibco, Thermo Fischer Scientific, Vienna, Austria) supplemented with 1% Penicillin/Streptomycin (Sigma-Aldrich, Austria) and 10% Fetal Calf Serum (FCS, Sigma-Aldrich, Vienna, Austria) in T25 polystyrene cell culture flasks for suspension cells (Sarstedt, Wiener Neudorf, Austria) at 37 °C and 5% CO_2. Cells were counted with a MOXI Z Mini Automated Cell Counter (ORFLO Technologies, Ketchum, ID, USA) using Moxi Z Type M Cassettes (ORFLO Technologies, Ketchum, ID, USA) and the number of seeded cells for the experiments calculated based of these results. For all experiments the cells were used in passages 22–26.

5.2. Differentiation with Phorbol 12-Myristate 13-Acetate (PMA) and Inflammatory Activation with Lipopolysaccharides (LPS)

All experiments were carried out in triplicates of LPS activation and control. For the proteomics and eicosadomics measurements 2×10^6 cells were seeded in T25 polystyrene cell culture flasks with cell growth surface for adherent cells (Sarstedt, Wiener Neudorf, Austria) with 5 mL fully supplemented media and 100 ng/mL PMA (Phorbol 12-myristate 13-acetate $\geq$ 99%, Sigma-Aldrich, Vienna, Austria) to induce differentiation. After 48 h incubation the medium was withdrawn and used for eicosanoid measurements. Three ml of new fully supplemented media was added either with 1 µg/mL LPS (Lipopolysaccharides from Escherichia coli 055:B5, γ-irradiated, BioXtra, Sigma-Aldrich, Vienna, Austria) or 1 µL PBS per 1 mL medium as control. After 24 h activation the medium was withdrawn again and used for eicosanoid measurements. The cells were gently washed twice with 5 mL phosphate buffered saline (PBS) and 3 mL new medium without FCS was added and incubated. After 4 h the supernatant was withdrawn and used for proteomics measurements. The cells were used for a subcellular fractionation as described before and cytoplasm and nuclear fraction were used for proteomics analysis [36].

5.3. Test of Different FCS Batches

Throughout the experiments different suppliers and batches of FCS were used. Additional details concerning the FCS batches are listed in Table 1. FCS batches A-C were heat inactivated at 56 °C for 30 min, batch D was already bought heat inactivated. Also different concentrations of HETEs and labelled arachidonic acid were supplemented and the respective controls treated with the same amount of LC-MS grade methanol (5 µL/3 mL medium). The experimental workflow of PMA differentiation and LPS activation was done for every condition similarly, only exchanging the FCS batch, supplier or eicosanoid. Additionally, for every condition 3 aliquots (3 × 3 mL) of the fully supplemented media were used for eicosanoid measurements to determine the default levels of eicosanoids present.

5.4. C13 Labelled Arachidonic Acid

For the investigation of the uptake and release of PUFAs an experiment was carried out with the supplementation of 1,2,3,4,5-^{13}C arachidonic acid (C13 AA, Cayman chemicals, Ann Arbor, Michigan, USA). Whenever C13 AA was added, a control experiment was supplemented with the same concentration of unlabeled arachidonic acid. For the first experiment the C13 AA was added at a concentration of 1.6 µM (used for all AA supplementations) to the fully supplemented medium during the 48 h PMA differentiation step. This concentration is around double that of the endogenous arachidonic acid, thus the supplementation tripled the concentration of biologically active arachidonic acid. Afterwards, the now adherent cells were washed three times with PBS and medium without supplemented AA was added together with or without LPS for 24 h. For the second experiment the cells were differentiated with PMA in standard medium, washed three times with PBS and medium supplemented with C13 AA was added together with and without LPS for 24 h. The eicosanoids were collected and measured in the supplemented medium without incubation (t0), after 48 h PMA differentiation and after 24 h LPS activation. The experimental setup is illustrated in Figure 3.

5.5. Proteomics of Supernatant (SN), Cytoplasm (CYT), and Nuclear Extract (NE)

For the proteomics sample preparation, the s-trap system (Protifi, Huntington, NY) was employed following the manufacturers protocol with slight modifications. The precipitated proteins were dissolved in lysis buffer (8 M Urea, 0.05 M triethylammonium bicarbonate (TEAB) and 5% sodium dodecyl sulfate SDS) and diluted to obtain a protein concentration of about 1 µg/µL. Twenty µg of protein was used for each digestion. First, the sample was reduced with 20 µL dithiothreitol DTT (Sigma-Aldrich) at a final concentration of 32 mM for 10 min at 95 °C. Afterwards, 5 µL iodoacetamide IAA (Sigma-Aldrich) was added to a final concentration of 54 mM and incubated for 30 min at 30 °C in the dark. After adding 4.5 µL 12% ortho-phosphoric acid (Sigma-Aldrich) and 297 µl S- Trap buffer (90% Methanol (v/v) in H_2O and 0.1 M TEAB) the sample was loaded onto the S-Trap Filter. The S-trap filters were centrifuged at 4000 xg for 1 min to pass through all the sample and trap the proteins onto the resin and afterwards washed four times with 150 µL S-Trap buffer. Twenty µg aliquots of Trypsin/Lys-C (MS grade; Promega Corporation, Madison, WI, USA) were dissolved in 400 µL 50 mM TEAB and 20 µg of this solution was added directly onto the resin of the filter (corresponding to 1 µg Trypsin/Lys-C per sample) and incubated for 1 h at 47 °C. After finishing the digestion, the peptides were eluted with 40 µL of 50 mM TEAB followed by 40 µL of 0.2% formic acid (FA) in H_2O and 35 µL of 50% (v/v) acetonitrile (ACN) with 0.2% FA in H_2O. The peptides were dried for about 2 h with vacuum centrifugation and stored at −20 °C until LC-MS/MS measurement.

5.6. HPLC-MS/MS for Proteomics

For the HPLC-MS/MS analysis the peptides were resolved in 5 µL 30% formic acid and diluted with 40 µL of mobile phase A (97.9% H_2O, 2% acetonitrile, 0.1% formic acid). One µL for the supernatant samples and 5 µL of cytoplasmic and nuclear samples were injected into the Dionex UltiMate 3000 RSLCnano liquid chromatography (LC) system coupled to the QExactive Orbitrap MS (all Thermo Fisher Scientific, Austria). Peptides were trapped on a C18 2 cm × 100 µm precolumn and LC separation was performed on a 50 cm × 75 µm Pepmap100 analytical column (both Thermo Fisher Scientific, Austria). Separation was achieved applying a 43 min gradient from 7% to 40% mobile phase B (79.9% acetonitrile, 20% H_2O, 0.1% formic acid) for supernatant samples and 95 min gradients from 8% to 40% mobile phase B for cytoplasmic and nuclear samples, both at a flow rate of 300 nL/min, resulting in a total run time of 85 min and 135 min, respectively. Mass spectrometric settings were the same for all fractions. The resolution on the MS1 level was set to 70,000 (at m/z = 200) with a scan range from 400 to 1400 m/z. The top eight abundant peptide ions were chosen for fragmentation at 30% normalized collision energy and resulting fragments analyzed in the Orbitrap at a resolution of 17,500 (at m/z = 200).

5.7. Proteomics Data Analysis

Raw data were subjected to the freely available software MaxQuant (version 1.6.0.1) [37] utilizing the Andromeda search engine, which returns label free quantification (LFQ) values for each identified protein as subsequently used for further data evaluation. For the MaxQuant search, a minimum of two peptide identifications, at least one of them being a unique peptide, was required for valid protein identification. Digestion mode was set to "Specific" choosing Trypsin/P. The peptide mass tolerance was set to 50 ppm for the first search and to 25 ppm for the main search. The false discovery rate (FDR) was set to 0.01 both on peptide and protein level. The database applied for the search was the human Uniprot database (version 03/2018, reviewed entries only) with 20,316 protein entries. Further settings for the search included carbamidomethylation as fixed modification and oxidation of methionine and acetylation of the protein C terminus as variable modifications. Each peptide was allowed to have a maximum of two missed cleavages and two modifications, "Match between runs" was enabled and the alignment window set to 10 min, with the match time window of 1 min. Statistical evaluation was performed with Perseus software (version 1.6.0.2) [38] using LFQ intensities of the MaxQuant result file. After filtering potential contaminants, the LFQ values were Log(2)-transformed. Technical duplicates were averaged. Only proteins detected in three of three biological replicates in either control and/or treatment groups were considered for data evaluation. Permutation-based FDR was set to 0.05 for *t*-tests and provided significant protein expression changes corrected for multi-parameters (S0 = 0.1). The mass spectrometry proteomics data were deposited in the ProteomeXchange Consortium (http://proteomecentral.proteomexchange.org) via the PRIDE partner repository [39] with the dataset identifier PXD020617 and 10.6019/PXD020617.

5.8. Eicosanoid Sample Preparation

Cell supernatants were spiked with 5 µL of internal standards (Supplementary Table S1) and centrifuged at 726 g for 5 min to remove cells and debris. Three ml of the supernatant was mixed with 12 mL of ice cold ethanol and left at −20 °C overnight to precipitate the contained proteins. The samples were centrifuged for 30 min with 4536 xg at 4 °C and the supernatant transferred into a new 15 mL Falcon tube. Ethanol was evaporated via vacuum centrifugation at 37 °C until the original sample volume was restored. Samples were loaded on conditioned 30 mg/mL StrataX solid phase extraction (SPE) columns (Phenomenex, Torrance, CA, USA). Columns were washed with 2 mL MS grade water and eicosanoids were eluted with 500 µL methanol (MeOH abs.; VWR International, Vienna, Austria) containing 2% formic acid (FA; Sigma-Aldrich). MeOH was evaporated using N_2 stream at room temperature and reconstituted in 150 µL reconstitution buffer (H2O/ACN/MeOH + 0.2% FA—65:31,5:3,5), containing a second set of internal eicosanoid standards at a concentration of 10–100 nM (Supplementary Table S1).

5.9. UHPLC-MS/MS for Eicosanoid Measurements

Analytes were separated using a Thermo Scientific Vanquish (UHPLC) system and a Kinetex® C18-column (2.6 µm C18 100 Å, LC Column 150 × 2.1 mm; Phenomenex®). Applying a 20 min gradient flow method, starting at 35% B steadily increasing to 90% B (1–10 min), going up to 99% B in 0.25 min. Flow rate was kept at 200 µL/min, 20 µL injection volume and column oven temperature was set to 40 °C. Eluent A contains H_2O + 0.2% FA and eluent B ACN:MeOH (90:10) + 0.2% FA.

Mass Spectrometric analysis was performed with a Q Exactive HF Quadrupole-Orbitrap mass spectrometer (Thermo Fisher Scientific, Austria), equipped with a HESI source for negative ionization. Mass spectra were recorded operating from 250 to 700 m/z at a resolution of 60,000 @ 200 m/z on MS1 level. The two most abundant precursor ions were selected for fragmentation (HCD 24 normalized collision energy), preferentially molecules from an inclusion list which contained 32 *m/z* values specific for eicosanoids (Supplementary Table S2). MS2 was operated at a resolution of 15,000 @ 200 m/z. For negative ionization, a spray voltage of 2.2 kV and a capillary temperature of 253 °C were applied, with the sheath gas set to 46 and the auxiliary gas to 10 arbitrary units.

Generated raw files were analyzed manually using Thermo Xcalibur 4.1.31.9 (Qual browser), comparing reference spectra from the Lipid Maps depository library from July 2018 [40]. For peak integration and quantitative data analysis the software TraceFinder™ (version 4.1-Thermo Scientific, Austria) was used. For the quantification of arachidonic acid (Figure 3), a calibration curve was generated (Supplementary Figure S4).

5.10. Immunofluorescence

For fluorescence microscopy 8 × 10^4 cells in 400 µL were seeded in a µ-Slide 8 well (Ibitreat coating, ibidi GmbH Martinsried, Germany). Differentiation of the cells was induced with 100 ng/mL PMA for 48 h. Afterwards the cell supernatant was exchanged with fully supplemented medium without PMA for an additional 24 h. Sample preparation was performed as previously described with minor modifications [41]. Cells were fixed with pre-warmed formaldehyde (3.7%) for 15 min and permeabilized with Triton-X 100 (0.2%) for 10 min. Blocking was performed with Donkey serum (2% in PBS-A) for 1 h, room temperature (RT). Primary antibodies were incubated 2h at RT at dilution 1:500. After washing, specie-specific fluorescent-labelled secondary antibodies were added and slides incubated in a dark humidified chamber for 1.5 h. For our study, Anti PMP70 Antibody (Rabbit polyclonal, PA1-650) and Anti TOM20 (F-10, Mouse Monoclonal Sc-17764), Alexa Fluor 488 Donkey Anti Mouse (A21202_LOT2090565) and Alexa Fluor 568 Donkey Anti-Rabbit (A10042_LOT2136776) were used. The slides were washed and post-fixed with 3.7% formaldehyde (10 min, RT); at the end of the post-fixation, 100 mM glycine was used to mask reactive sites and slides were mounted and sealed with Roti-Mount FluoCare (Roth, Graz, Austria) with DAPI. SIM Images were acquired with a Confocal LSM Zeiss 710 equipped with ELYRA PS. 1 system. Structured Illumination Microscopy (SIM) images were obtained with (Plan Apochromat 63X/1.4 oil objective) grid 5 rotation. For the quantification of fluorescence intensities (Figure 4B), 30 optical fields/region of interest (ROI) were quantified for every experimental condition from at least 3 independent experiments.

5.11. Differentiation Status by Flow Cytometry

In order to confirm the differentiation status obtained via PMA treatment the cells were tested for the differentiation marker CD11b (ITGAM) using FACS analysis. Therefore, U937 cells were treated with 100 ng/mL PMA for 48 h using 2×10^5 cells per well in 6-well plates. After the incubation time, the cells were washed three times with PBS and put on ice. The differentiation status was assessed by labelling with an anti-CD11b antibody (APC clone D12, BD Bioscience) and subsequent evaluation of the CD11b+ population. Three biological replicates were analyzed per condition on an FACS Canto II cytometer (BD Bioscience).

Supplementary Materials: The following are available online at https://www.mdpi.com/2218-273X/11/1/113/s1, Figure S1: FACS analysis of PMA-treated U937 cells, Table S1: Eicosanoid Standards, Table S2: Inclusion list of eicosanoids as used for the MS/MS analysis, Table S3: Results from proteomic analyses of U937 cells, Table S4: Results from eicosanoid analyses of U937 cells, Table S5: Results from proteomic analyses of FCS batches, Table S6: Results from eicosanoid analyses of FCS batches.

Author Contributions: L.N. performed research, analyzed and interpreted data, B.N. performed research, analyzed and interpreted data, J.B. performed research and analyzed data, L.J. performed research and analyzed data, A.B. interpreted data and wrote manuscript, G.D.F. performed research and analyzed data, C.G. conceptualized the project, interpreted data and wrote the manuscript. All authors have read and agreed to the published version of the manuscript.

Funding: This research received no external funding.

Data Availability Statement: The mass spectrometry proteomics data were deposited in the ProteomeXchange Consortium (http://proteomecentral.proteomexchange.org) via the PRIDE partner repository [39] with the dataset identifier PXD020617 and 10.6019/PXD020617.

Acknowledgments: We acknowledge support by the Mass Spectrometry Center and the Core Facility Multimodal Imaging of the Faculty of Chemistry, University of Vienna, as well as support by the Joint Metabolome Facility of the University of Vienna and Medical University of Vienna, all members of the Vienna Life Science Instruments (VLSI).

Conflicts of Interest: The authors declare no conflict of interest.

Abbreviations

AA	Arachidonic acid
FCS	Fetal calf serum
FFA	Free fatty acids
CDM	Chemically defined medium
FBS	Fetal bovine serum
HETE	Hydroxyeicosatetraenoic Acid
LC	Liquid chromatography
LPS	Lipopolysaccharides
MS	Mass spectrometry
MS/MS	Tandem mass spectrometry
PMA	Phorbol 12-myristate 13-acetate
PUFA	Polyunsaturated fatty acids

References

1. Prinz, F.; Schlange, T.; Asadullah, K. Believe it or not: How much can we rely on published data on potential drug targets? *Nat. Rev. Drug Discov.* **2011**, *10*, 712. [CrossRef] [PubMed]
2. Baker, M. Reproducibility: Respect your cells! *Nature* **2016**, *537*, 433–435. [CrossRef] [PubMed]
3. Wolfensohn, S. A review of the contributions of cross-discipline collaborative European IMI/EFPIA research projects to the development of Replacement, Reduction and Refinement strategies. *Altern. Lab. Anim.* **2018**, *46*, 91–102. [CrossRef] [PubMed]
4. Del Favero, G.; Kraegeloh, A. Integrating Biophysics in Toxicology. *Cells* **2020**, *9*, 1282. [CrossRef] [PubMed]
5. Freedman, L.P.; Cockburn, I.M.; Simcoe, T.S. The Economics of Reproducibility in Preclinical Research. *PLoS Biol.* **2015**, *13*, e1002165. [CrossRef]
6. Ben-David, U.; Siranosian, B.; Ha, G.; Tang, H.; Oren, Y.; Hinohara, K.; Strathdee, C.A.; Dempster, J.; Lyons, N.J.; Burns, R.; et al. Genetic and transcriptional evolution alters cancer cell line drug response. *Nature* **2018**, *560*, 325–330. [CrossRef]
7. Liu, Y.; Yang, M.; Mueller, T.; Kreibich, S.; Williams, E.G.; Van Drogen, A.; Borel, C.; Frank, M.; Germain, P.; Bludau, I.; et al. Multi-omic measurements of heterogeneity in HeLa cells across laboratories. *Nat. Biotechnol.* **2019**, *37*, 314–322. [CrossRef]
8. Gstraunthaler, G.; Lindl, T.; van der Valk, J. A plea to reduce or replace fetal bovine serum in cell culture media. *Cytotechnology* **2013**, *65*, 791–793. [CrossRef]
9. van der Valk, J.; Bieback, K.; Buta, C.; Cochrane, B.; Dirks, W.; Fu, J.; Hickman, J.; Hohensee, C.; Kolar, R.; Liebsch, M.; et al. Fetal Bovine Serum (FBS): Past–Present–Future. *ALTEX* **2018**, *35*, 99–118. [CrossRef]
10. Sikora, M.J.; Johnson, M.D.; Lee, A.V.; Oesterreich, S. Endocrine Response Phenotypes Are Altered by Charcoal-Stripped Serum Variability. *Endocrinology* **2016**, *157*, 3760–3766. [CrossRef]
11. Dennis, E.A.; Norris, P.C. Eicosanoid storm in infection and inflammation. *Nat. Rev. Immunol.* **2015**, *15*, 511–523. [CrossRef] [PubMed]
12. Tahir, A.; Bileck, A.; Muqaku, B.; Niederstaetter, L.; Kreutz, D.; Mayer, R.L.; Wolrab, D.; Meier-Menches, S.M.; Slany, A.; Gerner, C. Combined Proteome and Eicosanoid Profiling Approach for Revealing Implications of Human Fibroblasts in Chronic Inflammation. *Anal. Chem.* **2017**, *89*, 1945–1954. [CrossRef] [PubMed]
13. Reichl, B.; Niederstaetter, L.; Boegl, T.; Neuditschko, B.; Bileck, A.; Gojo, J.; Buchberger, W.; Peyrl, A.; Gerner, C. Determination of a Tumor-Promoting Microenvironment in Recurrent Medulloblastoma: A Multi-Omics Study of Cerebrospinal Fluid. *Cancers* **2020**, *12*, 1350. [CrossRef] [PubMed]
14. Muqaku, B.; Pils, D.; Mader, J.C.; Stefanie, A.; Mangold, A.; Muqaku, L.; Slany, A.; Del Favero, G.; Gerner, C. Neutrophil Extracellular Trap Formation Correlates with Favorable Overall Survival in High Grade Ovarian Cancer. *Cancers* **2020**, *12*, 505. [CrossRef]
15. Sundstrom, C.; Nilsson, K. Establishment and characterization of a human histiocytic lymphoma cell line (U-937). *Int. J. Cancer* **1976**, *17*, 565–577. [CrossRef] [PubMed]
16. Haque, M.A.; Jantan, I.; Harikrishnan, H. Zerumbone suppresses the activation of inflammatory mediators in LPS-stimulated U937 macrophages through MyD88-dependent NF-kappaB/MAPK/PI3K-Akt signaling pathways. *Int. Immunopharmacol.* **2018**, *55*, 312–322. [CrossRef] [PubMed]
17. Tian, X.; Xie, G.; Ding, F.; Zhou, X. LPS-induced MMP-9 expression is mediated through the MAPKs-AP-1 dependent mechanism in BEAS-2B and U937 cells. *Exp. Lung Res.* **2018**, *44*, 217–225. [CrossRef]
18. Kim, P.K.; Mullen, R.T.; Schumann, U.; Lippincott-Schwartz, J. The origin and maintenance of mammalian peroxisomes involves a de novo PEX16-dependent pathway from the ER. *J. Cell Biol.* **2006**, *173*, 521–532. [CrossRef]
19. Yao, C.; Narumiya, S. Prostaglandin-cytokine crosstalk in chronic inflammation. *Br. J. Pharmacol.* **2019**, *176*, 337–354. [CrossRef] [PubMed]
20. Zhang, L.; Chen, Y.; Li, G.; Chen, M.; Huang, W.; Liu, Y.; Li, Y. TGF-beta1/FGF-2 signaling mediates the 15-HETE-induced differentiation of adventitial fibroblasts into myofibroblasts. *Lipids Health Dis.* **2016**, *15*, 2. [CrossRef] [PubMed]

21. Zhang, L.; Ma, J.; Shen, T.; Wang, S.; Ma, C.; Liu, Y.; Ran, Y.; Wang, L.; Liu, L.; Zhu, D. Platelet-derived growth factor (PDGF) induces pulmonary vascular remodeling through 15-LO/15-HETE pathway under hypoxic condition. *Cell. Signal.* **2012**, *24*, 1931–1939. [CrossRef] [PubMed]
22. van der Vusse, G.J. Albumin as fatty acid transporter. *Drug Metab. Pharmacokinet.* **2009**, *24*, 300–307. [CrossRef]
23. Saifer, A.; Goldman, L. Free Fatty Acids Bound to Human Serum Albumin. *J. Lipid Res.* **1961**, *2*, 268–270. [CrossRef]
24. Vorum, H.; Pedersen, A.O.; Honore, B. Fatty-Acid and Drug-Binding to a Low-Affinity Component of Human Serum-Albumin, Purified by Affinity-Chromatography. *Int. J. Pept. Prot. Res.* **1992**, *40*, 415–422. [CrossRef] [PubMed]
25. Umamaheswaran, S.; Dasari, S.K.; Yang, P.; Lutgendorf, S.K.; Sood, A.K. Stress, inflammation, and eicosanoids: An emerging perspective. *Cancer Metastasis Rev.* **2018**, *37*, 203–211. [CrossRef]
26. Harizi, H.; Gualde, N. The impact of eicosanoids on the crosstalk between innate and adaptive immunity: The key roles of dendritic cells. *Tissue Antigens* **2005**, *65*, 507–514. [CrossRef]
27. Hammock, B.D.; Wang, W.; Gilligan, M.M.; Panigrahy, D. Eicosanoids: The Overlooked Storm in Coronavirus Disease 2019 (COVID-19)? *Am. J. Pathol.* **2020**, *190*, 1782–1788. [CrossRef]
28. Dieter, P.; Ambs, P.; Fitzke, E.; Hidaka, H.; Hoffmann, R.; Schwende, H. Comparative studies of cytotoxicity and the release of TNF-alpha, nitric oxide, and eicosanoids of liver macrophages treated with lipopolysaccharide and liposome-encapsulated MTP-PE. *J. Immunol.* **1995**, *155*, 2595–2604.
29. Kuhn, H.; O'Donnell, V.B. Inflammation and immune regulation by 12/15-lipoxygenases. *Prog. Lipid Res.* **2006**, *45*, 334–356. [CrossRef]
30. Powell, W.S.; Rokach, J. Biochemistry, biology and chemistry of the 5-lipoxygenase product 5-oxo-ETE. *Prog. Lipid Res.* **2005**, *44*, 154–183. [CrossRef]
31. Powell, W.S.; Rokach, J. Biosynthesis, biological effects, and receptors of hydroxyeicosatetraenoic acids (HETEs) and oxoeicosatetraenoic acids (oxo-ETEs) derived from arachidonic acid. *Biochim. Biophys. Acta* **2015**, *1851*, 340–355. [CrossRef] [PubMed]
32. Christofides, A.; Konstantinidou, E.; Jani, C.; Boussiotis, V.A. The role of Peroxisome Proliferator-Activated Receptors (PPAR) in immune responses. *Metabolism* **2020**, 154338. [CrossRef]
33. Fransen, M.; Nordgren, M.; Wang, B.; Apanasets, O. Role of peroxisomes in ROS/RNS-metabolism: Implications for human disease. *Biochim. Biophys. Acta* **2012**, *1822*, 1363–1373. [CrossRef]
34. Sato, T.; Morita, M.; Nomura, M.; Tanuma, N. Revisiting glucose metabolism in cancer: Lessons from a PKM knock-in model. *Mol. Cell. Oncol.* **2018**, *5*, e1472054. [CrossRef] [PubMed]
35. Zaabi, N.A.; Kendi, A.; Al-Jasmi, F.; Takashima, S.; Shimozawa, N.; Al-Dirbashi, O.Y. Atypical PEX16 peroxisome biogenesis disorder with mild biochemical disruptions and long survival. *Brain Dev.* **2019**, *41*, 57–65. [CrossRef] [PubMed]
36. Kreutz, D.; Bileck, A.; Plessl, K.; Wolrab, D.; Groessl, M.; Keppler, B.K.; Meier-Menches, S.M.; Gerner, C. Response Profiling Using Shotgun Proteomics Enables Global Metallodrug Mechanisms of Action To Be Established. *Chem. A Eur. J.* **2017**, *23*, 1881–1890. [CrossRef] [PubMed]
37. Cox, J.; Mann, M. MaxQuant enables high peptide identification rates, individualized p.p.b.-range mass accuracies and proteome-wide protein quantification. *Nat. Biotechnol.* **2008**, *26*, 1367–1372. [CrossRef] [PubMed]
38. Tyanova, S.; Temu, T.; Sinitcyn, P.; Carlson, A.; Hein, M.Y.; Geiger, T.; Mann, M.; Cox, J. The Perseus computational platform for comprehensive analysis of (prote)omics data. *Nat. Methods* **2016**, *13*, 731–740. [CrossRef]
39. Vizcaino, J.A.; Deutsch, E.W.; Wang, R.; Csordas, A.; Reisinger, F.; Ríos, D.; Dianes, J.A.; Sun, Z.; Farrah, T.; Bandeira, N.; et al. ProteomeXchange provides globally coordinated proteomics data submission and dissemination. *Nat. Biotechnol.* **2014**, *32*, 223–226. [CrossRef]
40. Fahy, E.; Sud, M.; Cotter, D.; Subramaniam, S. LIPID MAPS online tools for lipid research. *Nucleic Acids Res.* **2007**, *35*, W606–W612. [CrossRef]
41. Del Favero, G.; Hohenbichler, J.; Mayer, R.M.; Rychlik, M.; Marko, D. Mycotoxin Altertoxin II Induces Lipid Peroxidation Connecting Mitochondrial Stress Response to NF-kappaB Inhibition in THP-1 Macrophages. *Chem. Res. Toxicol.* **2020**, *33*, 492–504. [CrossRef] [PubMed]

biomolecules

MDPI

Article

Epithelial Cell Line Derived from Endometriotic Lesion Mimics Macrophage Nervous Mechanism of Pain Generation on Proteome and Metabolome Levels

Benjamin Neuditschko [1,2,†], Marlene Leibetseder [1,†], Julia Brunmair [1], Gerhard Hagn [1], Lukas Skos [1], Marlene C. Gerner [3], Samuel M. Meier-Menches [1,2,4], Iveta Yotova [5] and Christopher Gerner [1,4,*]

1 Department of Analytical Chemistry, Faculty of Chemistry, University of Vienna, Waehringer Straße 38, 1090 Vienna, Austria; benjamin.neuditschko@univie.ac.at (B.N.); marlene_leibetseder@gmx.at (M.L.); julia.brunmair@univie.ac.at (J.B.); gerhard.hagn@univie.ac.at (G.H.); lukas.skos@univie.ac.at (L.S.); samuel.meier@univie.ac.at (S.M.M.-M.)
2 Department of Inorganic Chemistry, Faculty of Chemistry, University of Vienna, Waehringer Straße 42, 1090 Vienna, Austria
3 Division of Biomedical Science, University of Applied Sciences, FH Campus Wien, Favoritenstraße 226, 1100 Vienna, Austria; marlene.gerner@fh-campuswien.ac.at
4 Joint Metabolome Facility, Faculty of Chemistry, University of Vienna, Waehringer Straße 38, 1090 Vienna, Austria
5 Department of Obstetrics and Gynaecology, Medical University of Vienna, Spitalgasse 23, 1090 Vienna, Austria; iveta.yotova@meduniwien.ac.at
* Correspondence: christopher.gerner@univie.ac.at
† Authors contributed equally.

check for updates

Citation: Neuditschko, B.; Leibetseder, M.; Brunmair, J.; Hagn, G.; Skos, L.; Gerner, M.C.; Meier-Menches, S.M.; Yotova, I.; Gerner, C. Epithelial Cell Line Derived from Endometriotic Lesion Mimics Macrophage Nervous Mechanism of Pain Generation on Proteome and Metabolome Levels. *Biomolecules* **2021**, *11*, 1230. https://doi.org/10.3390/biom11081230

Academic Editors: Peter Roman Jungblut and Vladimir N. Uversky

Received: 20 July 2021
Accepted: 13 August 2021
Published: 17 August 2021

Abstract: Endometriosis is a benign disease affecting one in ten women of reproductive age worldwide. Although the pain level is not correlated to the extent of the disease, it is still one of the cardinal symptoms strongly affecting the patients' quality of life. Yet, a molecular mechanism of this pathology, including the formation of pain, remains to be defined. Recent studies have indicated a close interaction between newly generated nerve cells and macrophages, leading to neurogenic inflammation in the pelvic area. In this context, the responsiveness of an endometriotic cell culture model was characterized upon inflammatory stimulation by employing a multi-omics approach, including proteomics, metabolomics and eicosanoid analysis. Differential proteomic profiling of the 12-Z endometriotic cell line treated with TNFα and IL1β unexpectedly showed that the inflammatory stimulation was able to induce a protein signature associated with neuroangiogenesis, specifically including neuropilins (NRP1/2). Untargeted metabolomic profiling in the same setup further revealed that the endometriotic cells were capable of the autonomous production of 7,8-dihydrobiopterin (BH2), 7,8-dihydroneopterin, normetanephrine and epinephrine. These metabolites are related to the development of neuropathic pain and the former three were found up-regulated upon inflammatory stimulation. Additionally, 12-Z cells were found to secrete the mono-oxygenated oxylipin 16-HETE, a known inhibitor of neutrophil aggregation and adhesion. Thus, inflammatory stimulation of endometriotic 12-Z cells led to specific protein and metabolite expression changes suggesting a direct involvement of these epithelial-like cells in endometriosis pain development.

Keywords: endometriosis; inflammation; metabolomics; multi-omics; proteomics

1. Introduction

Endometriosis is a chronic inflammatory disease describing the abnormal growth of uterine tissue outside of the uterine cavity in the pelvic area [1]. Endometriotic cells are characterized by invasive phenotypes. They successfully attach to pelvic organs and cause pelvic inflammation [2,3]. Studies estimate that the disease is affecting about 1 in 10 women worldwide. The clinical symptoms include dysmenorrhea, dyspareunia, chronic pelvic pain and infertility. To date, there exists no curative treatment [1]. As one of the cardinal symptoms, pain strongly affects the patients' quality of life and can only be treated symptomatically so far. A large fraction of the published studies focused on phenotypic investigations and the molecular mechanisms, especially those associated with pain development, remain to be fully elucidated [4].

According to the rAFS/ASRM system, the extent of the disease does not correlate with the pain level experienced by individual patients [5]. The severity of the pain sensation seems to be connected to a mixture of neuropathy, neurogenic inflammation, nociception and hyperalgesia [6]. It is known that the lesion's surrounding nerves are infiltrated and compressed by endometriotic cells [7]. Moreover, there is evidence that nascent nerve cells can be attracted by the endometriotic lesion. Their inflammatory activated state may also directly cause and transmit pain stimuli to the central nervous system [8]. Some nerve cells may even release proinflammatory factors, which ultimately lead to neurogenic inflammation and increased local vascular permeability, enhancing migration of ectopic endometrial cells [9–11]. Furthermore, it is well accepted that endometriosis-associated pain is directly linked to dorsal root ganglion neurons [12].

Immunological dysregulation seems to represent a main pathogenesis driver in endometriosis [13]. Normally, cell-mediated immune responses contribute to the elimination of immune invaders and clearance of ectopic endometrial tissue. In endometriosis, the clearance of endometrial tissue in the peritoneal cavity is abolished due to an impaired immune response at the site of implantation [14,15]. Deregulated T-cell immunity and a suppressed activity of macrophages and NK cells were found to contribute to this process [16]. The activation of an inflammatory response in the peritoneum of women with the disease leads to local production of cytokines, chemokines and growth factors that enhance the growth of the ectopic endometrial tissue by inhibiting normal apoptotic processes and promoting local angiogenesis and neurogenesis [17]. The macrophage-nervous axis in endometriosis is commonly accepted to be the main cause for disease-associated pain [18]. Indeed, nerve infiltration is positively correlated with high density of tissue-resident macrophages in the lesion [19]. Alternatively, these nerve fibres are attracted by the action of semaphorins. As semaphorins normally regulate axon migration, axonal growth and guidance, altered semaphorin secretion may lead to aberrant nervous innervation in endometriosis [18]. Infiltrating the endometriotic lesions, they secrete neuroangiogenic factors and create a neovasculative environment [20]. Herein, especially VEGF secretion plays an important role in axonal outgrowth functioning as a neurotrophic factor [21,22]. Once neuroangiogenesis and infiltration in nerve fibres was initiated by aberrant inflammatory signaling, the endometriotic lesion may be create its own altered microenvironment [23].

Macrophages secrete tumour necrosis factor α (TNFα) and interleukin-1β (IL1β) that contribute to disease progression [24,25]. These inflammatory cytokines can mediate neurogenic inflammation and secretion of further neuroangiogenic factors [26,27]. TNFα signaling increases the transient receptor potential vanilloid 1 (TRPV1) nociception in the dorsal root neurons which contributes to hyperalgesia and neuropathic pain sensation [28,29]. It was also found to sensitize sensory nerves to a constant induction of the action potential via TRPV1 in patients, mainly through overexpression of voltage-gated sodium channels [27]. TRPV1 expression is also increased in uterosacral ligaments in endometriosis patients [30] and it has been shown that pain is often driven by dorsal root ganglion neurons, in association with TRPV1 [12]. Neurogenesis seems to be at least partially responsible for neuropathic pain experiences [31,32]. Therefore, the molecular mechanisms of neurogenesis in endometriosis need to be comprehensively characterized to contribute towards novel therapy options in endometriosis pain management.

During the last years proteomics and metabolomics analysis were applied to uncover markers for early detection of endometriosis and to understand the molecular changes associated with its pathophysiology [33–35]. However, the use of these omics techniques is still sparse in endometriosis and multi-omics profiling was not yet applied, to the best of our knowledge. Combining proteomic with metabolomic profiling is especially attractive, since it may support a functional interpretation of the involved pathways. In addition, signaling lipids are key players during inflamation and act in a concerted fashion with proteins [36,37]. Thus, we have applied an in-depth proteome, metabolome and eicosanoid profiling of the endometriotic epithelial cell line 12-Z to investigate and characterize their responsiveness towards inflammatory signals. The cell line was isolated from a patient with peritoneal endometriosis and immortalized by Starzinski-Powitz [38]. 12-Z was characterized as epithelial-like (cytokeratin-positive/E-cadherin negative) and using a matrigel assay it was shown that the cell line was highly invasive [38]. To simulate the inflammatory macrophage signaling, the cells were treated with the cytokines TNFα or IL1β, which are typically upregulated in endometriotic lesions of patients [39]. We provide proteomic and metabolic evidence that the endometriotic 12-Z cells may independently express key mediators of neuroangiogenesis and neuropathic pain.

2. Material & Methods

2.1. Cell Culture

The 12-Z cell line was a generous gift of Dr. Anna Starzinski-Powitz (Goethe-Universität Frankfurt) [38]. Human epithelial endometriotic cell line 12-Z was cultivated in DMEM-F12 phenolredfree (Gibco, Thermo Fisher Scientific, Vienna, Austria) with 10% (v/v) heat inactivated fetal calf serum (FCS, SigmaAldrich, Vienna, Austria) and 1% (v/v) penicillin and streptomycin (Sigma-Aldrich, Vienna, Austria). Cultivation was done in humidified incubators at 37 °C and 5% CO_2. The 12-Z cell line was grown in T75 polystyrene cell culture flasks with cell growth surface for adherent cells (Sarstedt, Wiener Neudorf, Austria). Cells were sub-cultured every 3–4 days at a 1:3 ratio. Cells were cultivated until they reached a concentration of 80% confluency. Cell counting was performed using a MOXI Z mini automated cell counter (ORFLO Technologies, Ketchum, ID, USA) using Moxi Z Type M cassettes (ORFLO Technologies, Ketchum, ID, USA). Cells were routinely checked for mycoplasma contamination using MycoAlert™ mycoplasma detection kit (Szabo-Scanidc, Vienna, Austria). Cells were seeded on a 6-well plate with cell growth surface for adherent cells (Sarstedt, Wiener Neudorf, Austria) at a density of 300.000 cells/well in 3 mL growth medium. Inflammatory stimulation was applied for 48 h at a concentration of 10 ng mL^{-1} with either TNFα or IL1β (both Sigma-Aldrich).

2.2. Proteomics

After the indicated treatment, the cells were washed twice with PBS and fractionation was performed as previously described [40]. Isotonic lysis buffer (1 mL of 10mM HEPES/NaOH, pH 7.4, 0.25 M sucrose, 10 mM NaCl, 3 mM $MgCl_2$, 0.5% Triton × 100, protease and phosphatase inhibitor cocktail (Sigma-Aldrich)) was added to the cells, which were scraped using a cell scraper. Cell lysis was achieved using mechanical shear stress employing a syringe and a needle. After centrifugation at 2270× g for 5 min supernatant containing cytoplasmic proteins was precipitated in 4 mL of cold EtOH overnight.

Protein sample preparation: The ethanolic protein suspension was centrifuged at 4536g for 30 min (4 °C). The supernatant was discarded, while the protein pellet was dried and resuspended in lysis buffer (8 M urea, 50 mM TEAB, 5% SDS). Protein concentration was determined using the bicinchoninic acid assay. An aliquot of the samples containing 20 µg protein was digested using a modified Protifi protocol [41]. In short, the samples were diluted to a concentration of 1 µg μL^{-1}. The diluted sample (20 µg in 20 µL) were pipetted into a 1.5 mL Eppendorf microcentrifuge tube and 20 µL of DTT (64 mM) was added. The samples were heated for 10 min at 95 °C under constant shaking (300 rpm). The samples were cooled to room temperature, treated with iodacetamide (5 µL of 486 mM solution, and incubated in the dark for 30 min at 30 °C and 300 rpm. Afterwards, phosphoric acid (4.5 µL of 12%) was added, resulting in 1.2% final concentration of phosphoric acid. S-Trap buffer (297 µL, 90% MeOH (v/v), 0.1 M TEAB) was added to the solution. The sample was loaded on the Protifi S-Trap column and washed 4 times with S-trap buffer (150 µL). Trypsin/LysC (MS grade; Promega Corporation, Madison, WI, USA) was added in a 1:40

ratio (0.5 μg for 20 μg protein). After digestion for 2 h at 37 °C peptides were eluted, dried using a SpeedVac and stored until further analysis.

Data acquisition: Dried peptides were reconstituted in 5 μL 30% formic acid, containing four synthetic peptides, for quality control [42]. The samples were further diluted with 40 μL mobile phase A (97.9% H_2O, 2% acetonitrile, 0.1% formic acid). Peptides were analyzed with a Dionex UltiMate 3000 Nano LC system coupled to a Q Exactive Orbitrap mass spectrometer (Thermo Fisher Scientific, Vienna, Austria) using a previously published method [40]. In short, peptides were separated on a 50 cm × 75 μm PepMap100 analytical column (Thermo Fisher Scientific), at a flow rate of 300 nL min^{-1}. Gradient elution of the peptides was achieved by increasing the mobile phase B (79.9% acetonitrile, 20% H_2O, 0.1% formic acid) from 8% to 40%, with a total chromatographic run time of 135 min including washing and equilibration. Mass spectrometric resolution on the MS1 level was set to 70,000 (at *m/z* 200) with a scan range from *m/z* 400–1400. The 12 most abundant peptide ions were selected for fragmentation (Top12) at 30% normalized collision energy and analyzed in the Orbitrap at a resolution of 17,500 (at *m/z* 200).

Data analysis: Data was analyzedanalyzed in settings as previously described [42]. Briefly, raw data was submitted to the freely available software MaxQuant (version 1.6.6.0) [43] utilizing the Andromeda search engine. A minimum of two peptide identifications, at least one of them being a unique peptide, was required for valid protein identification. The false discovery rate (FDR) was set to 0.01 on both peptide and protein level. Uniprot database (Human, version 03/2018, reviewed entries only, 20,316 protein entries) was used to generate the fasta file used for the search. For statistical data evaluation MaxQuant companion software Perseus (version 1.6.1.0) was used. Reverse sequences and potential contaminants as well as proteins identified only by site were removed. Label-free quantification (LFQ) values were converted to Log2(x) and technical replicates averaged. Proteins were filtered for valid values, keeping only proteins that were identified in at least three measurements of one sample group. Evaluation of regulatory events between different samples groups was achieved by two-sided t-tests using a FDR < 0.05 calculated by permutation-based test [44,45]. Significantly regulated proteins were further analyzed using the Cytoscape [46] plugin ClueGo [47] with default settings. Gene ontology for Biological Processes (GOBP) was used as search space with medium network specificity. Statistical options were set to two-sided hypergeometriy test with Bonferroni step down p-value correction. String [48] analysis was further used to display protein network connections of regulated proteins. STRING protein query of species homo sapiens was used with a confidence (score) cut-off of 0.4 and with 0 additional interactors allowed. The mass spectrometry proteomics data have been deposited to the ProteomeXchange Consortium via the PRIDE [49] partner repository with the dataset identifier PXD022354 and 10.6019/PXD022354.

2.3. *Eicosanoid Analysis*

Eicosanoid sample collection and preparation: Supernatants from cell culture experiments or medium (3 mL) was added to cold ethanol (12 mL, EtOH, abs. 99%, −20 °C; AustroAlco) containing an internal standard mixture of 12S-HETE-d8, 15S-Hete-d8, 5-Oxo-ETE-d7, 11.12-DiHETrE-d11, PGE-d4 and 20-HETE-d6 (each 100 nM; Cayman Europe, Tallinn, Estonia). The samples were stored over-night at −20 °C. After centrifugation (30 min, 5000 rpm, 4 °C), the supernatant was transferred to a new 15 mL Falcon™ tube and EtOH was evaporated via vacuum centrifugation (37 °C) until the original sample volume was accomplished. Samples were loaded on preconditioned StrataX solid phase extraction (SPE) columns (30 mg mL^{-1}; Phenomenex, Torrance, CA, USA) using Pasteur pipettes. SPE columns were washed with MS grade water (3 mL) and elution of eicosanoids was performed with ice-cold methanol (500 μL, MeOH abs.; VWR International, Vienna, Austria) containing 2% formic acid (FA; Sigma-Aldrich). MeOH was evaporated under a gentle stream of nitrogen at room temperature and the samples were reconstituted in 150 μL reconstitution buffer (H_2O:ACN:MeOH + 0.2% FA–vol% 65:31.5:3.5), including a second mixture of internal standards, including 5S-HETE-d8, 14.15-DiHETrE-d11 and 8-iso-PGF2a-d4 (10–100 nM; Cayman Europe, Tallinn, Estonia). Reconstituted samples were stored at +4 °C and measured subsequently via LC-MS/MS.

Data acquisition: Separation of eicosanoids was performed on a Thermo Scientific™ Vanquish™ (UHPLC) system equipped with a Kinetex® C18-column (2.6 µm C18 100 Å, LC Column 150 × 2.1 mm; Phenomenex®). All samples were analyzed in technical duplicates. The injection volume was 20 µL and the flow rate was kept at 200 µL min^{-1}. The UHPLC method included a gradient flow profile (mobile phase A: H_2O + 0.2% FA, mobile phase B: ACN:MeOH (vol% 90:10) + 0.2% FA) starting at 35% B and increasing to 90% B (1–10 min), further increasing to 99% B within 0.5 min and held for 5 min. Afterwards solvent B was decreased to the initial level of 35% within 0.5 min and the column was equilibrated for 4 min, resulting in a total run time of 20 min. The column oven temperature was set to 40 °C. The UHPLC system was coupled to a Q Exactive™ HF Quadrupole-Orbitrap™ mass spectrometer (Thermo Fisher Scientific, Austria), equipped with a HESI source for negative ionization to perform the mass spectrometric analysis. The resolution on the MS1 level was set to 60,000 (at *m/z* 200) with a scan range from *m/z* 250–700. The two most abundant precursor ions were picked for fragmentation (HCD 24 normalized collision energy), preferentially from an inclusion list containing *m/z* 31 values specific for eicosanoids and their precursor molecules. Resulting fragments were analyzed on the MS2 level at a resolution of 15,000 (at *m/z* 200). Operating in negative ionization mode, a spray voltage of 2.2 kV and a capillary temperature of 253 °C were applied. Sheath gas was set to 46 and the auxiliary gas to 10 (arbitrary units).

Data analysis: Data interpretation of raw files generated by the Q Exactive™ HF Quadrupole-Orbitrap™ mass spectrometer was performed manually using Thermo Xcalibur™ 4.1.31.9 (Qual browser). Spectra were compared with reference spectra from the Lipid Maps depository library from July 2018 [50]. Peaks were integrated using the TraceFinder™ software package (version 4.1—Thermo Scientific, Vienna, Austria).

2.4. Metabolomics

Metabolomic sample collection and preparation: For metabolomics analysis, cells were seeded at 10^6 cells per T25 flask in complete medium (5 mL) and left to adhere over-night. They were then treated with IL1β or TNFα for 48 h similarly to the proteomic experiment. Thereafter, the medium was removed and centrifuged (1100 rpm, 2 min, 4 °C). An aliquot (200 µL) of each medium sample was precipitated in cold MeOH (100%, 800 µL) and stored at −80 °C. The methanolic solution contained dopamine-d4, melatonin-d4 (both Santa Cruz Biotechnology, Dallas, TX, USA) and *N*-acetyl-serotonin-d3 (Toronto Research Chemicals BIOZOL) as internal standards at concentrations of 120 pg µL^{-1}. The cell samples were washed once with PBS (3 mL) and metabolites were extracted with cold MeOH (80%, containing stds, 1 mL). The 80% methanolic solution contained dopamine-d4, melatonin-d4 (both Santa Cruz Biotechnology, Dallas, TX, USA) and *N*-acetyl-serotonin-d3 (Toronto Research Chemicals BIOZOL) as internal standards at concentrations of 100 pg µL^{-1}. Each flask was processed at a time and immediately snap-frozen in liquid nitrogen. Three replicates per condition were then thawed together and the cells were scraped, transferred into labelled Eppendorf tubes and were stored at −80 °C. Samples were dried and reconstituted in 120 µL of 1% methanol and 0.2% formic acid and 1 pg µL^{-1} caffeine-(trimethyl-D9) (Sigma Aldrich) and transferred into HPLC vials equipped with a 200 µL V-shape glass insert (both Macherey-Nagel GmbH Co. KG) suitable for LC-MS/MS analysis. Caffeine-(trimethyl-d9) (1 pg µL^{-1}) was used as an additional internal standard. Again, the experiment was carried out in biological triplicates.

Data acquisition: Samples were separated on a reversed phase Kinetex XB-C18 column (2.6 µm, 150 × 2.1 mm, 100 Å, Phenomenex Inc., Torrance, CA, USA) using a Vanquish UHPLC System (Thermo Fisher Scientific). Mass spectrometric analysis was performed on a Q Exactive HF orbitrap (Thermo Fisher Scientific). Mobile phase A consisted of water with 0.2% formic acid, mobile phase B of methanol with 0.2% formic acid and the following gradient program was run: 1 to 5% B in 0.5 min, 5 to 40% B from 0.5–5 min, then 40 to 90% B from 5–8 min, followed by a wash phase at 90% B for 2.5 min and then an equilibration phase at 1% B for 2 min, yielding a total run-time of 12.5 min. Flow rate was 0.5 mL min^{-1}, injection volume was 5 µL and the column temperature was set to 40 °C. The injection needle was washed in between runs with 10% methanol. All samples were analyzed in technical replicates. Samples were analyzed in positive, as well as in negative ionization mode. Scan range was from *m/z* 100–1000 and resolution was set to 60,000 (at *m/z* 200) for MS1 and 15,000 (at *m/z* 200) for MS2. The four most abundant ions of the full scan were selected for further fragmentation in the HCD collision cell applying 30 eV normalized collision energy. Dynamic exclusion was applied for 6 s. Instrument control was performed using Xcalibur 4.0 Qual browser (Thermo Fisher Scientific).

Data analysis: Raw files generated by the Q Exactive HF were loaded into the Compound Discoverer Software 3.1 (Thermo Fisher Scientific). Compounds were identified in Compound Discoverer with a user workflow tree. A maximum retention time shift of 0.1 min was allowed for aligning features and using a maximal mass tolerance of 5 ppm. Metabolites were matched against mzcloud (Copyright © 2021–2020 HighChem LLC, Slovakia mzCloud is a trademark of HighChem LLC, Bratislava, Slovakia). Compounds with a match factor ≥80 were manually checked. This was performed with Xcalibur 4.0 Qual browser (ThermoFisher Scientific). For peak integration and calculation of peak areas, the Tracefinder Software 4.1 (ThermoFisher Scientific) was used. The generated batch table was exported and further processed with Microsoft Excel, GraphPad Prism (Version 6.07) and the Perseus software (Version 1.6.12).

2.5. Cell Cycle Analysis

Flow cytometry was performed to determine the cell cycle distribution with and without inflammatory stimulation. Therefore, BD Cycletest™ Plus DNA Kit (BD Biosciences, Vienna, Austria) was used according to the manufacturer's protocol to prepare the cells and measured on a CytoFLEX Flow Cytometer (Beckman&Coulter, Vienna, Austria) in the PE-channel. Statistical significance was evaluated with a bidirectional student t-test and three biological replicates.

3. Results

Macrophages can stimulate endometriotic cells by secreting cytokines such as TNFα and IL1β [24,25]. Here, an in-vitro model of endometriosis was investigated to evaluate the effects of such inflammatory stimuli on endometriotic cells by means of a multi-omics approach, including untargeted shotgun proteomics, metabolomics and eicosanoid analysis (Figure S1). The endometriotic 12-Z cell line was treated with TNFα or IL1β at 10 ng mL^{-1} for 48 h (Figure 1A). Flow cytometry analysis showed a slight increase in the tetraploid G2/M phase, as well as S-Phase and a corresponding decrease of cells in G0/G1 phase (Figure S2).

3.1. Eicosanoid Analysis Reveals the Uptake of Eicosanoid Precursors by 12-Z Cells from the Growth Medium

Eicosanoids from the supernatants of control and inflammatory stimulated 12-Z cells were enriched by a solid-phase extraction protocol and subsequently analyzed by mass spectrometry. A total of 49 eicosanoids and polyunsaturated fatty acids (PUFAs) were detected in the supernatants. The composition of the fully supplemented medium was additionally verified. The epithelial-like 12-Z cells efficiently depleted the growth medium of the eicosanoid precursors arachidonic acid (AA), docosahexaenoic acid (DHA) and eicosapentaenoic acid (EPA) irrespective of treatment condition (Figure S2). Strikingly, EPA was not detectable after culturing the cells for 48 h. In contrast, the mono-oxygenated hydroxyeicosatetraenoic acids 16-HETE and 18-HETE were not detected in the growth medium, but only in the presence of the cultured cells. Inflammatory stimulation had little impact on the differential expression of eicosanoids.

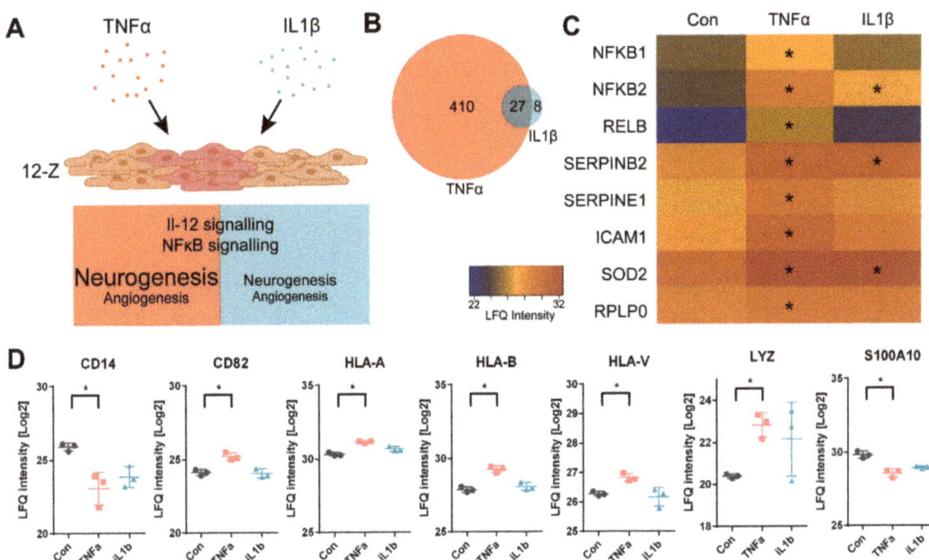

Figure 1. (**A**) Schematic representation of endometriotic 12-Z cells treated with TNFα (red) or IL1β (blue). IL-12 and NF-κB signaling was induced in both treatments while neuroangiogenesis was much more pronounced for TNFα stimulation. (**B**) Venn-Diagram showing the number of significantly regulated proteins (FDR = 0.05, S0 = 0.1) for TNFα (red) and IL1β (blue) compared to the control state. Twenty-seven protein groups were significantly regulated in both treatments. (**C**) Heatmap highlighting proteins involved in NF-κB and Il-12 signaling and downstream targets. (**D**) Protein signature characteristic for the phenotype of intermediate monocytes upon treatment of epithelial-like 12-Z cells with TNFα (red) and not with IL1β (blue). Asterisks (*) show multi-parameter corrected significant regulations of protein abundance (FDR = 0.05, S0 = 0.1) compared to untreated controls (Con).

3.2. TNFα- and IL1β-Stimulated 12-Z Cells Show Enhanced Proliferation and Activation of IL-12 and NF-κB Signaling Pathways

Proteomic profiling was performed on the cytoplasmic (soluble) fraction and resulted in the identification of 3684 protein groups. Label-free quantification (LFQ) was used to compare inflammatory stimulated with untreated conditions. In LFQ proteomics, the unlabeled peptides are quantified on the MS1 level and the intensities were adjusted to the overall intensity of all quantified peptides. Multi-parameter corrected statistical analysis (FDR = 0.05, S0 = 0.1) revealed 437 and 35 significantly regulated proteins after treatment with TNFα or IL1β, respectively, while 27 protein groups were regulated in both treatments (Figure 1B). Thus, inflammatory stimulation of 12-Z cells by TNFα led to 10-fold higher number of significantly regulated proteins compared to stimulation by IL1β. The most prominent regulations confirmed a successful inflammatory stimulation of the cells by upregulating IL-12 and NF-κB signaling pathways (Figure 1A). Upon TNFα treatment the classical (NFKB1) and alternative (NFKB2, RELB) NF-κB pathways were significantly upregulated while IL1β triggered partly the alternative pathway [51]. Both treatments, however, significantly induced the expression of downstream IL-12 and NF-κB targets (i.e., SOD2, SERPINB2) as exemplified in the heatmap in Figure 1C. Intermediate monocytes are a subpopulation of monocytes characterized by antigen presentation and transendothelial migration [52]. In contrast to classical monocytes, intermediate monocytes feature lower CD14 levels, but increased antigen presentation (HLA), lysozyme, S100A8 and S100A10 as identified by transcriptomic profiling. We found a strikingly similar protein signature corresponding to this intermediate monocyte state for the TNFα-treated, but not the IL1β-treated 12-Z cells (Figure 1D).

3.3. TNFα Induces the Expression of Proteins Involved in Neuroangiogenesis in 12-Z Cells

TNFα stimulation showed the upregulation of several proteins involved in neuroangiogenesis. For example, regulated protein groups revealed dorsal root ganglion morphogenesis, sensory neuron axon guidance, neuron projection extension/guidance, semaphorin signaling and positive regulation of sprouting angiogenesis (Figure 2A). ClueGo analysis of significantly regulated proteins further revealed a network corresponding to angiogenesis, including amongst others the VEGF pathway (Figure 2B). Furthermore, the neuropilin receptors NRP1/NRP2 and RPL10 were found to be significantly upregulated [53]. STRING network analysis revealed a strong interconnection of proteins associated with neuroangiogenesis, which were significantly upregulated upon treatment with TNFα, including angiogenesis promoters ICAM1, VCAM1 [54], and ITGA5 [55] (Figure 2C). Furthermore, downstream targets of vascular endothelial growth factor (VEGF) signaling have been found upregulated upon stimulation with TNFα (e.g., BCRA1, ITGAV). A significant upregulation of proteins involved in semaphorin signaling was observed (e.g., OPTN, EPHA4, DHRS3). While none of the described proteins were significantly upregulated by IL1β treatment, they showed a similar trend, which highlighted the differential responses of these endometriotic cells to distinct inflammatory stimulations.

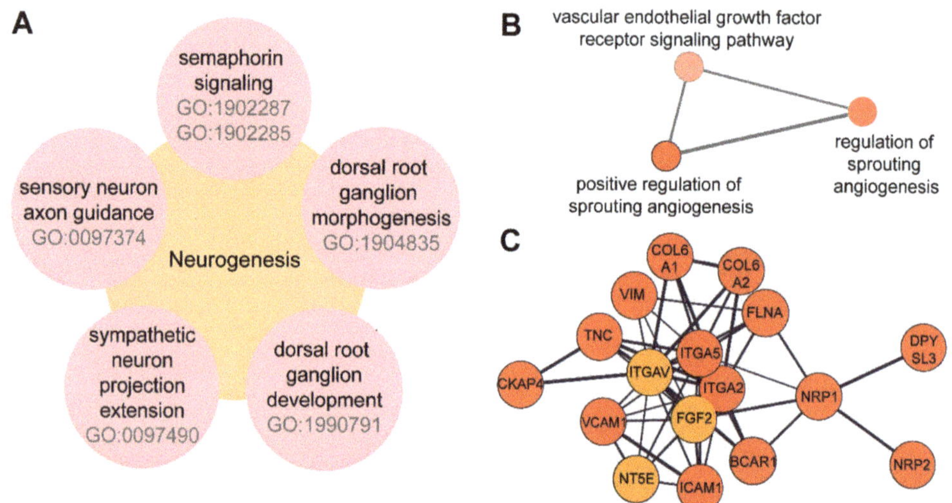

Figure 2. (**A**) Gene Ontology terms for biological processes (GOBP) associated with neurogenesis. (**B**) ClueGo network for angiogenesis. (**C**) STRING network analysis for proteins involved in neurogenesis and angiogenesis. Red indicates multi-parameter corrected significant regulation while yellow-colored proteins show higher expression but are not significant.

3.4. Untargeted Metabolomics Reveals the Upregulation of 7,8-Dihydroneopterin, 7,8-Dihydrobiopterin (BH2) and Normetanephrine in 12-Z Cells

The signature of upregulated proteins involved in neuroangiogenesis and neuropathic pain motivated the investigation of pain-associated signaling molecules on the level of metabolites. For this purpose, an untargeted metabolomics assay was carried out by collecting whole cell lysates and supernatants of control and inflammatory stimulated 12-Z cells. Additionally, the fully supplemented medium was analyzed using the same method to determine the composition of the metabolic background similarly to the eicosanoid analysis. The experiment included a database search based on MS^2 fragment spectra which resulted in the identification of 29,607 features, from which the software annotated 633 compounds with a match factor $\geq$80 (Table S3). After manual review, 63 metabolites were selected and their abundances were quantified on MS^1 level as area under the curve based on accurate masses and retention time (Table S2). Multi-parameter corrected statistical analysis (FDR = 0.05, S0 = 0.1) revealed 15 and 4 significantly regulated metabolites for TNFα and IL1β in the whole cell lysates, respectively (Figure 3A). An analogous analysis of the supernatants revealed 3 and 26 significantly regulated metabolites for TNFα and IL1β treatments, respectively (Figure 3B). Strikingly, 7,8-dihydroneopterin, 7,8-dihydrobiopterin (BH2), epinephrine (identity of the molecule verified with external standards) and normetanephrine (annotated based on MS^2) were detected in endometriotic 12-Z cells (Figures 3 and 4). These metabolites were not detected in the fully supplemented cell culture medium (Figure S3), but only in whole cell extracts and supernatants only in the presence of the 12-Z cells (Figure 4A). The recorded fragmentation spectra matched well the reference spectra from mzcloud database and corroborated the identification of these molecules (Figure 4B). In fact, BH2 and normetanephrine were significantly upregulated in the cellular interior during inflammatory stimulation with TNFα or IL1β, while 7,8-dihydroneopterin was upregulated upon TNFα stimulation only (Figures 3 and 4). The induction of BH2, 7,8-dihydroneopterin and normetanephrine was more pronounced upon activation with TNFα compared to IL1β. Interestingly, the enzymes involved in the biosynthesis of these metabolites remained largely constant upon inflammatory stimulation (e.g., dihydrofolate reductase DHFR or catechol O-methyltransferase COMT), with the exception of sepiapterin reductase (SPR), which was down-regulated by TNFα treatment (Figure 4). Epinephrine and normetanephrine were unexpectedly detected in 12-Z cells, as they were not yet associated with these endometriotic epithelial-like cells.

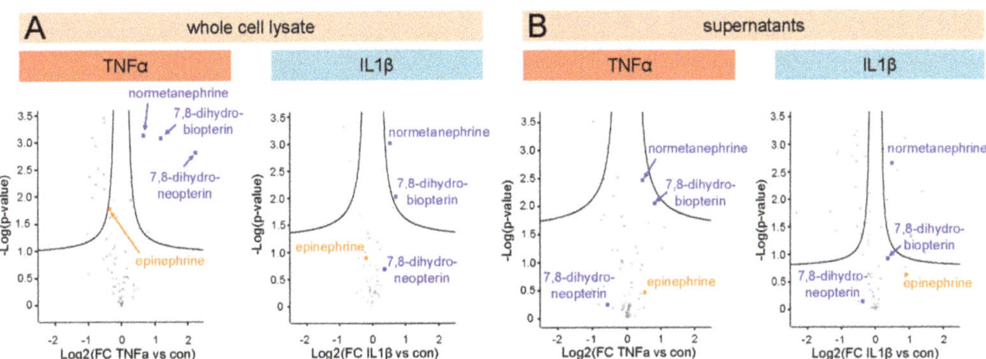

Figure 3. Volcano plots comparing metabolite profiles of control experiments with TNFα and IL1β treatments in whole cell lysates (**A**) and supernatants (**B**). X-axis displays the calculated difference of treatment-control an a log2-scale and y-axis show the -log p-value for each molecule. Metabolites above significance curves represent multi-parameter corrected significantly regulated metabolites (FDR = 0.05, S0 = 0.1).

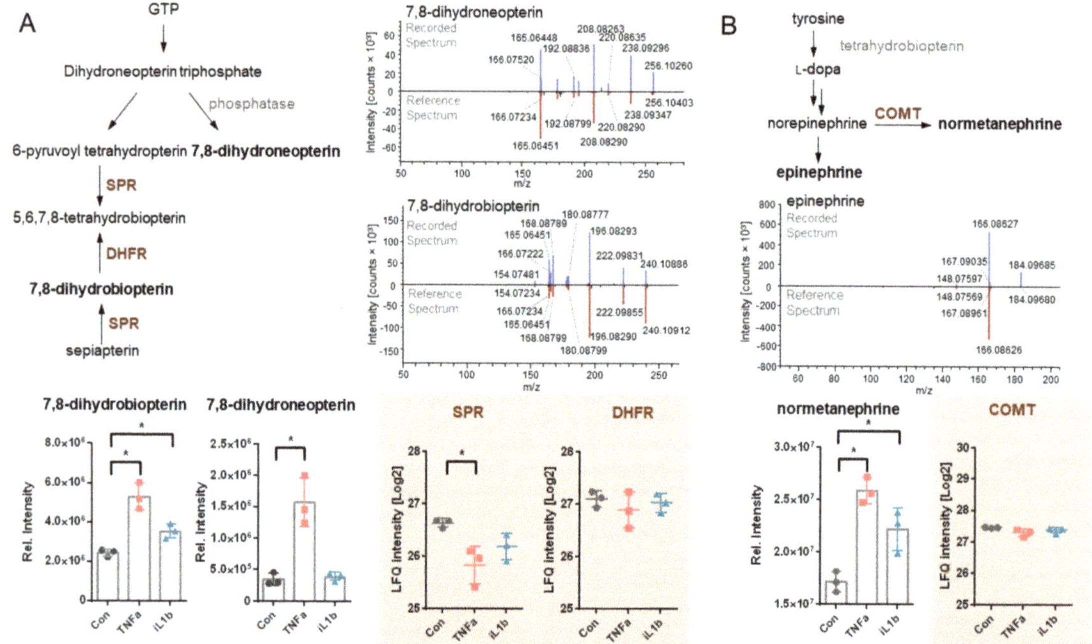

Figure 4. (**A**) The metabolic pathway for the synthesis of 7,8-dihydroneopterin and 7,8-dihydrobiopterin is depicted together with the MS^2 spectra of the measured metabolites compared to the reference spectrum from the mzcloud database. Intensity values of the metabolite and protein levels (orange background) in control, as well as TNFα– and IL1β-treatment, are given below. (**B**) Biosynthetic pathway of epinephrine derivatives. Normetanephrine is obtained from norepinephrine by the catechol-*O*-methyl transferase (COMT). MS^2 spectra of normetanephrine compared to the reference spectrum from the mzcloud database. Intensity values of the metabolite and protein levels (orange background) in control, as well as TNFα– and IL1β-treatment, are given below. None of the metabolites were detected in the fully supplemented medium (Figure S3). Asterisks (*) show multi-parameter corrected significant regulations of metabolite intensities compared to untreated controls (FDR = 0.05, S0 = 0.1). The orange shadows distinguish the abundance changes of proteins from those of the metabolites.

4. Discussion

Although endometriosis is affecting the quality of life of millions of women worldwide, representing a clear unmet medical need, the underlying molecular mechanism of this disease remains largely unknown. As pain sensation is among the most prevalent symptoms, investigating molecular mechanisms responsible for the development of pain may be key to identify useful therapeutic approaches. The interplay among endometriotic cells, macrophages and nerve cells in the ectopic lesions of endometriosis is of special interest for the origin of pain. We performed a multi-omics analysis, including proteomics, metabolomics and eicosanoid analysis, of the epithelial-like 12-Z endometriotic cell line in order to characterize the responses of these cells to inflammatory stimulation and their potential involvement in the development of pain. The 12-Z endometriotic cells were previously characterized as a proliferating and invasive cell line [38]. We found that inflammatory stimulation with TNAα or IL1β did not greatly affect the cell cycle distribution compared to untreated cells. In accordance, the eicosanoid precursors AA, DHA and EPA were efficiently depleted from growth medium irrespective of the inflammatory stimulus and were probably incorporated in the membranes of 12-Z cells. The mono-oxygenated 16– and 18-HETE are cytochrome P450 metabolic products of AA and were released from 12-Z cells. Importantly, 16-HETE is typically generated by exposure of resting neutrophils to AA [56] and represents an endogenous inhibitor of neutrophil activation [57] and thus exhibits anti-inflammatory effects. In our setup, the extent of 16-HETE release was independent of treatment condition.

The proteomic data suggested that the 12-Z endometriotic cells, when stimulated with TNFα, may mimic an intermediate monocytic phenotype, which is generally characterized by transendothelial migration [52], and is actively involved in forming and sustaining a neuroangiogenic microenvironment characteristic for endometriotic lesions. Endometriotic cells were previously shown to exhibit enhanced migratory properties upon exposure to proinflammatory factors [9–11]. Moreover, dysfunction in macrophage-mediated phagocytosis of endometrial cells that undergo retrograde transport to the peritoneal cavity is considered an important factor in the development of endometriosis. In fact, this mimicry phenotype of the 12-Z cells may contribute towards the dysregulated immune clearance observed in endometriosis [14,15]. Generally, the 12-Z cells seem more susceptible towards stimulation by TNFα compared to IL1β. Inflammatory stimulation led to the upregulation of proteins involved in neuronal interactions as well as dorsal root ganglion morphogenesis and axon guidance. It is known that the pathology of endometriosis features neuronal interactions and especially neurogenesis [22]. Furthermore, a previous study already showed differential expression of semaphorins and neuropilin receptors correlating to dysmenorrhea [58]. Semaphorins are a group of evolutionarily highly conserved surface or locally secreted nerve repellent factors that can regulate axon migration, axonal growth and guidance [59–61]. The potential role of semaphorin 3A and its receptor (NRP1) in the regulation of aberrant sympathetic innervation in peritoneal endometriosis have been previously described [58]. Neuropilin receptors are prominent neurogenesis promoters, which function as axon guidance signaling receptors, as well as angiogenesis activation [31]. Our study shows that stimulation of 12-Z cells with TNFα upregulates the levels of NRP1, NRP2, DPYSL3, OPTN, EPHA4 and DHRS3 proteins suggesting an active involvement of endometriotic epithelial cells in semaphorin signaling. Normally, the process of nervous generation is a conserved feature present during embryonic development [62]. Although proteins like RPL10 and NRP1/2 are mostly associated with embryonic developmental signaling, they have been found significantly overexpressed upon TNFα stimulation in 12-Z in this study and suggest an unrecognized functional plasticity of these cells, which may contribute towards an increased understanding of this pathology.

We further combined the proteome profiling with untargeted metabolomic analysis, investigating whether metabolites of 12-Z cells may be able to contribute to neuronal interaction and signaling. Especially, the capability of the production of 7,8-dihydrobiopterin (BH2), 7,8-dihydroneopterin, epinephrine and normetanephrine by endometriotic epithelial cells was striking. The differential expression of these metabolites was not correlated with the corresponding enzymes in their biosynthetic pathways (Figure 4). 7,8-Dihydroneopterin is an accepted metabolic inflammation marker normally generated by macrophages and has been related to impaired phagocytosis in endometriosis patients [14,63]. Epinephrine is a neurotransmitter secreted by the adrenal medulla. It is required for the vagus-mediated modulation of the nociceptive threshold and acts as inflammatory mediator induced in hyperalgesia [64]. Norepinephrine, the epinephrine precursor, from sympathetic nerve fibers is known to bind the oestradiol ß2 receptor on macrophages, leading to activation of PKA signaling and thus stimulating TNFα mediated inflammation [27]. Norepinephrine is generally involved in inflammation as well as endometriosis pathology [18,65]. This mechanism was previously described as an interaction between nerve cells and macrophages. The deregulation of epinephrine and semaphoring/NRP1 signaling pathways in the nerve cells of endometriosis lesion has been shown to support macrophage polarization [66–68]. Our data, however, suggest that epithelial endometriotic cells might themselves be capable of producing these metabolites, subsequently leading to enhanced TNFα secretion by polarized macrophages.

The present model proposes not only a potential influence of endometriosis-associated epithelial cells on macrophages but on nerve cells as well. The significant upregulation of neurogenesis-related proteins demonstrated that the 12-Z cells may be capable of independently modulating neuronal mechanisms. Here again, norepinephrine and its metabolites might play an important role in the activation of pain [69]. It has been shown that epinephrine activates unmyelinated afferents in lesioned nerves [70]. Since endometriotic lesions contain large amounts of unmyelinated nerve fibres [71], there might be a connection between epinephrine secretion and the induction of pain sensation. The detection of epinephrine and normetanephrine in cell lysates and supernatants of IL1β-activated endometriotic cells is unprecedented, since this metabolite is normally produced by adrenal glands. Thus, TNFα and IL1β activation might be important to perpetuate the disease by affecting proteome, metabolome and eicosanoid levels differentially. Finally, the conversion of tetrahydrobiopterin (BH4) into BH2 is involved in the biosynthesis of norepinephrine as the initial hydroxylation step from tyrosine [72]. BH4 application in vivo has been shown to cause heat hypersensitivity and increased pain sensation through TRPV1 [73]. TRPV1 receptor is overexpressed in ectopic endometriosis implants, as well as in dorsal root ganglia of rats with endometriosis [12,74].

5. Conclusions

In summary, the data presented in this work highlights the proteomic, metabolomic and eicosanoid alterations upon inflammatory stimulation of the endometriotic epithelial-like cell line 12-Z. Besides the expected activation of inflammatory signaling cascades upon cytokine stimulation, these cells displayed an unexpected protein signature related to neuroangiogenesis which clearly underlined their capability to support neurogenesis in the lesion. Putative novel mediators in endometriosis pathology and pain development were discovered on the protein, metabolite and eicosanoid levels. This study indicates that 12-Z endometriotic cells may mimic an intermediate monocytic phenotype and actively participate in the crosstalk of the macrophage-nervous network within the lesion on the protein and metabolite levels. Thus, inflammatory stimulation of endometriotic cells by TNFα and IL1β seem to play an important role in the perpetuation of the characteristic inflammatory phenotype. They further seem to create factors enhancing the pain sensation through neurogenic inflammation. The actual interaction of endometriotic cells with macrophages and nerve cells requires further investigation but the presented data provided experimental evidence that they might be capable of hijacking immune cell functions in order to support the development and growth of an endometriotic lesion outside the uterine cavity.

Supplementary Materials: The following supplementary information is available online at https://www.mdpi.com/article/10.3390/biom11081230/s1: Figure S1: Schematic experimental setup, Figure S2: Cell cycle analysis and selected eicosanoids, Figure S3: Extracted ion chromatograms of selected metabolites; Table S1: Results from proteomic analysis, Table S2: Results from metabolomics analysis.

Author Contributions: Conceptualisation, B.N., M.L., I.Y. and C.G.; Investigation, B.N., M.L., J.B., G.H., L.S. and M.C.G.; Formal analysis, B.N., M.L., J.B., G.H. and M.C.G.; Resources, I.Y. and C.G.; Validation, S.M.M.-M., C.G.; Visualisation, B.N., M.L. and S.M.M.-M.; Writing—original draft, B.N., M.L.; Writing—review & editing, S.M.M.-M., I.Y., C.G. All authors have read and agreed to the published version of the manuscript.

Funding: Open Access Funding by the University of Vienna.

Data Availability Statement: The mass spectrometry proteomics data were deposited in the ProteomeXchange Consortium (http://proteomecentral.proteomexchange.org) via the PRIDE partner repository [39] with the dataset identifier PXD022354 and 10.6019/PXD022354.

Acknowledgments: The authors are grateful to the Core Facility of Mass Spectrometry at the Faculty of Chemistry, University of Vienna and the Joint Metabolome Facility, University of Vienna and Medical University of Vienna. Both facilities are members of the Vienna Life-Science Instruments (VLSI).

Conflicts of Interest: The authors declare no conflict of interest.

References

1. Sonavane, S.K.; Kantawala, K.P.; Menias, C.O. Beyond the boundaries-endometriosis: Typical and atypical locations. *Curr. Probl. Diagn. Radiol.* **2011**, *40*, 219–232. [CrossRef] [PubMed]
2. Renner, S.P.; Strissel, P.L.; Beckmann, M.W.; Lermann, J.; Burghaus, S.; Hackl, J.; Fasching, P.A.; Strick, R. Inhibition of adhesion, proliferation, and invasion of primary endometriosis and endometrial stromal and ovarian carcinoma cells by a nonhyaluronan adhesion barrier gel. *Biomed. Res. Int.* **2015**, *2015*, 450468. [CrossRef] [PubMed]
3. McKinnon, B.D.; Kocbek, V.; Nirgianakis, K.; Bersinger, N.A.; Mueller, M.D. Kinase signalling pathways in endometriosis: Potential targets for non-hormonal therapeutics. *Hum. Reprod. Update* **2016**, *22*, 382–403. [CrossRef] [PubMed]
4. Imanaka, S.; Maruyama, S.; Kimura, M.; Nagayasu, M.; Kobayashi, H. Towards an understanding of the molecular mechanisms of endometriosis-associated symptoms (Review). *World Acad. Sci. J.* **2020**, *2*, 12. [CrossRef]
5. Haas, D.; Shebl, O.; Shamiyeh, A.; Oppelt, P. The rASRM score and the Enzian classification for endometriosis: Their strengths and weaknesses. *Acta Obstet. Gynecol. Scand.* **2013**, *92*, 3–7. [CrossRef]
6. Morotti, M.; Vincent, K.; Becker, C.M. Mechanisms of pain in endometriosis. *Eur. J. Obstet. Gynecol. Reprod. Biol.* **2017**, *209*, 8–13. [CrossRef]
7. Zhang, X.; Yao, H.; Huang, X.; Lu, B.; Xu, H.; Zhou, C. Nerve fibres in ovarian endometriotic lesions in women with ovarian endometriosis. *Hum. Reprod.* **2010**, *25*, 392–397. [CrossRef]
8. Walch, K.; Kernstock, T.; Poschalko-Hammerle, G.; Gleiss, A.; Staudigl, C.; Wenzl, R. Prevalence and severity of cyclic leg pain in women with endometriosis and in controls—Effect of laparoscopic surgery. *Eur. J. Obstet. Gynecol. Reprod. Biol.* **2014**, *179*, 51–57. [CrossRef]
9. Porpora, M.G.; Koninckx, P.R.; Piazze, J.; Natili, M.; Colagrande, S.; Cosmi, E.V. Correlation between endometriosis and pelvic pain. *J. Am. Assoc. Gynecol. Laparosc.* **1999**, *6*, 429–434. [CrossRef]
10. Bloski, T.; Pierson, R. Endometriosis and Chronic Pelvic Pain: Unraveling the Mystery Behind this Complex Condition. *Nurs. Womens Health* **2008**, *12*, 382–395. [CrossRef]
11. Klemmt, P.A.B.; Starzinski-Powitz, A. Molecular and Cellular Pathogenesis of Endometriosis. *Curr. Womens Health Rev.* **2018**, *14*, 106–116. [CrossRef] [PubMed]
12. Lian, Y.L.; Cheng, M.J.; Zhang, X.X.; Wang, L. Elevated expression of transient receptor potential vanilloid type 1 in dorsal root ganglia of rats with endometriosis. *Mol. Med. Rep.* **2017**, *16*, 1920–1926. [CrossRef]
13. Izumi, G.; Koga, K.; Takamura, M.; Makabe, T.; Satake, E.; Takeuchi, A.; Taguchi, A.; Urata, Y.; Fujii, T.; Osuga, Y. Involvement of immune cells in the pathogenesis of endometriosis. *J. Obstet. Gynaecol. Res.* **2018**, *44*, 191–198. [CrossRef]
14. Chuang, P.C.; Wu, M.H.; Shoji, Y.; Tsai, S.J. Downregulation of CD36 results in reduced phagocytic ability of peritoneal macrophages of women with endometriosis. *J. Pathol.* **2009**, *219*, 232–241. [CrossRef]
15. Herington, J.L.; Bruner-Tran, K.L.; Lucas, J.A.; Osteen, K.G. Immune interactions in endometriosis. *Expert Rev. Clin. Immunol.* **2011**, *7*, 611–626. [CrossRef] [PubMed]
16. Dmowski, W.P.; Gebel, H.M.; Braun, D.P. The role of cell-mediated immunity in pathogenesis of endometriosis. *Acta Obstet. Gynecol. Scand. Suppl.* **1994**, *159*, 7–14.
17. Nanda, A.; Thangapandi, K.; Banerjee, P.; Dutta, M.; Wangdi, T.; Sharma, P.; Chaudhury, K.; Jana, S.K. Cytokines, Angiogenesis, and Extracellular Matrix Degradation are Augmented by Oxidative Stress in Endometriosis. *Ann. Lab. Med.* **2020**, *40*, 390–397. [CrossRef] [PubMed]
18. Liang, Y.; Xie, H.; Wu, J.; Liu, D.; Yao, S. Villainous role of estrogen in macrophage-nerve interaction in endometriosis. *Reprod. Biol. Endocrinol.* **2018**, *16*, 122. [CrossRef]
19. Greaves, E.; Temp, J.; Esnal-Zufiurre, A.; Mechsner, S.; Horne, A.W.; Saunders, P.T. Estradiol is a critical mediator of macrophage-nerve cross talk in peritoneal endometriosis. *Am. J. Pathol.* **2015**, *185*, 2286–2297. [CrossRef] [PubMed]
20. Capobianco, A.; Rovere-Querini, P. Endometriosis, a disease of the macrophage. *Front. Immunol.* **2013**, *4*, 9. [CrossRef]
21. Sondell, M.; Sundler, F.; Kanje, M. Vascular endothelial growth factor is a neurotrophic factor which stimulates axonal outgrowth through the flk-1 receptor. *Eur. J. Neurosci.* **2000**, *12*, 4243–4254. [CrossRef] [PubMed]
22. Sheveleva, T.; Bejenar, V.; Komlichenko, E.; Dedul, A.; Malushko, A. Innovative approach in assessing the role of neurogenesis, angiogenesis, and lymphangiogenesis in the pathogenesis of external genital endometriosis. *Gynecol. Endocrinol.* **2016**, *32*, 75–79. [CrossRef] [PubMed]
23. Vallve-Juanico, J.; Houshdaran, S.; Giudice, L.C. The endometrial immune environment of women with endometriosis. *Hum. Reprod. Update* **2019**, *25*, 564–591. [CrossRef] [PubMed]
24. Weiss, G.; Goldsmith, L.T.; Taylor, R.N.; Bellet, D.; Taylor, H.S. Inflammation in reproductive disorders. *Reprod. Sci.* **2009**, *16*, 216–229. [CrossRef]
25. Wang, X.M.; Ma, Z.Y.; Song, N. Inflammatory cytokines IL-6, IL-10, IL-13, TNF-alpha and peritoneal fluid flora were associated with infertility in patients with endometriosis. *Eur. Rev. Med. Pharmacol. Sci.* **2018**, *22*, 2513–2518. [CrossRef]
26. Casazza, A.; Laoui, D.; Wenes, M.; Rizzolio, S.; Bassani, N.; Mambretti, M.; Deschoemaeker, S.; Van Ginderachter, J.A.; Tamagnone, L.; Mazzone, M. Impeding macrophage entry into hypoxic tumor areas by Sema3A/Nrp1 signaling blockade inhibits angiogenesis and restores antitumor immunity. *Cancer Cell* **2013**, *24*, 695–709. [CrossRef] [PubMed]
27. Wu, J.; Xie, H.; Yao, S.; Liang, Y. Macrophage and nerve interaction in endometriosis. *J. Neuroinflamm.* **2017**, *14*, 53. [CrossRef] [PubMed]

28. Brawn, J.; Morotti, M.; Zondervan, K.T.; Becker, C.M.; Vincent, K. Central changes associated with chronic pelvic pain and endometriosis. *Hum. Reprod. Update* **2014**, *20*, 737–747. [CrossRef]
29. Vitonis, A.F.; Vincent, K.; Rahmioglu, N.; Fassbender, A.; Buck Louis, G.M.; Hummelshoj, L.; Giudice, L.C.; Stratton, P.; Adamson, G.D.; Becker, C.M.; et al. World Endometriosis Research Foundation Endometriosis Phenome and Biobanking Harmonization Project: II. Clinical and covariate phenotype data collection in endometriosis research. *Fertil. Steril.* **2014**, *102*, 1223–1232. [CrossRef]
30. Song, N.; Leng, J.H.; Lang, J.H. Expression of transient receptor potentials of vanilloid subtype 1 and pain in endometriosis. *Zhonghua Fu Chan Ke Za Zhi* **2012**, *47*, 333–336.
31. Hey-Cunningham, A.J.; Peters, K.M.; Zevallos, H.B.; Berbic, M.; Markham, R.; Fraser, I.S. Angiogenesis, lymphangiogenesis and neurogenesis in endometriosis. *Front. Biosci.* **2013**, *5*, 1033–1056. [CrossRef]
32. Morotti, M.; Vincent, K.; Brawn, J.; Zondervan, K.T.; Becker, C.M. Peripheral changes in endometriosis-associated pain. *Hum. Reprod. Update* **2014**, *20*, 717–736. [CrossRef] [PubMed]
33. Irungu, S.; Mavrelos, D.; Worthington, J.; Blyuss, O.; Saridogan, E.; Timms, J.F. Discovery of non-invasive biomarkers for the diagnosis of endometriosis. *Clin. Proteom.* **2019**, *16*, 1–16. [CrossRef]
34. Hudson, Q.J.; Perricos, A.; Wenzl, R.; Yotova, I. Challenges in uncovering non-invasive biomarkers of endometriosis. *Exp. Biol. Med.* **2020**, *245*, 437–447. [CrossRef] [PubMed]
35. Ortiz, C.N.; Torres-Reverón, A.; Appleyard, C.B. Metabolomics in endometriosis: Challenges and perspectives for future studies. *Reprod. Fertil.* **2021**, *2*, R35–R50. [CrossRef]
36. Khanapure, S.; Garvey, D.; Janero, D.; Gordon Letts, L. Eicosanoids in Inflammation: Biosynthesis, Pharmacology, and Therapeutic Frontiers. *Curr. Top. Med. Chem.* **2007**, *7*, 311–340. [CrossRef] [PubMed]
37. Medzhitov, R. Origin and physiological roles of inflammation. *Nature* **2008**, *454*, 428–435. [CrossRef] [PubMed]
38. Zeitvogel, A.; Baumann, R.; Starzinski-Powitz, A. Identification of an invasive, N-cadherin-expressing epithelial cell type in endometriosis using a new cell culture model. *Am. J. Pathol.* **2001**, *159*, 1839–1852. [CrossRef]
39. Malutan, A.M.; Drugan, T.; Costin, N.; Ciortea, R.; Bucuri, C.; Rada, M.P.; Mihu, D. Pro-inflammatory cytokines for evaluation of inflammatory status in endometriosis. *Cent. Eur. J. Immunol.* **2015**, *40*, 96–102. [CrossRef] [PubMed]
40. Neuditschko, B.; Janker, L.; Niederstaetter, L.; Brunmair, J.; Krivanek, K.; Izraely, S.; Sagi-Assif, O.; Meshel, T.; Keppler, B.K.; Del Favero, G.; et al. The Challenge of Classifying Metastatic Cell Properties by Molecular Profiling Exemplified with Cutaneous Melanoma Cells and Their Cerebral Metastasis from Patient Derived Mouse Xenografts. *Mol. Cell. Proteom.* **2020**, *19*, 478–489. [CrossRef]
41. Zougman, A.; Selby, P.J.; Banks, R.E. Suspension trapping (STrap) sample preparation method for bottom-up proteomics analysis. *Proteomics* **2014**, *14*, 1006–1010. [CrossRef] [PubMed]
42. Mayer, R.L.; Schwarzmeier, J.D.; Gerner, M.C.; Bileck, A.; Mader, J.C.; Meier-Menches, S.M.; Gerner, S.M.; Schmetterer, K.G.; Pukrop, T.; Reichle, A.; et al. Proteomics and metabolomics identify molecular mechanisms of aging potentially predisposing for chronic lymphocytic leukemia. *Mol. Cell. Proteom.* **2018**, *17*, 290–303. [CrossRef]
43. Cox, J.; Mann, M. MaxQuant enables high peptide identification rates, individualized p.p.b.-range mass accuracies and proteome-wide protein quantification. *Nat. Biotechnol.* **2008**, *26*, 1367–1372. [CrossRef] [PubMed]
44. Gerner, M.C.; Niederstaetter, L.; Ziegler, L.; Bileck, A.; Slany, A.; Janker, L.; Schmidt, R.L.J.; Gerner, C.; Del Favero, G.; Schmetterer, K.G. Proteome analysis reveals distinct mitochondrial functions linked to interferon response patterns in activated CD4+ and CD8+ T cells. *Front. Pharmacol.* **2019**, *10*, 727. [CrossRef] [PubMed]
45. Bileck, A.; Kreutz, D.; Muqaku, B.; Slany, A.; Gerner, C. Comprehensive assessment of proteins regulated by dexamethasone reveals novel effects in primary human peripheral blood mononuclear cells. *J. Proteome Res.* **2014**, *13*, 5989–6000. [CrossRef] [PubMed]
46. Shannon, P.; Markiel, A.; Ozier, O.; Baliga, N.S.; Wang, J.T.; Ramage, D.; Amin, N.; Schwikowski, B.; Ideker, T. Cytoscape: A software environment for integrated models of biomolecular interaction networks. *Genome Res.* **2003**, *13*, 2498–2504. [CrossRef]
47. Bindea, G.; Mlecnik, B.; Hackl, H.; Charoentong, P.; Tosolini, M.; Kirilovsky, A.; Fridman, W.H.; Pages, F.; Trajanoski, Z.; Galon, J. ClueGO: A Cytoscape plug-in to decipher functionally grouped gene ontology and pathway annotation networks. *Bioinformatics* **2009**, *25*, 1091–1093. [CrossRef] [PubMed]
48. Szklarczyk, D.; Gable, A.L.; Lyon, D.; Junge, A.; Wyder, S.; Huerta-Cepas, J.; Simonovic, M.; Doncheva, N.T.; Morris, J.H.; Bork, P.; et al. STRING v11: Protein-protein association networks with increased coverage, supporting functional discovery in genome-wide experimental datasets. *Nucleic Acids Res.* **2019**, *47*, D607–D613. [CrossRef] [PubMed]
49. Perez-Riverol, Y.; Csordas, A.; Bai, J.; Bernal-Llinares, M.; Hewapathirana, S.; Kundu, D.J.; Inuganti, A.; Griss, J.; Mayer, G.; Eisenacher, M.; et al. The PRIDE database and related tools and resources in 2019: Improving support for quantification data. *Nucleic Acids Res.* **2019**, *47*, D442–D450. [CrossRef]
50. Fahy, E.; Sud, M.; Cotter, D.; Subramaniam, S. LIPID MAPS online tools for lipid research. *Nucleic Acids Res.* **2007**, *35*, W606–W612. [CrossRef]
51. Karin, M.; Greten, F.R. NF-kappaB: Linking inflammation and immunity to cancer development and progression. *Nat. Rev. Immunol.* **2005**, *5*, 749–759. [CrossRef]
52. Kapellos, T.S.; Bonaguro, L.; Gemünd, I.; Reusch, N.; Saglam, A.; Hinkley, E.R.; Schultze, J.L. Human Monocyte Subsets and Phenotypes in Major Chronic Inflammatory Diseases. *Front. Immunol.* **2019**, *10*, 2035. [CrossRef] [PubMed]

53. Brooks, S.S.; Wall, A.L.; Golzio, C.; Reid, D.W.; Kondyles, A.; Willer, J.R.; Botti, C.; Nicchitta, C.V.; Katsanis, N.; Davis, E.E. A novel ribosomopathy caused by dysfunction of RPL10 disrupts neurodevelopment and causes X-linked microcephaly in humans. *Genetics* **2014**, *198*, 723–733. [CrossRef] [PubMed]
54. Laschke, M.W.; Menger, M.D. Basic mechanisms of vascularization in endometriosis and their clinical implications. *Hum. Reprod. Update* **2018**, *24*, 207–224. [CrossRef] [PubMed]
55. Somanath, P.R.; Malinin, N.L.; Byzova, T.V. Cooperation between integrin alphavbeta3 and VEGFR2 in angiogenesis. *Angiogenesis* **2009**, *12*, 177–185. [CrossRef]
56. Shoieb, S.M.; El-Sherbeni, A.A.; El-Kadi, A.O.S. Subterminal hydroxyeicosatetraenoic acids: Crucial lipid mediators in normal physiology and disease states. *Chem. Biol. Interact.* **2019**, *299*, 140–150. [CrossRef]
57. Kraemer, R.; Bednar, M.M.; Hatala, M.A.; Mullane, K.M. A neutrophil-derived cytochrome P450-dependent metabolite of arachidonic acid modulates neutrophil behavior. *Am. J. Pathol.* **1987**, *128*, 446–454. [PubMed]
58. Liang, Y.; Wang, W.; Huang, J.; Tan, H.; Liu, T.; Shang, C.; Liu, D.; Guo, L.; Yao, S. Potential Role of Semaphorin 3A and Its Receptors in Regulating Aberrant Sympathetic Innervation in Peritoneal and Deep Infiltrating Endometriosis. *PLoS ONE* **2015**, *10*, e0146027. [CrossRef]
59. Koncina, E.; Roth, L.; Gonthier, B.; Bagnard, D. Role of semaphorins during axon growth and guidance. *Adv. Exp. Med. Biol.* **2007**, *621*, 50–64. [CrossRef]
60. Markovinovic, A.; Cimbro, R.; Ljutic, T.; Kriz, J.; Rogelj, B.; Munitic, I. Optineurin in amyotrophic lateral sclerosis: Multifunctional adaptor protein at the crossroads of different neuroprotective mechanisms. *Prog. Neurobiol.* **2017**, *154*, 1–20. [CrossRef] [PubMed]
61. Hu, S.; Zhu, L. Semaphorins and Their Receptors: From Axonal Guidance to Atherosclerosis. *Front. Physiol.* **2018**, *9*, 1236. [CrossRef] [PubMed]
62. Malik, S.; Vinukonda, G.; Vose, L.R.; Diamond, D.; Bhimavarapu, B.B.; Hu, F.; Zia, M.T.; Hevner, R.; Zecevic, N.; Ballabh, P. Neurogenesis continues in the third trimester of pregnancy and is suppressed by premature birth. *J. Neurosci.* **2013**, *33*, 411–423. [CrossRef] [PubMed]
63. Shchepetkina, A.A.; Hock, B.D.; Miller, A.; Kennedy, M.A.; Gieseg, S.P. Effect of 7,8-dihydroneopterin mediated CD36 down regulation and oxidant scavenging on oxidised low-density lipoprotein induced cell death in human macrophages. *Int. J. Biochem. Cell Biol.* **2017**, *87*, 27–33. [CrossRef]
64. Levine, J.D.; Khasar, S.G.; Green, P.G. Neurogenic inflammation and arthritis. *Ann. N. Y. Acad. Sci.* **2006**, *1069*, 155–167. [CrossRef] [PubMed]
65. Gosain, A.; Jones, S.B.; Shankar, R.; Gamelli, R.L.; DiPietro, L.A. Norepinephrine modulates the inflammatory and proliferative phases of wound healing. *J. Trauma* **2006**, *60*, 736–744. [CrossRef]
66. Grailer, J.; Haggadone, M.; Ward, P. Catecholamines promote an M2 macrophage activation phenotype (P1299). *J. Immunol.* **2013**, *190*, 63–67.
67. Grailer, J.J.; Haggadone, M.D.; Sarma, J.V.; Zetoune, F.S.; Ward, P.A. Induction of M2 regulatory macrophages through the beta2-adrenergic receptor with protection during endotoxemia and acute lung injury. *J. Innate Immun.* **2014**, *6*, 607–618. [CrossRef]
68. Qin, J.F.; Jin, F.J.; Li, N.; Guan, H.T.; Lan, L.; Ni, H.; Wang, Y. Adrenergic receptor beta2 activation by stress promotes breast cancer progression through macrophages M2 polarization in tumor microenvironment. *BMB Rep.* **2015**, *48*, 295–300. [CrossRef]
69. Pertovaara, A. Noradrenergic pain modulation. *Prog. Neurobiol.* **2006**, *80*, 53–83. [CrossRef]
70. Habler, H.J.; Janig, W.; Koltzenburg, M. Activation of unmyelinated afferents in chronically lesioned nerves by adrenaline and excitation of sympathetic efferents in the cat. *Neurosci. Lett.* **1987**, *82*, 35–40. [CrossRef]
71. Medina, M.G.; Lebovic, D.I. Endometriosis-associated nerve fibers and pain. *Acta Obstet. Gynecol. Scand.* **2009**, *88*, 968–975. [CrossRef] [PubMed]
72. Nichol, C.A.; Smith, G.K.; Duch, D.S. Biosynthesis and Metabolism of Tetrahydrobiopterin and Molybdopterin. *Annu. Rev. Biochem.* **1985**, *54*, 729–764. [CrossRef]
73. Latremoliere, A.; Costigan, M. GCH1, BH4 and pain. *Curr. Pharm. Biotechnol.* **2011**, *12*, 1728–1741. [CrossRef] [PubMed]
74. Rocha, M.G.; e Silva, J.C.; Ribeiro da Silva, A.; Candido Dos Reis, F.J.; Nogueira, A.A.; Poli-Neto, O.B. TRPV1 expression on peritoneal endometriosis foci is associated with chronic pelvic pain. *Reprod. Sci.* **2011**, *18*, 511–515. [CrossRef] [PubMed]

 *biomolecules* 

Article

LPS Tolerance Inhibits Cellular Respiration and Induces Global Changes in the Macrophage Secretome

Joseph Gillen [1,†], Thunnicha Ondee [2,†,‡], Devikala Gurusamy [3], Jiraphorn Issara-Amphorn [2], Nathan P. Manes [1], Sung Hwan Yoon [1], Asada Leelahavanichkul [2,4] and Aleksandra Nita-Lazar [1,*]

1 Functional Cellular Networks Section, Laboratory of Immune System Biology, National Institute of Allergy and Infectious Diseases, National Institutes of Health, Bethesda, MD 20892-1892, USA; joseph.gillen@nih.gov (J.G.); nathan.manes@nih.gov (N.P.M.); sunghwan.yoon@nih.gov (S.H.Y.)
2 Department of Microbiology, Faculty of Medicine, Chulalongkorn University, Bangkok 10330, Thailand; thunnichaon@yahoo.com (T.O.); jiraphorn298@gmail.com (J.I.-A.); aleelahavanit@gmail.com (A.L.)
3 Surgery Branch, National Cancer Institute, Bethesda, MD 20892-1892, USA; devikala.gurusamy@nih.gov
4 Translational Research in Inflammation and Immunology Research Unit (TRIRU), Department of Microbiology, Chulalongkorn University, Bangkok 10330, Thailand
* Correspondence: nitalazarau@niaid.nih.gov
† These authors contributed equally to this work.
‡ Current address: Department of Preventive and Social Medicine, Faculty of Medicine, Chulalongkorn University, Bangkok 10330, Thailand.

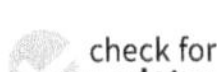

Citation: Gillen, J.; Ondee, T.; Gurusamy, D.; Issara-Amphorn, J.; Manes, N.P.; Yoon, S.H.; Leelahavanichkul, A.; Nita-Lazar, A. LPS Tolerance Inhibits Cellular Respiration and Induces Global Changes in the Macrophage Secretome. *Biomolecules* **2021**, *11*, 164. https://doi.org/10.3390/biom11020164

Academic Editors: Peter Jungblut and Michelle Hill

Received: 1 December 2020
Accepted: 23 January 2021
Published: 27 January 2021

Publisher's Note: MDPI stays neutral with regard to jurisdictional claims in published maps and institutional affiliations.

Abstract: Inflammatory response plays an essential role in the resolution of infections. However, inflammation can be detrimental to an organism and cause irreparable damage. For example, during sepsis, a cytokine storm can lead to multiple organ failures and often results in death. One of the strongest triggers of the inflammatory response is bacterial lipopolysaccharides (LPS), acting mostly through Toll-like receptor 4 (TLR4). Paradoxically, while exposure to LPS triggers a robust inflammatory response, repeated or prolonged exposure to LPS can induce a state of endotoxin tolerance, a phenomenon where macrophages and monocytes do not respond to new endotoxin challenges, and it is often associated with secondary infections and negative outcomes. The cellular mechanisms regulating this phenomenon remain elusive. We used metabolic measurements to confirm differences in the cellular metabolism of naïve macrophages and that of macrophages responding to LPS stimulation or those in the LPS-tolerant state. In parallel, we performed an unbiased secretome survey using quantitative mass spectrometry during the induction of LPS tolerance, creating the first comprehensive secretome profile of endotoxin-tolerant cells. The secretome changes confirmed that LPS-tolerant macrophages have significantly decreased cellular metabolism and that the proteins secreted by LPS-tolerant macrophages have a strong association with cell survival, protein metabolism, and the metabolism of reactive oxygen species.

Keywords: host-pathogen interactions; proteomics; secretome; macrophages

1. Introduction

Macrophages and monocytes are innate immune cells playing an important role in orchestrating the initial response to bacterial infection and tissue damage [1]. During Toll-like receptor (TLR) stimulation, macrophages are activated and produce pro-inflammatory cytokines and chemokines to recruit other cells to the site of infection [1,2]. In sepsis, lipopolysaccharides (LPS), an outer membrane component of Gram-negative bacteria, are considered to be a major activator of macrophages, triggering an inflammatory response [3]. However, in response to a second or prolonged LPS stimulation, macrophages are initially activated but produce lower amounts of pro-inflammatory cytokines. This phenomenon is called "LPS tolerance" or "endotoxin tolerance" and has been known since the 1940s [4–6]. While the lower cytokine production during LPS tolerance prevents a severe "cytokine storm" response and lethal effects in the host, decreased cytokine levels might not be sufficient to maintain an effective defense against pathogens. Indeed, LPS tolerance has been reported to be associated with the immune suppression stage known as immune exhaustion [6]. A concept of innate immunity bearing a memory of past insults termed "trained immunity" encompasses endotoxin tolerance, and its exploration may result in discoveries of new immunotherapies [7].

The mechanisms inhibiting the LPS response and moving cells into a tolerant state have still not been completely elucidated [8]. Findings from several groups emphasize the roles in this process of epigenetic reprogramming [4], microRNA [9,10], alteration of gene expression patterns [11], sometimes by specific transcription factors such as hypoxia-inducible factor 1-alpha (HIF-1α) [12], non-coding RNAs [13], and energy depletion [14]. The metabolic changes in LPS-challenged macrophages after treatment with LPS have been indicated by several recent studies, with varying experimental designs focusing on a specific protein [15], pathway [16], or general phenotype [17]. Regulation of cellular signaling leads to changes in multiple secreted proteins that are responsible for the immune response during TLR stimulation (e.g., interleukin (IL)-6 and tumor necrosis factor (TNF)-α). These proteins act as autocrine, paracrine, or chemoattracting signaling molecules for communication with other immune cells [18]. We have recently demonstrated the role of secreted lipocalin 2 (Lcn2) in the reduction in macrophage cytokine release in LPS-tolerant cells [15], but a comprehensive secretome analysis of LPS tolerance has not been previously reported. Investigating the secretome during tolerance induction could provide directions for explaining the phenomenon of immune tolerance and exhaustion.

In this study, we used metabolic measurements to confirm differences in the cellular metabolism of naïve macrophages and that of either macrophages responding to LPS stimulation or macrophages in the LPS-tolerant state. Next, we used mass spectrometry-based proteomics to thoroughly investigate, for the first time, the changes in the extracellular proteome (secretome) following the induction of LPS tolerance. Furthermore, we investigated the secretome profile during the induction of LPS tolerance to identify possible regulators of cellular metabolism and the production of proteins. In our analysis, we confirmed that LPS-tolerant macrophages have significantly decreased cellular metabolism and that the proteins secreted by LPS-tolerant macrophages have a strong association with cell survival, protein metabolism, and reactive oxygen species metabolism.

2. Materials and Methods

2.1. Cell Culture and Reagents

RAW 264.7 mouse macrophage cells were cultured in Dulbecco's Modified Eagle's Medium (DMEM) supplemented with 10% fetal bovine serum(FBS), 1 × glutamine, and 20 mM 4-(2-hydroxyethyl)-1-piperazineethanesulfonic acid (HEPES) buffer, referred to as complete DMEM (cDMEM). For stable isotope labeling by amino acids in cell culture (SILAC), the cells were cultured in DMEM for SILAC purchased from Thermo Fisher Scientific (Waltham, MA, USA), supplemented with 10% FBS, 1 × glutamine, 20 mM HEPES buffer, and isotopically labeled lysine and arginine purchased from Cambridge Isotope Laboratories, Inc. (Tewksbury, MA, USA). The cells were cultured in the labeled media for five passages prior to analysis to allow for > 95% incorporation of the labeled amino acids. Lipopolysaccharide (LPS) from *Salmonella minnesota* R595 was purchased from Enzo Life Sciences, Inc. (Farmingdale, NY, USA).

2.2. Quantification of Secreted Cytokines

Secreted TNF-α, IL-6, and IL-10 were quantified with ELISA kits (Thermo-Scientific, Rockford, IL, USA) following the manufacturer's protocols.

2.3. Extracellular Flux Analysis

The energy metabolism profiles of macrophages can be used to estimate glycolysis and mitochondrial oxidative phosphorylation on the basis of the extracellular acidification rate (ECAR) and the oxygen consumption rate (OCR), which were measured using Seahorse

XF Analyzers (Agilent, Santa Clara, CA, USA). RAW 264.7 cells in different experimental groups, namely untreated (NT/NT or Con), LPS-responsive (NT/LPS or LR), and LPS-tolerant (LPS/LPS or LT), were dispersed into monolayers for measurement. A RAW mitochondrial stress test and a glucose stress test were performed at 37 °C using the Seahorse XFe96 bioanalyzer (Agilent, Santa Clara, CA, USA). RAW 264.7 cells in various treatment groups were collected and washed in 1× PBS. Cells seeded at 4×10^5 cells per well of the Seahorse analysis plates were centrifuged at 400 rpm with acceleration and deceleration set to 1 for 5 min to achieve an even monolayer of cells for accurate measurement. OCR and ECAR for the mitochondrial stress test were measured in xeno-free (XF) media (containing 25 mM glucose, 2 mM L-glutamine, and 1 mM sodium pyruvate) under basal conditions and in response to 2 μM oligomycin, 1.5 μM fluoro-carbonyl cyanide phenylhydrazone (FCCP), and 0.5 μM rotenone and antimycin A (Sigma-Aldrich, St. Louis, MO, USA). For the glucose stress test, the cells were cultured in XF media (containing 2 mM L-glutamine), and the ECAR readout was obtained at basal conditions and in response to 10 mM glucose, 1 μM oligomycin, and 10 mM 2-deoxy-glucose (2-DG).

2.4. Collection of Secreted Proteins

For the secretome analysis by quantitative liquid chromatography–tandem mass spectrometry (LC-MS/MS), we used the method we established earlier [19]. Briefly, prior to stimulation, 1×10^6 RAW 264.7 cells were seeded in a well of a 12-well plate and grown at 37 °C for 24 h. To decrease the amount of non-specific protein in the media prior to stimulation, the cDMEM was removed from the cell culture wells and replaced with cDMEM lacking FBS. We extensively evaluated cell death in this method and found it to be negligible [19]. To test the effect of multiple stimulations of the innate immune system, the RAW 264.7 cells were treated in one of three ways. The first group contained control cells that received no LPS (NT/NT or Con) and were grown in cDMEM labeled with Arg^0 and Lys^0. The second group received a single stimulation with 100 ng/mL LPS 6 h prior to the collection of the media (NT/LPS or LR) and were grown in cDMEM labeled with Arg^{+6} and Lys^{+4}. The third group received two stimulations with 100 ng/mL LPS separated by 24 h, with the second stimulation being 6 h prior to the sample collection (LPS/LPS or LT), and were grown in cDMEM labeled with Arg^{+10} and Lys^{+8}. After the stimulations, the media were collected and equal parts of the three groups (v/v, as in [19–21]) were combined into a single 1.5-mL tube. Any cellular debris or detached cells were separated from the media by filtration using a 0.22-μm polysaccharide filter, and then, the medium was centrifuged at 400× g for 5 min. Finally, the supernatant was transferred to a 1.5-mL tube and the proteins were concentrated in a vacuum centrifuge (SpeedVac, Thermo Fisher Scientific, Waltham, MA) to dryness. Overall, this method was repeated twice with two biological replicates each time to produce four biological replicates.

2.5. In-Gel Digestion of Secreted Proteins

The dried proteins were resuspended in 2 × NuPAGE loading buffer, and then, the proteins were denatured by boiling for 10 min. The proteins were separated using a 10% Bis-Tris NuPAGE gel (Invitrogen, 8 × 8 cm) with 3-(N-morpholino)propanesulfonic acid (MOPS) buffer and run with 200 V for 40 min to ensure that there were no significant visual differences in the band patterns between samples. The gel was fixed using 47.5% methanol and 5% glacial acetic acid for 30 min at room temperature and then washed three times with ddH_2O. The fixed proteins were stained with PageBlue protein staining solution (Thermo Fisher Scientific, Waltham, MA, USA) for 1 h at room temperature and then destained with ddH_2O overnight at 4 °C. Following destaining, the lanes were cut from the gel using razor blades, sectioned into five equal units to avoid processing excess gel in one sample, and cubed into approximately 1-mm^3 pieces. The gel pieces from each section were collected into 1.5-mL microcentrifuge tubes and then processed according to a previously published protocol [22].

In brief, 500 μL of acetonitrile (ACN) was added to the gel pieces and the tubes were incubated at room temperature for 10 min before a brief centrifugation and removal of the supernatant. Next, 50 μL of 10 mM dithriothreitol (DTT, Sigma-Aldrich, St. Louis, MO, USA) in 100 mM ammonium bicarbonate (ABC) was added to the gel pieces and the tubes were incubated at 56 °C for 30 min followed by a second incubation with ACN. Next, 50 μL of 55 mM 2-chloroacetamide (CA, Sigma-Aldrich, St. Louis, MO, USA) in 100 mM of ABC was added to the gel pieces and the tubes were incubated at room temperature in the dark for 20 min followed by a third incubation with ACN. Then, 100 μL of 50% ACN, 50 mM ABC was added to the gel pieces and the tubes were incubated at room temperature with occasional vortexing, followed by a fourth incubation with ACN. The gel pieces were saturated with 13 ng/μL sequence-grade modified trypsin (Promega; Madison, WI, USA) in 10 mM ABC, 10% ACN and the tubes were incubated at 37 °C overnight. To extract the peptides, 100 μL of a 1%:25% mix of formic acid:acetonitrile was added to the gel pieces and the tubes were incubated for 15 min in a 37 °C shaker. The tubes were centrifuged briefly and the supernatant was collected in 1.5-mL tubes. At this point, the peptides from the gel sections were recombined to make one sample per lane and the peptides were concentrated in a vacuum centrifuge (SpeedVac, Thermo Fisher Scientific, Waltham, MA, USA). Lastly, the samples were mixed with formic acid and ACN to generate peptide samples with a final concentration of 0.1% formic acid, 2% ACN.

2.6. Mass Spectrometry

The Thermo Orbitrap Q-Exactive HF (Thermo Fisher Scientific, Bremen, Germany) and the Thermo UltiMate 3000 systems (Thermo Fisher Scientific, Bremen, Germany) were used for LC-MS/MS experiments. Peptides were trapped on an Acclaim C18 PepMap 100 trap column (5 μm, 100 Å, 300 μm i.d. × 5 mm, Thermo Fisher Scientific, Pittsburgh, PA, USA) and separated on a PepMap RSLC C18 column (2 μm, 100 Å, 75 μm i.d. × 50 cm, Thermo Fisher Scientific, Pittsburgh, PA, USA) at 40 °C. Peptides were eluted with a linear gradient of 2.5% to 5% mobile phase B (0.1% formic acid in ACN) for 15 min and then 5% to 35% mobile phase B over 90 min. Gradient changes were followed at 105 min to 35% mobile phase B and then increased to 99% mobile phase B at 110 min. The gradient was changed back to 2.5% mobile phase B at 125 min to equilibrate for 20 minutes prior to the next injection. Eluted peptides were ionized in positive ion polarity at a 2.3-kV spraying voltage. MS1 full scans were recorded in the range of m/z 400 to 1600 with a resolution of 60,000 at 200 m/z using the Orbitrap mass analyzer. Automatic gain control was set at 1×10^6 with 40 ms of maximum injection time. The top 20 data-dependent acquisition mode was used to maximize the number of MS2 spectra from each cycle. Higher-energy collision-induced dissociation (HCD) was used to fragment selected precursor ions with a normalized collision energy of 27%. Each biological replicate was analyzed twice to create two technical replicates. The mass spectrometry-based proteomics data have been deposited to the ProteomeXchange Consortium via the PRIDE partner repository with the dataset identifier PXD021925 [23].

2.7. Analysis of MS Results

The RAW MS files were processed with MaxQuant software (version 1.6.5.0, Max Planck Institute, Munich, Germany) [24] and searched with the Andromeda search engine [25] against a mouse UniProt FASTA database (download date: 26.03.2019, 22,325 entries) supplemented with common contaminants and reverse sequences of all entries [26]. The Andromeda search engine parameters were: type = three labels—light (Arg0, Lys0), medium—(Arg6, Lys4), and heavy—(Arg10, Lys8); fixed modification = carbamidomethylation of cysteine; variable modifications = oxidation of methionine, acetylation of lysine, and acetylation of protein N-terminus; minimum peptide length = 7; and max missed cleavages = 2. The false-discovery rate was set to 0.01 at the peptide spectrum matches (PSM), peptide, and protein levels.

The protein group abundance data were filtered to remove possible protein contaminants. In addition, at least 2 identified peptides were required for each protein (if only one peptide was identified, at least 12 valid abundance values were required). This resulted in an estimated protein group false discovery rate of 1.54%. If a SILAC triplet contained one or two missing values, they were imputed by randomly generating a value between 10% and 100% of the minimum protein intensity value (equally distributed; separately for each LC-MS dataset). The abundance ratios were log2-transformed, and mean values were calculated across the technical replicates. InfernoRDN software v1.1.7626.35996 (https://omics.pnl.gov/software/infernordn) [27] was used to perform *t*-tests and to calculate post-hoc *q*-values.

3. Results

3.1. LPS-Tolerant Macrophages Decrease Cellular Respiration

LPS tolerance is the decreased response of immune cells following secondary or prolonged stimulation with LPS. This process is typified by the decreased secretion of cytokines (Figure 1A–C), but the causes of tolerance and the processes that maintain it are not completely elucidated. This decreased response, although useful as a mechanism preventing a lethal outcome, can have tragic consequences for patients as the decreased secretion of cytokines might often lead to an increase in secondary infections. To examine the causes and regulation of LPS tolerance in macrophages, we used purified LPS to stimulate RAW 264.7 cells as an in vitro model, a methodology successfully used previously for secretome analysis with conclusions on changes in innate immune pathways and cellular metabolism by us [19] and others (for example, [21,28]). We found that while a single LPS stimulation enhanced cytokine release (LPS-Responding (LR)), two sequential LPS stimulations over a 24-h period induced decreased cytokine levels following a 6-h incubation (LPS-Tolerant (LT)) (Figure 1A–C). These results established the conditions required to induce LPS tolerance in RAW cells.

While the decrease in secretion could be due to many factors such as a lack of available amino acids to build proteins, inhibition of vesicle transport, or increased turnover of specific mRNAs, we hypothesized that LPS-tolerant cells would display changes in their metabolic functions. The glycolytic and mitochondrial functions of control (Con or NT), LR, and LT cells were determined by measuring the ECAR and OCR using the Seahorse XF Extracellular Flux Analyzer. Both functions were impaired in LT cells compared to Con or LR cells (Figure 2A–E). Hence, the lower macrophage cytokine production in LPS-tolerant cells compared with control cells might be associated with the low cell energy.

3.2. Variations in LPS Treatment Lead to Variations in the Secretome

To examine the conditions that contributed to the decreased respiration of LT cells, we analyzed the media collected from cells in each condition. By using SILAC metabolic labeling to mark each of the conditions prior to mass spectrometric analysis, as we have done in an earlier analysis of the TLR ligand-induced secretomes [19], we could simultaneously process and quantify the relative amounts of the proteins secreted by the cells in each condition (Figure 3A). We have reliably identified and quantified 1189 proteins across all conditions. Using a *t*-test to compare the intensities of the protein signals identified in the LR or LT samples to the Con samples, we found that several proteins had a two-fold or higher change in relative quantity and a significant change (p-value ≤ 0.05) versus the control (Figure 3B,C). In total, we found 56 and 107 proteins with significantly different levels in LR and LT cellular media, respectively.

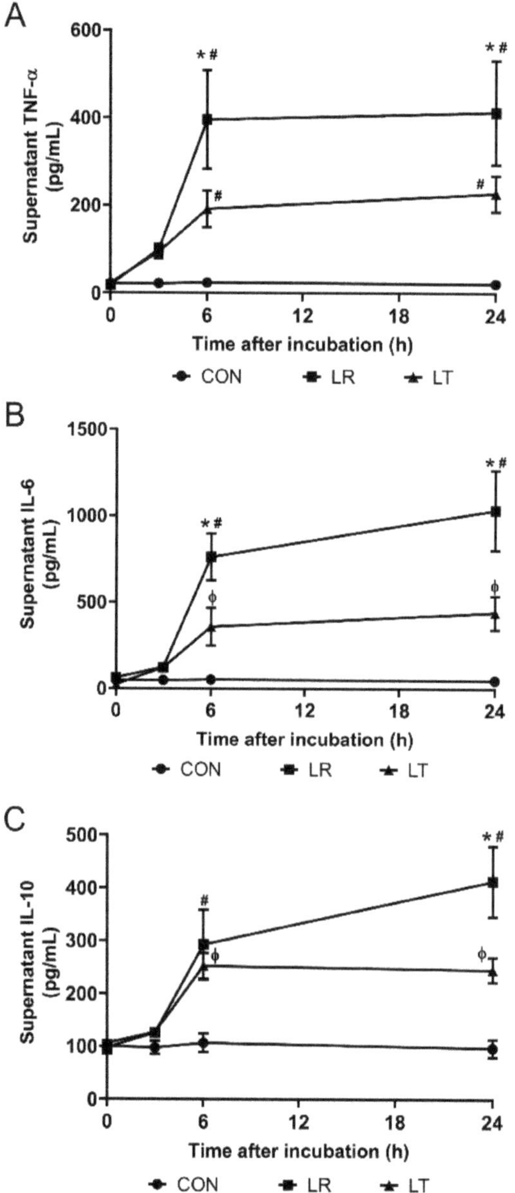

Figure 1. Induction of lipopolysaccharide (LPS) tolerance inhibits cytokine production by RAW 264.7 cells. Inflammatory cytokines in the supernatants (**A–C**) (n = 4/time point) of macrophages treated either once (LPS-Responding (LR)) or twice with LPS stimulation (LPS-Tolerant (LT)) and untreated control samples (Con), measured using ELISA kits, show significantly inhibited secretion of tumor necrosis factor (TNF)-α (**A**), interleukin (IL)-6 (**B**), and IL-10 (**C**) from LT cells compared to LR cells. *, p-value < 0.05 vs. LR; #, p-value < 0.001 vs. Con; ϕ, p-value < 0.05 vs. Con.

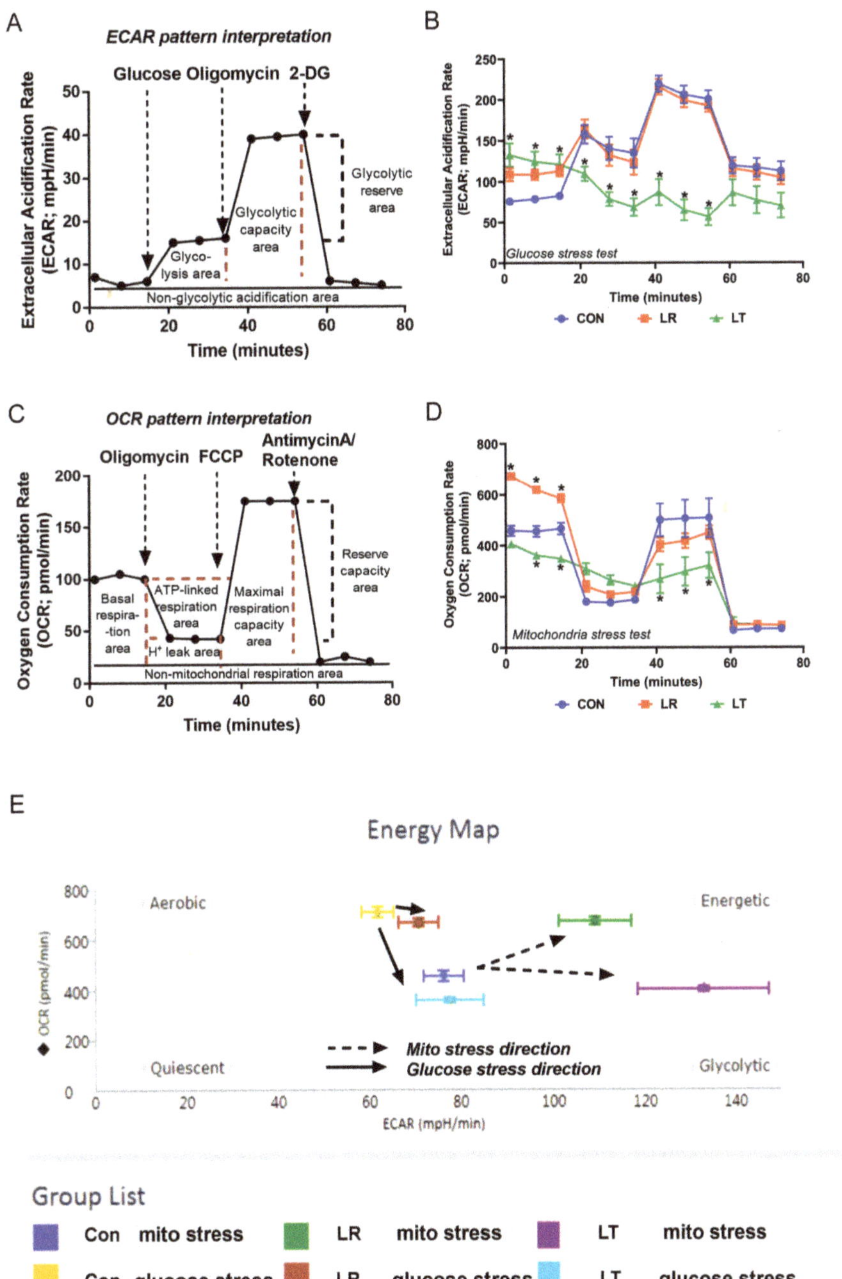

Figure 2. LPS-tolerant RAW 264.7 cells have significantly decreased cellular metabolism compared to LPS-responding and unstimulated cells. The general pattern of the estimation of glycolysis and mitochondrial functions through extracellular acidification rate (ECAR) and oxygen consumption rate (OCR), respectively (**A**,**C**); the pattern of macrophages treated with LPS either once (LPS-Responding (LR, red)) or twice (LPS-Tolerant (LT, green)) and untreated control samples (Con, blue) (**B**,**D**) (combination from triplicate experiments for **B**,**D**), and the energy map calculated using the Seahorse XF Extracellular Flux Assay (**E**). * = *p*-value < 0.001.

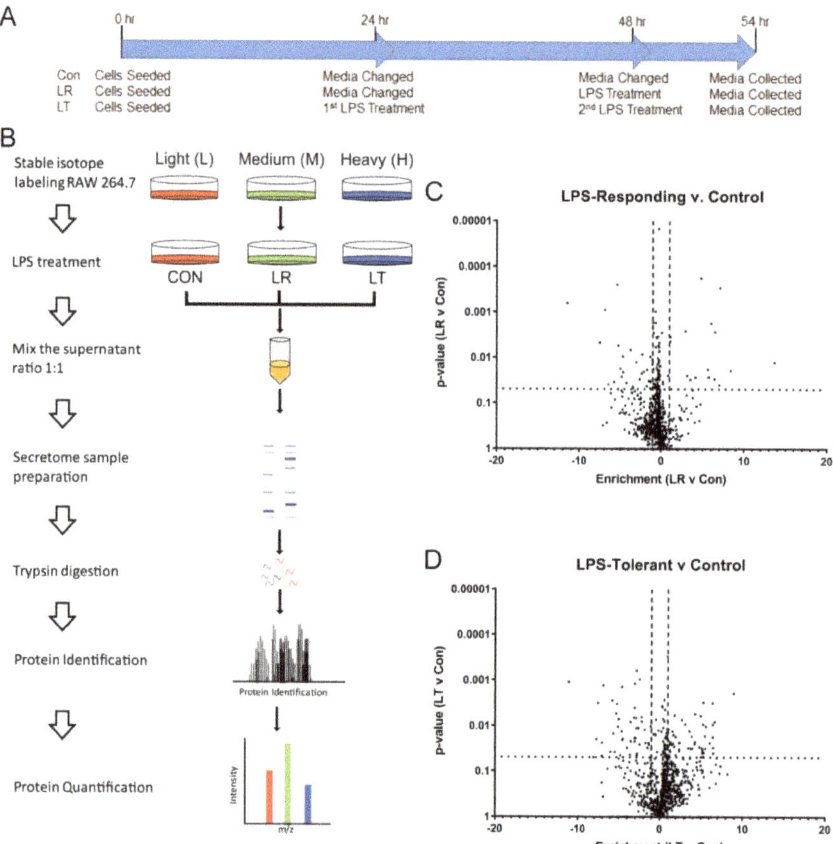

Figure 3. LPS-Responding and LPS-Tolerant RAW 264.7 cells secrete a wide variety of proteins. Schematic of the timing for LPS treatments of RAW 264.7 cells to induce the LPS response (LR) or LPS tolerance (LT) (**A**). MS analysis of the secretome of RAW 264.7 (**B**). Plot of *t*-test results comparing individual protein intensities calculated by MaxQuant (version 1.6.5.0) in LPS-Responding (LR) (**C**) or LPS-Tolerant (LT) (**D**) versus untreated control samples (NT/NT). Protein intensities of eight replicates were averaged and missing values for intrasample results were replaced with a random value between $1/2\times$ and $2\times$ the average of the 10 lowest values. The dotted lines indicate significance (p-value < 0.05) and the dashed lines indicate a onefold difference from the control.

We found that most proteins were found in both treatment groups but had different levels and directions of change compared to the control in both LR and LT conditions. To identify which changes in protein levels were specific to either LR or LT conditions, we plotted the p-values versus the control of each protein in the LR and LT datasets (Figure 4A). This plot identified four clear groups, identified as A through D. Group A proteins had a significant difference (p-value ≤ 0.05) in LR samples and included 33 proteins. Group B had a significant difference (p-value ≤ 0.05) in LT samples and included 84 proteins. Group C had a significant difference (p-value ≤ 0.05) in samples treated with either LR or LT samples and included 23 proteins. Group D had no significant difference following LPS treatment and included 608 proteins (Figure 4A).

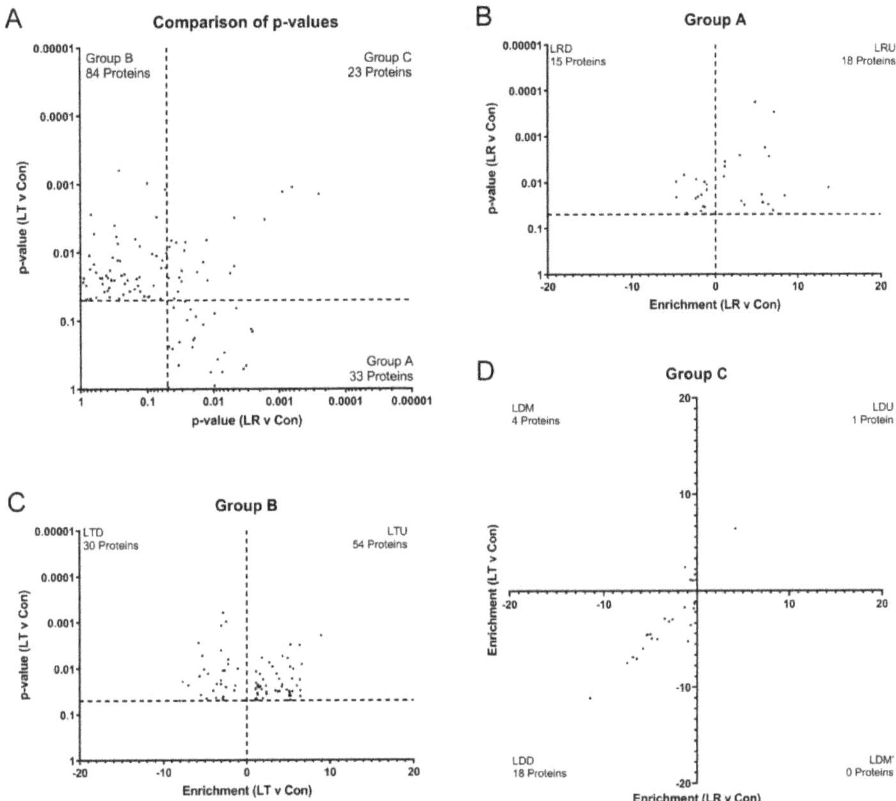

Figure 4. LPS-Tolerant and LPS-Responding RAW 264.7 cells have distinct secretomes. Plotting of *p*-values of LR vs. Con against *p*-values of LT vs. Con confirms that only 19.3% of proteins have significantly modified secretion in both conditions (**A**). Of the proteins with significantly modified secretion in only LR cells, over half of the proteins have increased secretion (**B**). Of the proteins with significantly modified secretion in only LT cells, over half of the proteins have increased secretion (**C**). A comparison of the enrichment of proteins secreted by the LR and LT cells with significantly modified secretion shows that the majority of these proteins were significantly decreased following treatment (**D**).

In addition to the significant difference from the control, the individual protein results could be further sorted by whether the intensity increased or decreased in comparison to the control (Figure 4B,C; Supplementary Table S1). The proteins that presented increased intensity following LPS treatments were termed subgroup one, while those with decreased intensity were termed subgroup two. Of the group A proteins, 18 were significantly increased (group "LPS-Responding Up" (LRU)) and 15 were significantly decreased (group "LPS-Responding Down" (LRD)). Group B proteins, while more numerous than those in group A, still had a bias towards increasing intensities, with 54 proteins from the group "LPS-Tolerant UP" (LTU) against 30 proteins from the group "LPS-Tolerant Down" (LTD). Lastly, for group C, significance following LPS treatment could lead to three possible outcomes: increased in both conditions (one protein), decreased in both conditions (18 proteins), or a discordant result with increased in one condition but decreased in the other condition (four proteins).

Amongst the proteins in the LRU group were several cytokines and chemokines (Supplementary Table S1), including C-C motif chemokine 4 (Ccl4), tumor necrosis factor (TNF), C-X-C motif chemokine 10 (Cxcl10), C-C motif chemokine 2 (Ccl2), and leukemia inhibitory factor (LIF), which all showed significant (*p*-value < 0.05) or highly significant

(p-value < 0.001) increases in their average intensity when compared to either the Con or the LT treatment group (Figure 4B, Supplementary Table S1). These data provide a perfect quality control for our dataset because these cytokines and chemokines are essential for the inflammatory response and are expected to be elevated in response to LPS.

The group LRD included three proteins (Beta-glucuronidase (Gusb), beta-hexosaminidase, subunit alpha (Hexb), and alpha-N-acetylglucosaminidase (Naglu)) that localize to the phagolysosome and are associated with the metabolism of carbohydrates [29]. These proteins, along with vinculin (Vin), Dipeptidyl peptidase 2 (Dpp7), and malate dehydrogenase mitochondrial (Mdh2), displayed between a 1.25- and 5.75-fold significant decrease in intensity with t-test p-values ranging from 0.007 to 0.0484 following the LPS treatment (Supplementary Table S1).

In contrast to the LR groups, the LT groups both contained many proteins typically found in the cytoplasm or other regions of the cell in addition to some secreted proteins (Supplementary Table S1). LTU proteins included 54 proteins with increases ranging from 48- to 2.78-fold versus the control sample and included osteopontin (Spp1), neutrophil gelatinase-associated lipocalin (Lcn2), sequestosome-1 (Sqstm1), and TAR DNA-binding protein 43 (Tarbp) (Figure 4C). The t-test p-values of each protein versus the control ranged from 0.0491 to 0.0001 (Supplementary Table S1). In contrast to the LTU proteins, nearly half (7/19) of the LTD proteins were associated with extracellular space. The 30 proteins from the LTD group showed between a 22- and 1.6-fold significant (p-values between 0.04 and 0.00001) decrease in overall intensity versus the control cells and included the urokinase-type plasminogen activator (Plau), sodium/potassium-transporting ATPase subunit gamma (Fxyd2), lysozyme C-2 (Lyz2), and cystatin-C (Cst3) (Supplementary Table S1).

The last and smallest group of proteins that showed significant differences versus the control depending on the treatment with LPS were group C proteins (LPS-Dependent (LD)). The inclusion of the second treatment group leads to three possible results: both increase (LDU), both decrease (LDD), or one increases and one decreases (mixed) (LDM). In our analysis, we found only one LDU protein, plasminogen activator inhibitor 1 (Serpine1), and four LDM proteins, Talin-1 (Tln), MARCKS-related protein (Marcksl1), cytosolic non-specific dipeptidase (Cndp2), and eukaryotic initiation factor 4A-I (Eif4a1), with increases in at least one treatment group by 29- to 1.6-fold versus the control set (Supplementary Table S1). The last group of 18 proteins identified in our analysis were the proteins with significant decreases in intensity (between 2000- and 1.5-fold) versus the control in both treatment conditions (group LDD), such as gelsolin (Gsn), low-density lipoprotein receptor-related protein 1 (Lrp1), macrophage colony-stimulating factor 1 receptor (Csf1r), and fibronectin (Fn1) (Supplementary Table S1).

3.3. Pathway Analysis of Critical Groups

Because either increasing or decreasing secretion of a signaling protein could have profound effects on the condition of cells, we analyzed all proteins with significant changes using the Ingenuity Pathway Analysis (IPA) software suite (Qiagen) (Figure 5). This analysis allowed us to identify several patterns, including pathways or functions enriched in either both or only one dataset.

The canonical process associated with LPS treatment is the inflammatory response. While both LR and LT groups had highly significant effect changes in the inflammatory response (both p-values < 0.01), the LR group showed a strong increase (z-score of 1.908) and the LT group had a smaller increase (z-score of 0.204) (visualized in Figure 6A, Supplementary Table S2). When we focus on the myeloid cell responses, the differences in the LR and LT groups become even more striking. While the "Immune Response of Myeloid Cells" is significantly affected in either condition (p-values of <0.001), the LR condition had an increased response (z-score 1.134) but the LT condition had a decreased response (z-score −0.348) (visualized in Figure 6B, Supplementary Table S2). By examining a heatmap of the proteins measured from each condition, it was found that while the LR group had several signaling molecules, including CXCL3, CXCL10, and TNF, the LT group had decreased recovery of these signaling molecules along with decreased secretion (compared to the untreated control) of the urokinase-type plasminogen activator (PLAU) (Figure 6B), a secreted enzyme that activates plasmin, a protein that is critical for the complement system [30]. These results confirm that either type of LPS treatment induces the inflammatory response, but the response after sequential LPS treatment is significantly reduced.

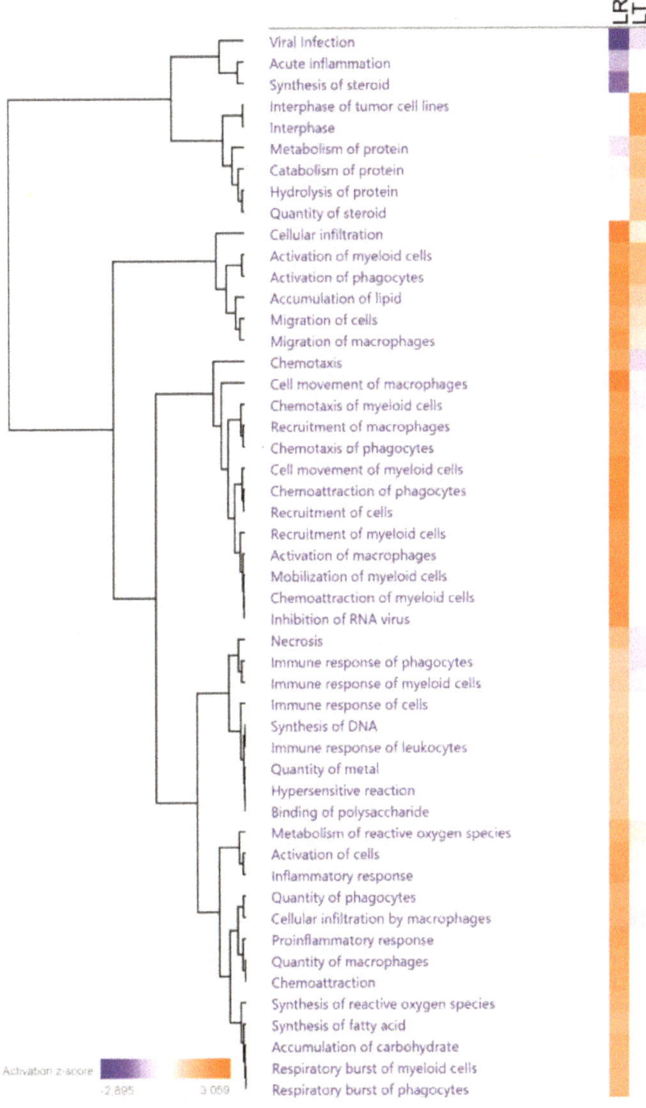

Figure 5. LPS-Responding RAW 264.7 cells have secretomes strongly associated with the immune response in contrast to LPS-Tolerant RAW 264.7 cells. Comparison of the Ingenuity Pathway Analysis of the proteins with significantly changed secretion in either LR or LT cells shows that while LR cells secreted proteins that strongly relate to the immune response and chemotaxis, LT cells secreted proteins that strongly relate to metabolism and cellular survival. Prepared using the Ingenuity Pathway Analysis program suite from QIAGEN (Germantown, MD, USA).

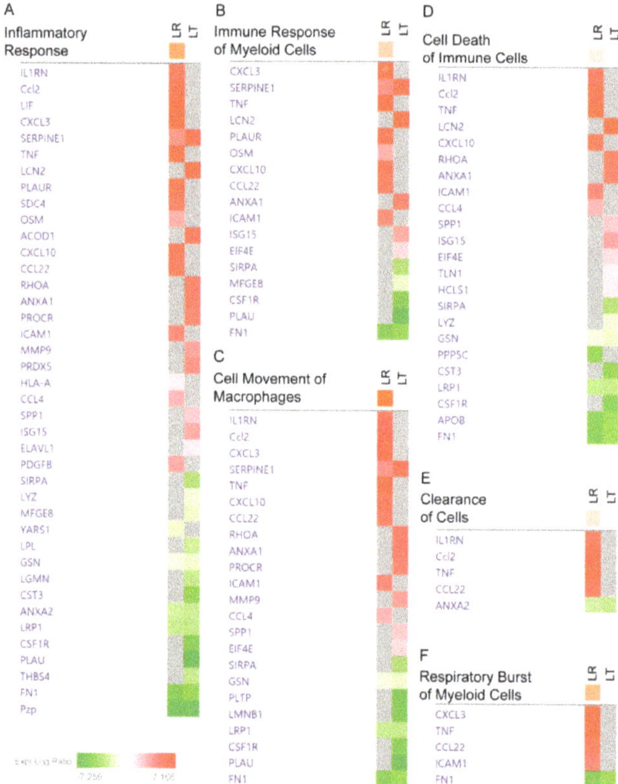

Figure 6. LPS-Responding RAW 264.7 cells' secretomes include cytokines and signaling proteins strongly related to the inflammatory response and cell motility, while LPS-Tolerant RAW 264.7 cells' secretomes include proteins strongly related to cell survival. Fold changes of the proteins associated by Ingenuity Pathway Analysis to Inflammatory Response (**A**); Immune Response of Myeloid Cells (**B**); Cell Movement by Macrophages (**C**); Cell Death of Immune Cells (**D**); Clearance of Cells (**E**), and Respiratory Burst of Myeloid Cells (**F**). Prepared using the Ingenuity Pathway Analysis program suite from QIAGEN (Germantown, MD, USA).

Another biological function associated with all three sets and with the LPS response was cellular motility. Due to the variety of cells and mechanisms of movement, most analysis platforms include both general terms and specific pathways. In the LR group, "Cell Movement of Macrophages" was significantly increased (p-value < 0.001, z-score 2.829), and while the LT group had a highly significant increase (p-value < 0.001), the overall degree of migration was lower (z-score −0.290). In our comparison, both datasets were associated with migration and contained at least five significantly elevated or decreased proteins (Figure 6C). This suggests that both treatments lead to cellular migration, but the overall effect was much higher in the LPS-responsive group.

While the processes of inflammation and movement are critical for the immune response, cell survival has been the hypothetical goal of LPS tolerance. In support of this hypothesis, our results found significant inhibition of "Cell Death of Immune Cells" in the LT group (p-value < 0.01, z-score −0.254) (Supplementary Table S2). In contrast, the LR group had a highly significant increase in the "Cell Death of Immune Cells" (p-value < 0.001, z-score 0.565) (Supplementary Table S2). The difference in the recovery of cell-survival-associated proteins from the LT and LR groups suggests a connection between cell survival and LPS tolerance (Figure 6D).

3.4. Relationship between Repeated LPS Stimulation and Cellular Exhaustion

We have hypothesized that cellular exhaustion is related to suppression of the LPS response in sequential LPS treatments. Two pathways that relate to exhaustion are metabolism and the production of reactive oxygen species. By filtering the IPA comparative analysis results only for processes related to metabolism or reactive oxygen species, we found distinct differences between the LR and LT groups (Figure 7). Overall, the LT group has a wide variety of affected processes, with both increased and decreased rates predicted.

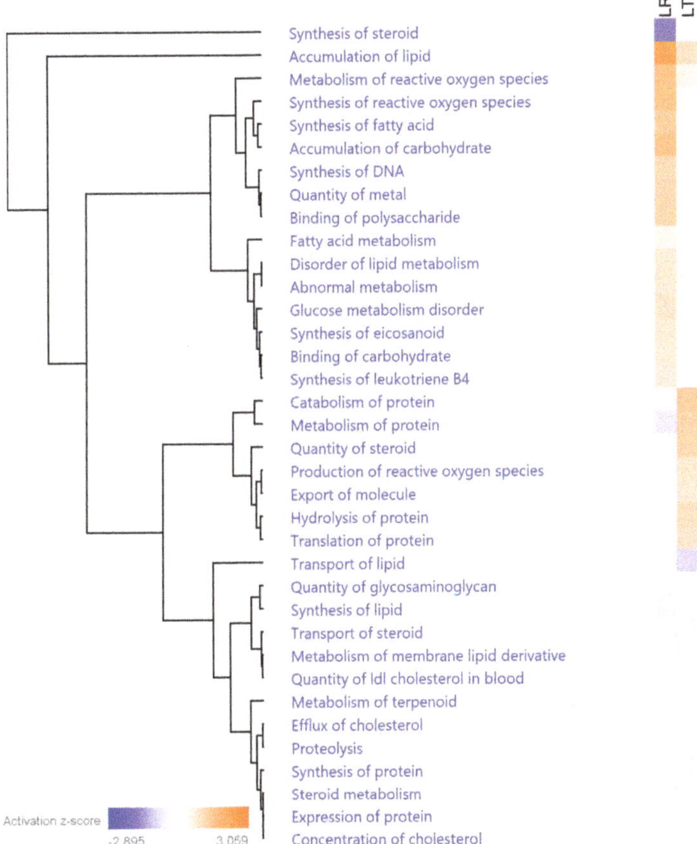

Figure 7. LPS-Tolerant RAW 264.7 cell secretomes include proteins related to initiation of protein metabolism. Comparison of the Ingenuity Pathway Analysis of proteins with significantly changed secretion in either LR or LT cells shows that while LR cells secreted proteins strongly relate to carbohydrate metabolism, LT cells secreted proteins that strongly relate to protein and reactive oxygen species metabolism. Prepared using the Ingenuity Pathway Analysis program suite from QIAGEN (Germantown, MD, USA).

Metabolism can be further defined by the class of molecule targeted, such as protein, lipid, or carbohydrate. The two classes that exhibited the clearest differences between the LR and LT groups were the processes related to carbohydrate and protein metabolism. In carbohydrate metabolism, the overall effect is that the LPS response induced increased carbohydrate metabolism, including the binding, accumulation, and metabolism of polysaccharides (Figure 8B). In contrast to the carbohydrate results, an examination of the processes related to protein metabolism showed increased association between the LT group and protein metabolism. Overall, protein metabolism appears to lean towards the accumulation of new proteins, with the increased z-score of overall protein metabolism and protein synthesis coinciding with decreases in protein catabolism and proteolysis (Figure 8C). The last aspect of metabolism with distinct differences between the LR and LT groups is the pathways related to reactive oxygen species. Overall, the LT group results were linked to lower metabolism and synthesis of ROS compared to the LR group (Figure 8D, Supplementary Table S2). These results confirm the modifications in the cellular environment that occur during both the LPS response and LPS tolerance.

Based on the strong association of metabolism and reactive oxygen species with the previously shown effects of LPS tolerance on cellular respiration, we concluded that the induction and maintenance of LPS tolerance is dependent on the rates of cellular respiration, and further studies of the modifiers of cellular respiration and metabolic rates could lead to greater understanding of the regulation of LPS tolerance.

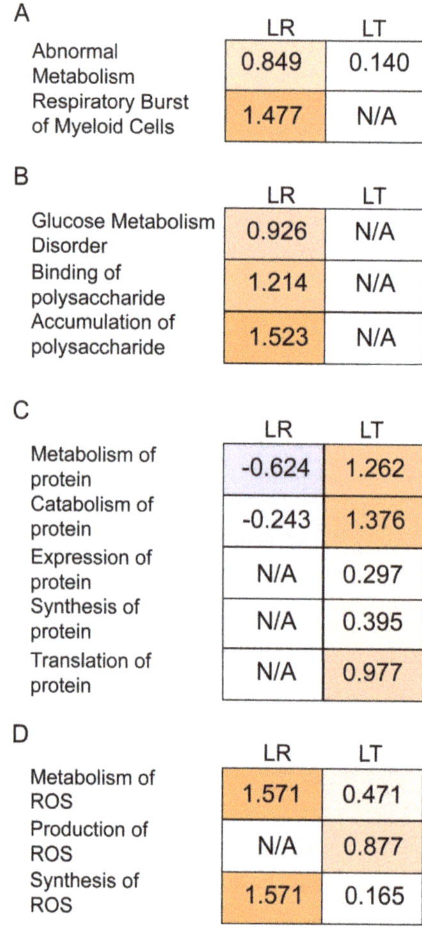

Figure 8. LPS-Tolerant RAW 264.7 cell secretomes include proteins related to global changes in cellular metabolism. z-scores of the metabolic pathway results generated by Ingenuity Pathway Analysis show abnormal metabolism and Respiratory Burst of Myeloid Cells in LR cells (**A**), increased carbohydrate metabolism by LR cells (**B**), increased protein metabolism along with decreased protein translation by LT cells (**C**), and increased maintenance of reactive oxygen species by both cells (**D**).

4. Discussion

LPS tolerance is a cellular condition defined by a lack of a typical immune response to LPS stimulation, originally characterized by decreased levels of secreted cytokines such as TNF-α, IL-6, and IL-10 (Figure 1). We have shown that LPS-tolerant RAW 264.7 cells secrete a wide variety of proteins, including several not typically found in the secretome, defined as proteins released from the cells as described by Koppenol-Raab et al. [19] (Figures 3 and 4). Similarly, while LPS-responding cells have basal metabolic rates the same as or above unstimulated control cells, LPS-tolerant cells show a significant decrease in their glycolytic and aerobic respirations (Figure 2).

4.1. Most Evident Protein Level Changes in the Secretome

Using MS analysis combined with SILAC labeling to allow for direct comparisons of the Con (NT), LR, and LT secretomes, we identified global changes in the secretome following the induction of either the LPS response or LPS tolerance (Figure 3). It is important to note that the experimental setup with serum-free media necessary to facilitate mass spectrometry-based proteomics may affect the cell response. We have established that the cells respond to TLR ligands for up to 24 h, with the secretion patterns of known inflammatory cytokines being the same as the cells in the complete media [19], but there is a probability that some elements of the response to the second LPS stimulation, although many controls are as predicted for the LPS-tolerant state, may be changed by this variable. A comparison of the LR and LT secretomes further confirmed the vast differences in the quantity and types of secreted proteins that had significantly enhanced secretion (Figure 4). Using pathway analysis of the secreted proteins, we found that the LR cell secretome is highly associated with the innate immune response (Figures 5 and 6). In contrast, the LT cell secretome is highly associated with cell survival and modulation of cellular metabolism (Figures 5–7). These modulations focus on many aspects of both protein metabolism and reactive oxygen species metabolism (Figure 8).

4.2. Potential Protein Regulators of LPS Tolerance

The clear differences between LR and LT cells raise the question of which signaling molecules induce and maintain LPS tolerance following multiple stimulations with LPS. Possible inducers or regulators of LPS tolerance could be secreted proteins (previously shown by either the protein itself, a closely related protein, or a homolog), specifically enriched in the LPS-tolerant cells, that have been previously linked to two or three of the critical functions we identified above (cell survival, protein metabolism, and maintenance of ROS). An examination of the LT group proteins identified several proteins that fulfill many of these requirements (Table 1). Three proteins that were linked to all three of the critical functions were superoxide dismutase 2 (SOD2, just below the statistical significance threshold but important to mention), sequestosome 1 (SQSTM1), and osteopontin 1 (SPP1). In addition, three secreted proteins were specifically enriched and involved in cell survival along with redox. The last group of seven proteins have been shown to be secreted, were specifically enriched, and were involved in cell survival along with protein metabolism.

One protein with a direct association with the metabolism of protein and reactive oxygen species is mitochondrial superoxide dismutase (SOD2), whose deficiency has been linked to inflammatory disorders [31]. SOD2 has been shown to be increased in the process of the macrophage protection from reactive oxygen species-induced cell death [32]. Interestingly, its upregulation was described together with the downregulation of PARP1, an enzyme adding ADP-ribose to many proteins, a modification which we have recently shown to be regulated by LPS in macrophages [33]. Since we have found many proteins involved in the inhibition of apoptosis and necrosis in LPS-tolerant cells, there may be crucial mechanisms affected by proteins within this group that can be targeted for tolerance induction.

Table 1. Potential regulators of LPS tolerance. Proteins with enhanced secretion by LPS-Tolerant RAW 264.7 cells vs. control and LPS-Responding cells that have associations with cell survival along with associations with either protein metabolism, reactive oxygen species metabolism, or both. "Secreted" results include "Yes" for proteins shown previously to be secreted in mice, "Yes (related)" for proteins shown previously to be secreted in humans, or "Yes (exosome)" for proteins shown previously to be secreted in exosomes in humans.

Uniprot	Protein Name	Entrez Gene Name	LT Enrichment/ LR Enrichment	Secreted	Cell Survival	Protein Metabolism	Redox
P09671	SOD2	superoxide dismutase 2	3.7913	Yes (exosome)	Yes	Yes	Yes
Q64337	SQSTM1	sequestosome 1	4.6730	Yes (exosome)	Yes	Yes	Yes
P10923	SPP1	secreted phosphoprotein 1	2.3679	Yes	Yes	Yes	Yes
Q62351	TFRC	transferrin receptor	4.7225	Yes	Yes	No	Yes
P60710	ACTB	actin beta	3.6136	Yes (related)	Yes	No	Yes
P99029	PRDX5	peroxiredoxin 5	5.3956	Yes (related)	Yes	No	Yes
Q3TCH7	CUL4A	cullin 4A	1.9345	Yes (exosome)	Yes	Yes	No
Q9CXW4	RPL11	ribosomal protein L7	1.1280	Yes (exosome)	Yes	Yes	No
P46471	PSMC2	proteasome 26S subunit, ATPase 2	1.6182	Yes (related)	Yes	Yes	No
P11438	LAMP1	lysosomal associated membrane protein 1	2.8676	Yes (exosome)	Yes	Yes	No
P22777	SERPINE1	serpin family E member 1	2.2662	Yes	Yes	Yes	No
P25085	IL1RN	interleukin 1 receptor antagonist	1.0980	Yes	Yes	Yes	No
P41245	MMP9	matrix metallopeptidase 9	3.7096	Yes	Yes	Yes	No

The second secreted protein that affects all three processes is sequestosome 1 (SQSTM1, or p62), a receptor for selective autophagy that is responsible for sequestering cytoplasmic components into an autophagosome [34] and which, by its role in regulation of autophagy, affects macrophage survival. Because of these roles in autophagy, its upregulation following LPS tolerance would be another indication of the switch to survival mode. Additionally, SQSTSM1/p62 has been proposed to act as an inflammatory signaling platform after activation by transforming growth factor beta-activated kinase 1 (TAK1) (one of the kinases essential in TLR4 signaling [35]), effectively disabling it as an autophagy receptor and inhibiting its own degradation [36].

The final secreted protein that affects all three processes is osteopontin 1 (SPP1), a secreted bone matrix glycoprotein protein that is essential for bone homeostasis and control of cell migration [37,38]. SPP1 has also been shown to be expressed by macrophages during tissue repair after myocardial infarction [39], indicating its function in the tissue homeostasis function of macrophages as opposed to the inflammatory function. The secreted proteins may also provide an autocrine signal to balance cytokine production, a main feature of LPS tolerance. For example, in LT, the IL-1 receptor antagonist (Table 1) might directly decrease cytokine production [40], while Lipocalin-2 counteracts LT through the induction of cytokine production [15]. Hence, the understanding of these proteins and complex feedback loops is fundamental to control LPS signaling and macrophage function.

The specificity of the secreted proteins associated with LPS tolerance does raise the question of the role of the regulation of protein signaling in leading to stimulation type-specific protein secretion in the initiation and maintenance of LPS tolerance. The role of post-translational modifications, especially phosphorylation, was pointed out as a regulatory mechanism in LPS tolerance nearly thirty years ago [41] and linked to crosstalk with other signaling pathways, for example, Fc gamma receptors (FcGRs) [42]. On the other hand, pathways intuitively associated with the regulation of the immune response may not be required for the induction of endotoxin tolerance, as shown for type 1 interferon signaling [43]. In addition to TLR4 signaling, NLRP3 inflammasome has been shown to play an important role in the response to LPS and has recently been shown to be regulated by specific lipid mediators [44]. These results may open another avenue for exploration of the mechanisms of LPS tolerance and for explanation of changes in the secretome. Cellular metabolism has recently emerged as a regulator of macrophage phenotype in general [45]. Unbiased secreted protein profiling and system-level characterization of changes in innate immune signaling and cellular metabolism, pointing to regulation at the post-transcriptional level, emphasize the importance of global studies that reach beyond gene expression analysis. Our study reveals the value of proteomics approaches that can explain rapid functional changes necessary for effective immune function.

In the clinic, macrophage LPS tolerance could be either beneficial or harmful to the host, depending on other factors. While well-controlled LPS tolerance reduces overwhelming cytokine production (cytokine storm) and attenuates sepsis severity [46], unhinged LPS tolerance immune exhaustion might be harmful [47]. Novel ways to inhibit cytokine secretion and controlled induction of LPS tolerance should therefore be considered as a future treatment of septic shock.

Supplementary Materials: The following are available online at https://www.mdpi.com/2218-273X/11/2/164/s1, Supplementary Table S1. Sheet 1: All of the proteins identified in all the analyses (protein identifiers: Column C). Sheet 2: Protein result filtering to identify proteins with two or more peptides and 12 or more valid values. Sheets 3–5: Imputation of missing values and conversion of data into Log2 Fold changes in protein intensity recorded. Sheet 6: Summary of protein quantification and *p*-values for all the proteins quantified. Supplementary Table S2. All of the biological processes and cellular functions examined in the LPS-Responding and LPS-Tolerant cells, with *p*-values and significance marked with asterisks (increasing significance is indicated with more asterisks).

Author Contributions: J.G.: Conceptualization, Data curation, Formal Analysis, Investigation, Methodology, Visualisation, Writing—Original draft, Writing—review&editing; T.O.: Investigation, Methodology, Writing—Original draft, Writing—review&editing; D.G.: Investigation, Methodology, Visualisation, Writing—Original draft, Writing—review&editing; J.I.-A.: Investigation, Methodology, Writing—Original draft, Writing—review&editing; N.P.M.: author (6) Formal Analysis, Methodology, Visualisation, Writing—review&editing; S.H.Y.: Investigation, Methodology, Writing—Original draft, Writing—review&editing; A.L.: Conceptualization, Funding acquisition, Methodology, Resources, Supervision, Writing—review&editing; A.N.-L.: Conceptualization, Funding acquisition, Methodology, Project Administration, Resources, Supervision, Writing—Original draft, Writing—review&editing. All authors have read and agreed to the published version of the manuscript.

Funding: This research was supported by the Intramural Research Program of NIAID, NIH, Thailand Government Fund (RSA-6080023), the Thailand Research Fund (RES_61_202_30_022), the Ratchadaphiseksomphot Endowment Fund 2017 (76001-HR), the Second Century Fund (C2F), Chulalongkorn University (to T. O.), and the Program Management Unit for Human Resources and Institutional Development Research and Innovation—CU (Global Partnership B16F630071 and Flagship B05F630073).

Institutional Review Board Statement: Not applicable.

Informed Consent Statement: Not applicable.

Data Availability Statement: The mass spectrometry-based proteomics data have been deposited to the ProteomeXchange Consortium via the PRIDE partner repository with the dataset identifier PXD021925.

Acknowledgments: This research was supported in part by the Intramural Research Program of NIAID, NIH.

Conflicts of Interest: The authors declare no conflict of interest.

References

1. Akira, S.; Uematsu, S.; Takeuchi, O. Pathogen recognition and innate immunity. *Cell* **2006**, *124*, 783–801. [CrossRef] [PubMed]
2. Kawai, T.; Akira, S. The role of pattern-recognition receptors in innate immunity: Update on Toll-like receptors. *Nat. Immunol.* **2010**, *11*, 373–384. [CrossRef] [PubMed]
3. Salomao, R.; Brunialti, M.K.; Rapozo, M.M.; Baggio-Zappia, G.L.; Galanos, C.; Freudenberg, M. Bacterial sensing, cell signaling, and modulation of the immune response during sepsis. *Shock* **2012**, *38*, 227–242. [CrossRef]
4. Seeley, J.J.; Ghosh, S. Molecular mechanisms of innate memory and tolerance to LPS. *J. Leukoc. Biol.* **2017**, *101*, 107–119. [CrossRef] [PubMed]
5. Hotchkiss, R.S.; Monneret, G.; Payen, D. Sepsis-induced immunosuppression: From cellular dysfunctions to immunotherapy. *Nat. Rev. Immunol.* **2013**, *13*, 862–874. [CrossRef] [PubMed]
6. Foster, S.L.; Medzhitov, R. Gene-specific control of the TLR-induced inflammatory response. *Clin. Immunol.* **2009**, *130*, 7–15. [CrossRef] [PubMed]
7. Netea, M.G.; Quintin, J.; van der Meer, J.W. Trained immunity: A memory for innate host defense. *Cell Host Microbe* **2011**, *9*, 355–361. [CrossRef] [PubMed]
8. Biswas, S.K.; Lopez-Collazo, E. Endotoxin tolerance: New mechanisms, molecules and clinical significance. *Trends Immunol.* **2009**, *30*, 475–487. [CrossRef]
9. Quinn, E.M.; Wang, J.; Redmond, H.P. The emerging role of microRNA in regulation of endotoxin tolerance. *J. Leukoc. Biol.* **2012**, *91*, 721–727. [CrossRef]
10. He, X.; Jing, Z.; Cheng, G. MicroRNAs: New regulators of Toll-like receptor signalling pathways. *Biomed. Res. Int.* **2014**, *2014*, 945169. [CrossRef]
11. Pena, O.M.; Hancock, D.G.; Lyle, N.H.; Linder, A.; Russell, J.A.; Xia, J.; Fjell, C.D.; Boyd, J.H.; Hancock, R.E. An Endotoxin Tolerance Signature Predicts Sepsis and Organ Dysfunction at Initial Clinical Presentation. *EBioMedicine* **2014**, *1*, 64–71. [CrossRef] [PubMed]
12. Shalova, I.N.; Lim, J.Y.; Chittezhath, M.; Zinkernagel, A.S.; Beasley, F.; Hernandez-Jimenez, E.; Toledano, V.; Cubillos-Zapata, C.; Rapisarda, A.; Chen, J.; et al. Human monocytes undergo functional re-programming during sepsis mediated by hypoxia-inducible factor-1alpha. *Immunity* **2015**, *42*, 484–498. [CrossRef] [PubMed]
13. Vergadi, E.; Vaporidi, K.; Tsatsanis, C. Regulation of Endotoxin Tolerance and Compensatory Anti-inflammatory Response Syndrome by Non-coding RNAs. *Front. Immunol.* **2018**, *9*, 2705. [CrossRef] [PubMed]
14. Lopez-Collazo, E.; del Fresno, C. Pathophysiology of endotoxin tolerance: Mechanisms and clinical consequences. *Crit. Care* **2013**, *17*, 242. [CrossRef]
15. Ondee, T.; Gillen, J.; Visitchanakun, P.; Somparn, P.; Issara-Amphorn, J.; Dang Phi, C.; Chancharoenthana, W.; Gurusamy, D.; Nita-Lazar, A.; Leelahavanichkul, A. Lipocalin-2 (Lcn-2) Attenuates Polymicrobial Sepsis with LPS Preconditioning (LPS Tolerance) in FcGRIIb Deficient Lupus Mice. *Cells* **2019**, *8*, 1064. [CrossRef]
16. Dominguez-Andres, J.; Novakovic, B.; Li, Y.; Scicluna, B.P.; Gresnigt, M.S.; Arts, R.J.W.; Oosting, M.; Moorlag, S.; Groh, L.A.; Zwaag, J.; et al. The Itaconate Pathway Is a Central Regulatory Node Linking Innate Immune Tolerance and Trained Immunity. *Cell Metab* **2019**, *29*, 211–220.e5. [CrossRef]
17. Widdrington, J.D.; Gomez-Duran, A.; Pyle, A.; Ruchaud-Sparagano, M.H.; Scott, J.; Baudouin, S.V.; Rostron, A.J.; Lovat, P.E.; Chinnery, P.F.; Simpson, A.J. Exposure of Monocytic Cells to Lipopolysaccharide Induces Coordinated Endotoxin Tolerance, Mitochondrial Biogenesis, Mitophagy, and Antioxidant Defenses. *Front. Immunol.* **2018**, *9*, 2217. [CrossRef]
18. Mosser, D.M.; Edwards, J.P. Exploring the full spectrum of macrophage activation. *Nat. Rev. Immunol.* **2008**, *8*, 958–969. [CrossRef]
19. Koppenol-Raab, M.; Sjoelund, V.; Manes, N.P.; Gottschalk, R.A.; Dutta, B.; Benet, Z.L.; Fraser, I.D.; Nita-Lazar, A. Proteome and Secretome Analysis Reveals Differential Post-transcriptional Regulation of Toll-like Receptor Responses. *Mol. Cell Proteom.* **2017**, *16*, S172–S186. [CrossRef]
20. Meissner, F.; Scheltema, R.A.; Mollenkopf, H.J.; Mann, M. Direct proteomic quantification of the secretome of activated immune cells. *Science* **2013**, *340*, 475–478. [CrossRef]
21. Eichelbaum, K.; Krijgsveld, J. Rapid temporal dynamics of transcription, protein synthesis, and secretion during macrophage activation. *Mol. Cell Proteom.* **2014**, *13*, 792–810. [CrossRef] [PubMed]
22. Shevchenko, A.; Tomas, H.; Havlis, J.; Olsen, J.V.; Mann, M. In-gel digestion for mass spectrometric characterization of proteins and proteomes. *Nat. Protoc.* **2006**, *1*, 2856–2860. [CrossRef] [PubMed]
23. Perez-Riverol, Y.; Csordas, A.; Bai, J.; Bernal-Llinares, M.; Hewapathirana, S.; Kundu, D.J.; Inuganti, A.; Griss, J.; Mayer, G.; Eisenacher, M.; et al. The PRIDE database and related tools and resources in 2019: Improving support for quantification data. *Nucleic Acids Res.* **2019**, *47*, D442–D450. [CrossRef] [PubMed]
24. Cox, J.; Mann, M. MaxQuant enables high peptide identification rates, individualized p.p.b.-range mass accuracies and proteome-wide protein quantification. *Nat. Biotechnol.* **2008**, *26*, 1367–1372. [CrossRef] [PubMed]
25. Cox, J.; Neuhauser, N.; Michalski, A.; Scheltema, R.A.; Olsen, J.V.; Mann, M. Andromeda: A peptide search engine integrated into the MaxQuant environment. *J. Proteome Res.* **2011**, *10*, 1794–1805. [CrossRef]
26. Tyanova, S.; Temu, T.; Cox, J. The MaxQuant computational platform for mass spectrometry-based shotgun proteomics. *Nat. Protoc.* **2016**, *11*, 2301–2319. [CrossRef]

27. Polpitiya, A.D.; Qian, W.J.; Jaitly, N.; Petyuk, V.A.; Adkins, J.N.; Camp, D.G., 2nd; Anderson, G.A.; Smith, R.D. DAnTE: A statistical tool for quantitative analysis of -omics data. *Bioinformatics* **2008**, *24*, 1556–1558. [CrossRef]
28. Motani, K.; Kosako, H. Activation of stimulator of interferon genes (STING) induces ADAM17-mediated shedding of the immune semaphorin SEMA4D. *J. Biol. Chem.* **2018**, *293*, 7717–7726. [CrossRef]
29. Hernandez, M.X.; Jiang, S.; Cole, T.A.; Chu, S.H.; Fonseca, M.I.; Fang, M.J.; Hohsfield, L.A.; Torres, M.D.; Green, K.N.; Wetsel, R.A.; et al. Prevention of C5aR1 signaling delays microglial inflammatory polarization, favors clearance pathways and suppresses cognitive loss. *Mol. Neurodegener.* **2017**, *12*, 66. [CrossRef]
30. Heissig, B.; Salama, Y.; Takahashi, S.; Osada, T.; Hattori, K. The multifaceted role of plasminogen in inflammation. *Cell Signal.* **2020**, *75*, 109761. [CrossRef]
31. Scheurmann, J.; Treiber, N.; Weber, C.; Renkl, A.C.; Frenzel, D.; Trenz-Buback, F.; Ruess, A.; Schulz, G.; Scharffetter-Kochanek, K.; Weiss, J.M. Mice with heterozygous deficiency of manganese superoxide dismutase (SOD2) have a skin immune system with features of "inflamm-aging". *Arch. Dermatol. Res.* **2014**, *306*, 143–155. [CrossRef] [PubMed]
32. Regdon, Z.; Robaszkiewicz, A.; Kovacs, K.; Rygielska, Z.; Hegedus, C.; Bodoor, K.; Szabo, E.; Virag, L. LPS protects macrophages from AIF-independent parthanatos by downregulation of PARP1 expression, induction of SOD2 expression, and a metabolic shift to aerobic glycolysis. *Free Radic. Biol. Med.* **2019**, *131*, 184–196. [CrossRef] [PubMed]
33. Daniels, C.M.; Kaplan, P.R.; Bishof, I.; Bradfield, C.; Tucholski, T.; Nuccio, A.G.; Manes, N.P.; Katz, S.; Fraser, I.D.C.; Nita-Lazar, A. Dynamic ADP-Ribosylome, Phosphoproteome, and Interactome in LPS-Activated Macrophages. *J. Proteome Res.* **2020**, *19*, 3716–3731. [CrossRef] [PubMed]
34. Lamark, T.; Svenning, S.; Johansen, T. Regulation of selective autophagy: The p62/SQSTM1 paradigm. *Essays BioChem.* **2017**, *61*, 609–624. [CrossRef]
35. Irie, T.; Muta, T.; Takeshige, K. TAK1 mediates an activation signal from toll-like receptor(s) to nuclear factor-kappaB in lipopolysaccharide-stimulated macrophages. *FEBS Lett.* **2000**, *467*, 160–164. [CrossRef]
36. Kehl, S.R.; Soos, B.A.; Saha, B.; Choi, S.W.; Herren, A.W.; Johansen, T.; Mandell, M.A. TAK1 converts Sequestosome 1/p62 from an autophagy receptor to a signaling platform. *EMBO Rep.* **2019**, *20*, e46238. [CrossRef]
37. Giachelli, C.; Bae, N.; Lombardi, D.; Majesky, M.; Schwartz, S. Molecular cloning and characterization of 2B7, a rat mRNA which distinguishes smooth muscle cell phenotypes in vitro and is identical to osteopontin (secreted phosphoprotein I, 2aR). *BioChem. Biophys. Res. Commun.* **1991**, *177*, 867–873. [CrossRef]
38. Miyauchi, A.; Alvarez, J.; Greenfield, E.M.; Teti, A.; Grano, M.; Colucci, S.; Zambonin-Zallone, A.; Ross, F.P.; Teitelbaum, S.L.; Cheresh, D.; et al. Recognition of osteopontin and related peptides by an alpha v beta 3 integrin stimulates immediate cell signals in osteoclasts. *J. Biol. Chem.* **1991**, *266*, 20369–20374. [CrossRef]
39. Murry, C.E.; Giachelli, C.M.; Schwartz, S.M.; Vracko, R. Macrophages express osteopontin during repair of myocardial necrosis. *Am. J. Pathol.* **1994**, *145*, 1450–1462.
40. Henricson, B.E.; Neta, R.; Vogel, S.N. An interleukin-1 receptor antagonist blocks lipopolysaccharide-induced colony-stimulating factor production and early endotoxin tolerance. *Infect. Immun.* **1991**, *59*, 1188–1191. [CrossRef]
41. Nakano, M.; Saito, S.; Nakano, Y.; Yamasu, H.; Matsuura, M.; Shinomiya, H. Intracellular protein phosphorylation in murine peritoneal macrophages in response to bacterial lipopolysaccharide (LPS): Effects of kinase-inhibitors and LPS-induced tolerance. *Immunobiology* **1993**, *187*, 272–282. [CrossRef]
42. Ondee, T.; Jaroonwitchawan, T.; Pisitkun, T.; Gillen, J.; Nita-Lazar, A.; Leelahavanichkul, A.; Somparn, P. Decreased Protein Kinase C-beta Type II Associated with the Prominent Endotoxin Exhaustion in the Macrophage of FcGRIIb-/- Lupus Prone Mice is Revealed by Phosphoproteomic Analysis. *Int. J. Mol. Sci* **2019**, *20*, 1354. [CrossRef] [PubMed]
43. Karimi, Y.; Poznanski, S.M.; Vahedi, F.; Chen, B.; Chew, M.V.; Lee, A.J.; Ashkar, A.A. Type I interferon signalling is not required for the induction of endotoxin tolerance. *Cytokine* **2017**, *95*, 7–11. [CrossRef] [PubMed]
44. Lopategi, A.; Flores-Costa, R.; Rius, B.; Lopez-Vicario, C.; Alcaraz-Quiles, J.; Titos, E.; Claria, J. Frontline Science: Specialized proresolving lipid mediators inhibit the priming and activation of the macrophage NLRP3 inflammasome. *J. Leukoc. Biol.* **2019**, *105*, 25–36. [CrossRef] [PubMed]
45. Saha, S.; Shalova, I.N.; Biswas, S.K. Metabolic regulation of macrophage phenotype and function. *Immunol. Rev.* **2017**, *280*, 102–111. [CrossRef] [PubMed]
46. Cavaillon, J.M.; Adib-Conquy, M. Bench-to-bedside review: Endotoxin tolerance as a model of leukocyte reprogramming in sepsis. *Crit. Care* **2006**, *10*, 233. [CrossRef] [PubMed]
47. Ondee, T.; Surawut, S.; Taratummarat, S.; Hirankarn, N.; Palaga, T.; Pisitkun, P.; Pisitkun, T.; Leelahavanichkul, A. Fc Gamma Receptor IIB Deficient Mice: A Lupus Model with Increased Endotoxin Tolerance-Related Sepsis Susceptibility. *Shock* **2017**, *47*, 743–752. [CrossRef] [PubMed]

 MDPI

Article

Omics Technologies to Decipher Regulatory Networks in Granulocytic Cell Differentiation

Svetlana Novikova †, Olga Tikhonova †, Leonid Kurbatov, Tatiana Farafonova, Igor Vakhrushev, Alexey Lupatov, Konstantin Yarygin and Victor Zgoda *

Orekhovich Institute of Biomedical Chemistry, Pogodinskaya 10, 119121 Moscow, Russia; novikova.s.e3101@gmail.com (S.N.); ovt.facility@gmail.com (O.T.); kurbatovl@mail.ru (L.K.); farafonova.tatiana@gmail.com (T.F.); vakhrunya@gmail.com (I.V.); alupatov@inbox.ru (A.L.); kyarygin@yandex.ru (K.Y.)

* Correspondence: victor.zgoda@gmail.com

† These authors contributed equally to this work.

Abstract: Induced granulocytic differentiation of human leukemic cells under all-*trans*-retinoid acid (ATRA) treatment underlies differentiation therapy of acute myeloid leukemia. Knowing the regulation of this process it is possible to identify potential targets for antileukemic drugs and develop novel approaches to differentiation therapy. In this study, we have performed transcriptomic and proteomic profiling to reveal up- and down-regulated transcripts and proteins during time-course experiments. Using data on differentially expressed transcripts and proteins we have applied upstream regulator search and obtained transcriptome- and proteome-based regulatory networks of induced granulocytic differentiation that cover both up-regulated (HIC1, NFKBIA, and CASP9) and down-regulated (PARP1, VDR, and RXRA) elements. To verify the designed network we measured HIC1 and PARP1 protein abundance during granulocytic differentiation by selected reaction monitoring (SRM) using stable isotopically labeled peptide standards. We also revealed that transcription factor CEBPB and LYN kinase were involved in differentiation onset, and evaluated their protein levels by SRM technique. Obtained results indicate that the omics data reflect involvement of the DNA repair system and the MAPK kinase cascade as well as show the balance between the processes of the cell survival and apoptosis in a p53-independent manner. The differentially expressed transcripts and proteins, predicted transcriptional factors, and key molecules such as HIC1, CEBPB, LYN, and PARP1 may be considered as potential targets for differentiation therapy of acute myeloid leukemia.

Keywords: acute myeloid leukemia; HL-60 cell line; ATRA; induced differentiation; transcriptome; proteome; transcription factors; key molecules; regulatory pathway modelling; SRM

Citation: Novikova, S.; Tikhonova, O.; Kurbatov, L.; Farafonova, T.; Vakhrushev, I.; Lupatov, A.; Yarygin, K.; Zgoda, V. Omics Technologies to Decipher Regulatory Networks in Granulocytic Cell Differentiation. *Biomolecules* **2021**, *11*, 907. https://doi.org/10.3390/biom11060907

Academic Editors: Christopher Gerner and Michelle Hill

Received: 23 April 2021
Accepted: 15 June 2021
Published: 18 June 2021

1. Introduction

Cell differentiation is a fundamental process of the development, growth, reproduction of multicellular organisms. Regulation of cell differentiation has been for decades and remains an important task for investigation due to its importance in cancer and many other diseases therapy. Leukemic cells that are induced to differentiate under all-*trans*-retinoid acid (ATRA) treatment make a convenient model for studying of cell maturation in vitro.

Normally, ATRA in physiological dosage binds and activates a heterodimer receptor RAR/RXR followed by release of histone deacetylases (HDACs) and transcription co-repressors (N-CoR or SMRT), and by recruitment of transcription co-activators (NcoA-1/SRC-1, CBP/p300, p/CIP, and ACTR) [1]. In turn, retinoic acid response element (RARE) containing genes, which are repressed by nonactive RAR/RXR, trigger the further cascade of molecular events leading to myeloid precursor's maturation into functional granulocytes. Various mutations impair granulocytic differentiation resulting in highly heterogeneous acute myeloid leukemia (AML), which could be cured by high dosage of ATRA. In the case of AML subtype M3 (French–American–British (FAB) classification), a.k.a acute promyelocytic leukemia (APL), deleterious mutation, namely balanced chromosomal translocation between chromosomes 15 and 17 t (15;17) (q24; q21), affects retinoic acid (RA) receptor gene *RARα* resulting in formation of dominant negative fusion protein PLM-RARA [2]. The NB4 cell line that harbors such a hallmark mutation, is used as a model to study of the APL cell biology [3]. In APL cells the transcription of RA-responsive genes is blocked due to the increased avidity of PLM-RARA/RXR for co-repressor molecules [4]. The treatment with high dosage of ATRA induces dissociation of co-repressors from PML-RARA and triggers fusion protein degradation via the ubiquitin-proteasome or autophagy pathway [5,6]. The ATRA-based regimens that are used as a first-line treatment of APL patients, induce complete remission rates of 90% [7]. Nevertheless, other types of AML are not that successfully treatable with the 5-years survival rates only about 40–45% [8]. Meanwhile, antileukemic effect of ATRA was also observed in AML (non-APL) cell models, including HL-60, THP-1, MOLM-14, HF-6, and U937 cell lines [9].

The HL-60 promyelocytic leukemia cell line is classified as AML with maturation, also referred to as AML subtype M2 by FAB classification [10]. These cells were isolated in 1977 from a patient with acute myeloid leukemia. Later it was found that, these promyelocytic cells could be induced to differentiate into granulocytes in vitro by ATRA [11]. The HL60 cell genome contains normal *RARα* gene, an amplified *c-myc* proto-oncogene and deficient of *p53* gene [12,13]. Notably, deletion in the *p53* gene occurs at a frequency of up to 10% in de novo AML (non-APL) cases and associated with exceedingly adverse prognosis regardless of the type of mutation (missense, nonsense, small insertions, and deletions, etc.) [8]. Being ATRA-responsive, the HL60 cell line has been used for decades as a convenient model object for cell differentiation [11,14].

Omics technologies represent powerful tools for a full-scale analysis of gene and protein expression that allow for gaining important molecular information about differentiation process, and acquiring the complete picture of the cell maturation. Thus, using HL-60 (AML) and NB4 (APL) cell lines as model systems, the complexity of differentiation processes and the diversity of pathways involved in induced differentiation at transcriptome [15–17] and proteome [18,19] levels have been demonstrated.

Despite the fact that proteomics and transcriptomics alone represent the powerful techniques for investigation of ATRA-induced differentiation, the systems approach is appealing to the elucidation of molecular mechanisms. In this respect, the systems study was performed on the NB4 promyelocytic cell line under ATRA treatment (alone or in combination with arsenic trioxide (ATO)) in a time-course manner. By applying microarray technology and 2D-gel electrophoresis followed by MALDI-TOF-TOF analysis, transcription factors (TFs) and co-factors responsible for global changes in transcriptional regulation and involved in stimulation of the IFN-pathway, cell cycle arrest, and activation of signal transduction have been unmasked [3].

However, even simultaneous analysis of proteome and transcriptome differences observed in the experiment is not always sufficient to unravel regulatory mechanisms. The up- or down-regulation of protein and transcript levels under ATRA treatment is often caused by previous regulatory events. Predicting transcription factors, responsible for altered gene expression, and revealing, in turn, their putative regulators, a hierarchical model of induced differentiation could be built. Therefore, a bioinformatics search for upstream regulators, including transcription factors [20], is an appropriate tool for proteome and transcriptome data interpretation. Identification and analysis of TFs and regulatory pathways responsible for altered gene or protein expression that result in the cell differentiation may contribute to identification of the mechanism(s) underlying this complex process.

2. Materials and Methods

2.1. Experimental Design

The time-course studying of induced granulocytic differentiation allows obtainment of the most accurate data on molecular perturbations under ATRA treatment. Previously, several schedules of HL-60 cell harvesting after ATRA treatment have been applied in the time-course experiments [3,21]. To perform transcriptomic and proteomic profiling, we selected 24 and 96 h time points, when the molecular perturbations are prominent. To reveal the molecular onset of cell maturation at transcriptome and proteome levels, we also added the 3 h time point. For proteomic experiment we also studied the time point 48 h of treatment; during this period HL-60 cells underwent two division cycles. In our preliminary mass-spectrometry experiments we did not observe any significant changes in the ATRA-induced cell proteome within the first 2 h (compared to 0 h) after ATRA induction or at 72 h (compared to 96 h) after treatment (data not shown).

For proteome analysis, we performed the ATRA-induced differentiation experiments in three independent biological replicates. HL-60 cells were harvested at 0, 3, 24, 48, and 96 h after ATRA treatment (overall 15 samples). For the transcriptome analysis, HL-60 cells were subjected to ATRA treatment in three biological replicates and were harvested at 0, 3, 24, and 96 h (overall 12 samples).

For the proteome analysis, the LC-MS/MS experiments were carried out in five technical replicates per time point, and the whole-genome transcriptome analysis was performed in three technical replicates per time point.

Cells harvested before ATRA treatment (time point 0 h) served as controls for both transcriptomic and proteomic profiling. The study workflow is shown in Figure 1.

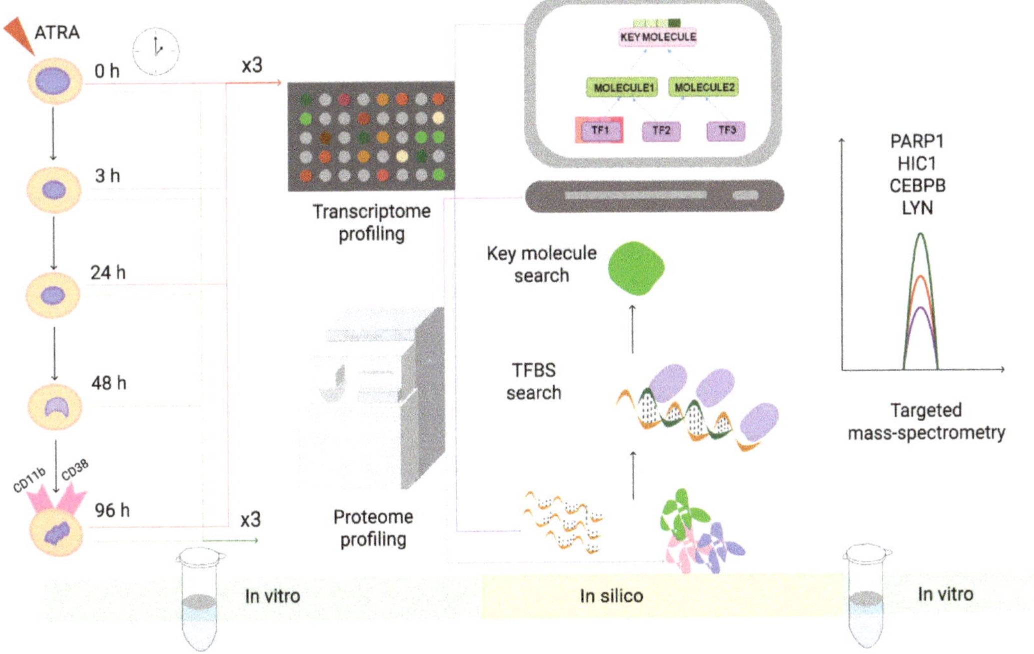

Figure 1. The study workflow. We applied a multi-disciplinary platform to study ATRA-induced granulocytic differentiation in a time-course manner using HL-60 cell line as a model. We combined LC-MS/MS analysis (0, 3, 24, 48, and 96 h after ATRA treatment, three bio repeats), whole-genome transcriptome analysis (0, 3, 24, and 96 h after ATRA treatment, three bio repeats), and bioinformatic search for transcription factor binding sites (TFBS) and for the key regulatory molecules. To verify the predicted regulatory networks the abundance of proteins HIC1, CEBPB, LYN, and PARP1, belonging to the designed model regulatory networks or involving in differentiation onset, were measured in time-course manner by selected reaction monitoring (SRM) using synthetic isotopically-labeled peptides as standard.

2.2. HL60 Cells Cultures

The HL-60 human promyelocytic leukemia cells (obtained from the cell culture bank Institute of Biomedical Chemistry (IBMC), Moscow, Russia) were grown in RPMI-1640 medium supplemented with 10% fetal bovine serum, 100 U/mL penicillin, 100 U/mL streptomycin and 2 mM L-glutamine (all Gibco™, Paisley, UK) in a CO_2 incubator under standard conditions (37 °C, 5% CO_2, 80% humidity). ATRA (Sigma-Aldrich, St. Louis, MO, USA) was dissolved in ethanol as a stock solution at 1 mM. HL-60 cells were treated with ATRA as described in [3] and control HL-60 cells were treated with an equal volume of the solvent (ethanol).

Cell differentiation was evaluated by the CD11b and CD38 expression measured by flow cytometry. At the selected time points, the cells were harvested, washed twice with PBS, transferred to 1.5-mL Eppendorf tubes, and pelleted by centrifugation at 3000× *g* for 15 min using an Eppendorf 5424R centrifuge (Eppendorf, Hamburg, Germany). After removing the supernatants, the cell pellets were frozen in liquid nitrogen and stored until transcriptomic and proteomic analysis.

2.3. Transcriptome Analysis

Total RNA was isolated from the cells using RNeasy Mini Kit (Qiagen, Hilden, Germany) at each time point studied. The quality of the extracted RNA was controlled using a Bioanalyzer 2100, RNA 6000 Nano LabChips, and the 2100 Expert standard software (all Agilent Technologies, Santa Clara, CA, USA). Approximately 0.5 µg of each RNA sample was used for cDNA preparation in the reaction of the reverse transcription performed using a Low RNA Input Linear Amp Kit (Agilent Technologies, Santa Clara, CA, USA) according to standard protocol. The cRNA samples for all time points were labeled with Cy5-CTP (Perkin Elmer, Waltham, MA, USA) and with Cy3-CTP (Perkin Elmer, Waltham, MA, USA) for the control sample (the time point 0 h). The cRNA fragmentations and hybridizations were performed using a standard protocol with an in situ Hybridization Kit Plus (Agilent Technologies, Santa Clara, CA, USA). Data acquisition was carried out using a DNA Microarray Scanner G2505C (Agilent Technologies, Santa Clara, CA, USA). The primary transcriptome data were processed using the Feature Extraction software (version 10.1.3.1; Agilent Technologies, Santa Clara, CA, USA).

Statistical data analysis by ANOVA with the *p*-value cut-off set at 0.05 was performed using the GeneSpring GX12.5 software (Agilent Technologies, Santa Clara, CA, USA). Thus, we prepared the lists of genes that showed more than two-fold expression difference at least at one time point studied.

2.4. Preparation of HL60 Cells Lysates and In-Solution Digestion with Trypsin

The cell samples were lysed using ice-cold buffer (150 µL) containing 3% sodium deoxycholate, 2.5 mM EDTA, 75 mM Tris-HCl (all Sigma-Aldrich, St. Louis, MO, USA), pH 8.5 and protease inhibitors cOmplete™ (Roche, Basel, Switzerland) with subsequent ultrasonication using the Bandelin Sonopuls probe ("BANDELIN electronic GmbH & Co. KG", Berlin, Germany). The cell lysates were centrifuged for 15 min at 5000× *g* using Eppendorf 5424R centrifuge. The supernatants were collected, and the pellets were dissolved in 100 µL of lysis buffer, and then subjected to the second round of protein solubilization as described above. The sample protein concentration was measured using a Pierce™ BCA Protein Assay Kit (Pierce, Rockford, IL, USA). Protein digestion was performed according to the protocol described in detail by Zgoda et al. [22]. Briefly, the protein sample (about 100 µg) was transferred into a clean tube and denaturation solution (5 M urea, 1% sodium deoxycholate, in a 50 mM triethylammonium bicarbonate buffer (TEAB) containing 20mM dithiothreitol (DTT) (all Sigma-Aldrich, St. Louis, MO, USA) 20 mM DTT) in volume of 20 µL was added to make the final concentration of total protein close to 5 mg/mL. Then the samples were heated for 60 min at 42 °C and, after cooling at room temperature, 25 µL of 15 mM 2-iodoacetamide in 50 mM TEAB was added. The alkylation reaction continued for 30 min at room temperature and the sample was then diluted up to 120 μL by 50 mM TEAB to decrease the final concentration of denaturation buffer compounds and dilute the final protein concentration close to 0.5 mg/mL. Trypsin (1 μg) was added to samples and incubated overnight at 37 °C. The hydrolysis was stopped by adding formic acid (to a final concentration of 5%). Samples were centrifuged for 10 min at 10 °C at 12,000× *g* to sediment deoxycholic acid. The supernatant was transferred into a clean tube. In the obtained supernatants, the total peptide concentration was determined by the colorimetric method using a Pierce™ Quantitative Colorimetric Peptide Assay kit (Thermo Scientific, Waltham, MA, USA) in accordance with the manufacturer's recommendations. The peptides were dried and dissolved in 0.1% formic acid to a final concentration of 1 μg/μL.

2.5. Shotgun Mass Spectrometry

The peptide samples obtained were analyzed using the Agilent HPLC system 1100 Series (Agilent Technologies, Santa Clara, CA, USA) connected to a hybrid linear ion trap LTQ Orbitrap Velos, equipped with a nanoelectrospray ion source (Thermo Scientific, Waltham, MA, USA). Peptide separations were carried out on a RP-HPLC Zorbax 300SB-C18 column (C18 3.5 μm, 75 μm inner diameter and 150 mm length, Agilent Technologies, Santa Clara, CA, USA) using a linear gradient from 95% solvent A (water, 0.1% formic acid) and 5% solvent B (water, 0.1% formic acid, and 80% acetonitrile) to 60% solvent B over 85 min at a flow rate of 0.3 μL/min.

Mass spectra were acquired in the positive ion mode using Orbitrap analyzer with a resolution of 30,000 (m/z = 400) for MS and 7500 (m/z = 400) for MS/MS scans. The AGC target was set at 2×10^5 and 1×10^5 with maximum ion injection time 50 ms and 100 ms for MS and MS/MS, respectively. Survey MS scan was followed by MS/MS spectra for five the most abundant precursors. The higher energy collisional dissociation (HCD) was used, and normalized collision energy was set to 35 eV. Signal threshold was set to 5000 for an isolation window of 2 m/z. The precursors fragmented were dynamically excluded from targeting with repeat count 1, repeat duration 10 s, and exclusion duration 60 s. Singly charged ions and those with not defined charge state were excluded from triggering the MS/MS scans.

2.6. Data Analysis

The mass spectrometry data were analyzed using SPIRE pipeline [23]. The raw mass spectrometry data were converted to the mzXML format with the RawToMzXML convertor and uploaded into the SPIRE server. The experimental data were assigned to five time points (0, 3, 24, 48, and 96 h); each point included three biological- with five technical replicates. The data obtained were searched by the in-built «Composite» search engine within SPIRE pipeline using the following parameters: enzyme specificity was set to trypsin, two missed cleavages were allowed. Carbamidomethylation of cysteines was set as fixed modification and methionine oxidation was set as variable modification for the peptide search. The mass tolerance for precursor ions was 10 ppm; the mass tolerance for fragment ions was 20 ppm. Human FASTA file (September 2015) was used as a protein sequence database. The spectra identified with 90% probability were assigned to peptides. The local false discovery rate for protein identification was set bellow 0.01 (locFDR < 0.01). locFDR was calculated in SPIRE utilizing randomized or decoy database searches [23].

Label-free quantitation was performed with the use of the SPIRE software by default settings. Expression ratios and p-values were calculated based on an over-dispersed Poisson model using an empirical Bayes correction [23]. The proteins with the expression fold change > 1.5, p-value < 0.05 and CV between biological repeats < 30%, were considered as differentially expressed. The imputation of missing data has not been applied to mass-spectrometric results.

The volcano plot was obtained using VolcaNoseR web app [24].

2.7. Functional Classification of Differentially Expressed Genes and Proteins

Functional analysis of differentially expressed genes/proteins was carried out using the «Functional classification» option of the geneXplain platform (http://platform.genexplain.com) with GO and PROTEOME Databases (BIOBASE) implemented as a module of the GeneXplain platform.

For the functional analysis of gene groups exhibiting altered expression at the selected time points of cell differentiation, the cut-off value for the probability of random gene allocation of a gene to a particular group (Adjusted p-value) was set at 5×10^{-4}. Only statistically significant classification of genes according to the GO categories, describing various biological processes in cells, was taken into consideration for the functional analysis.

The STRING database v.11.0 was used to retrieve the protein–protein interactions (PPIs) from the lists of DEGs of MCD group at 3, 24, and 96 h. A high confidence (0.9) score was applied. The active interaction sources were experiments and curated databases. The built-in functional enrichment analysis results according to the molecular function (GO), and KEGG pathways were used for visualization.

2.8. Search for Transcription Factors, Putatively Regulating Gene and Protein Expression during ATRA-Induced Differentiation of HL-60 Cells

The search for over-represented transcription factor binding sites (TFBS) was performed using geneXplain platform 2.0 software packages (http://platform.genexplain.com) and TRANSFAC® database [25]. The differentially expressed genes/proteins at different time points were considered as the test sets (Yes-sets). The gene/protein that did not show any expression changes after ATRA treatment were used as a background set (No-sets). The profile used for analysis contains a collection of vertebrate non-redundant transcription factor matrices. The promoter window was selected from −1000 to +100 from the transcription start site, and only the best-supported promoters of the genes analyzed were used. The cut-off values with a threshold of p-value < 0.005 were selected to obtain high-scoring binding sites. The matrices with high over-representation of site frequency in the promoters under study versus the background promoters (ratio > 1.4) were selected for further analysis. These matrices were converted to the set of the transcription factors (TFs), which can be responsible for expression changes in the group of genes/proteins under study.

2.9. Generation of Regulatory Networks

The identification of potential master regulators in the signal transduction network was performed using the «Regulator search» module of the geneXplain platform 2.0 software (http://platform.genexplain.com). The signal transduction network was provided by the manually curated database, TRANSPATH®. The algorithm starts from a set of TFs and performs a graph-topological search in the signal transduction network upstream of transcription factors to identify the "key nodes" that can play a crucial role in intracellular signaling from various receptors to the set of TFs identified. These key nodes may be considered as master regulators of the process studied. The following setting parameters were used: TRANSPATH® database, maximal search radius R = 10, Score cutoff = 0.2, FDR cutoff = 0.05 and Z-score cutoff = 1.0. Besides FDR, for each possible additional regulator the Score, Z-score and Ranks sum values were calculated. For the proteomic data analysis, the "Context genes" option was used for the search of key regulators. In this case, passing through the common network nodes, the nodes presented at the transcriptome data were preferentially selected. Among the overall list of regulators generated after the search, the statistically significant results were selected using the Ranks sum parameter. Thus, it was possible to find the molecules characterized by equally good "Score" and "Z-score" parameters. The "Score" parameter reflects how well a key molecule is associated with the other molecules in the database and how many molecules of the input TFs are present in the network for a given key molecule. The "Z-score" reflects how the proposed molecule corresponds to the input TFs set. The ranks sum is a combination of Score and Z-score. In other words, these "trivial" expected results attract interest as the well-known "nodes" in the network (Score) and more specific key molecules for the input sample, which are less likely to be detected as an important regulator in the case of the other TF sets used simultaneously.

2.10. Selected Reaction Monitoring (SRM)

The standard peptides for HIC1 (LEEAAPPSDPFR), CEBPB (VLELTAENER) LYN (TQPVPESQLLPGQR), and PARP1 (TLGDFAAEYAK) were obtained using the solid-phase peptide synthesis on the Overture™ Robotic Peptide Library Synthesizer (Protein Technologies, Manchester, UK) or Hamilton Microlab STAR devices according to the published method [26]. The isotopically labeled lysine ($^{13}C_6$,$^{15}N_2$), arginine ($^{13}C_6$,$^{15}N_4$) or serine ($^{13}C_3$,$^{15}N_1$) leucine ($^{13}C_6$,$^{15}N_1$) were used for isotopically labeled peptide synthesis instead of the unlabeled lysine (TLGDFAAEYA**K**), arginine (VLELTAENE**R**), leucine (TQPVPESQL**L**PGQR), or serine (LEEAAPP**S**DPFR), respectively. Concentrations of the synthesized peptides were measured by the method of amino acids analysis with fluorescent signal detection of amino acids derived after acidic hydrolysis of peptides as described in [27].

SRM experiments were performed in three biological replicates with five time points each (0, 3 h, 24 h, 48 h, and 96 h) and in five technical replicates for each time point studied. The digested samples were spiked with isotopically labeled peptide to the final concentration 50 fmol/µg of total protein. Peptide samples (2 µg) were separated on a RP-C18 column, (Zorbax 300SB-C18, 3.5 m, 150 mm × 0.075 mm, Agilent Technologies, Santa Clara, CA, USA) using the nanoflow UPLC DionexUltiMate 3000 RSLC nano System Series (Thermo Scientific, Waltham, MA, USA). Peptide separation was achieved using a linear gradient from 95% solvent A (0.1% formic acid) and 5% solvent B (80% acetonitrile, 0.1% formic acid) to 60% solvent A and 40% solvent B over 25 min at a flow rate of 0.4 µL/min. SRM analysis was performed on the QqQ TSQ Vantage (Thermo Scientific, Waltham, MA, USA) with capillary voltage set at 2100 V, isolation window was set to 0.7 Da. SRM transition details for all peptides are shown in Table S8. The results were processed using Skyline software v4.1.0 (MacCoss Lab Software, Seattle, WA, USA). The coefficient of variation (CV) of transition intensity did not exceed 25%, 12%, 12%, and 6% between technical replicates for LEEAAPPSDPFR, VLELTAENER TQPVPESQLLPGQR, and TLGDFAAEYAK, respectively.

3. Results

3.1. Transcriptome Analysis and Functional Annotation of Differentially Expressed Genes during ATRA-Induced Differentiation of HL-60 Cells

To validate HL-60 cell differentiation into neutrophils, expression of surface markers CD11b and CD38 was assessed by flow cytofluorometry at 96 h after ATRA treatment prior transcriptome/proteome analysis (Figure S1). Although measurement of CD11b is the most convenient way to evaluate granulocyte differentiation, to obtain more accurate data we have used additional marker CD38 that promotes induced myeloid maturation [28]. The mean fluorescence from HL-60 cells at 96 h after ATRA-treatment increased approximately 15-fold (CD38-from 171 to 2929; CD11b-from 112 to 1726) compared to untreated control. This indicates that the granulocyte differentiation of the HL60 cell line was successful.

To obtain the transcriptomic data, HL-60 cells were harvested at 3 h, 24 h, and 96 h after ATRA treatment followed by mRNA microarray profiling. A total of 14,543 gene expressions were detected at all the time points studied. Among them 159, 231, and 1449 genes with fold-change (FC) $\geq$2 were determined as differentially expressed genes (DEGs) at 3 h, 24 h, and 96 h after ATRA treatment, respectively (Supplemental Table S1).

Further, we focused on the bioinformatics reconstruction of putative regulatory pathways for DEGs that were involved in cell differentiation according to highly validated data. We annotated the altered expression genes by the Gene Ontology (GO) database category related to the biological processes (Figure 2).

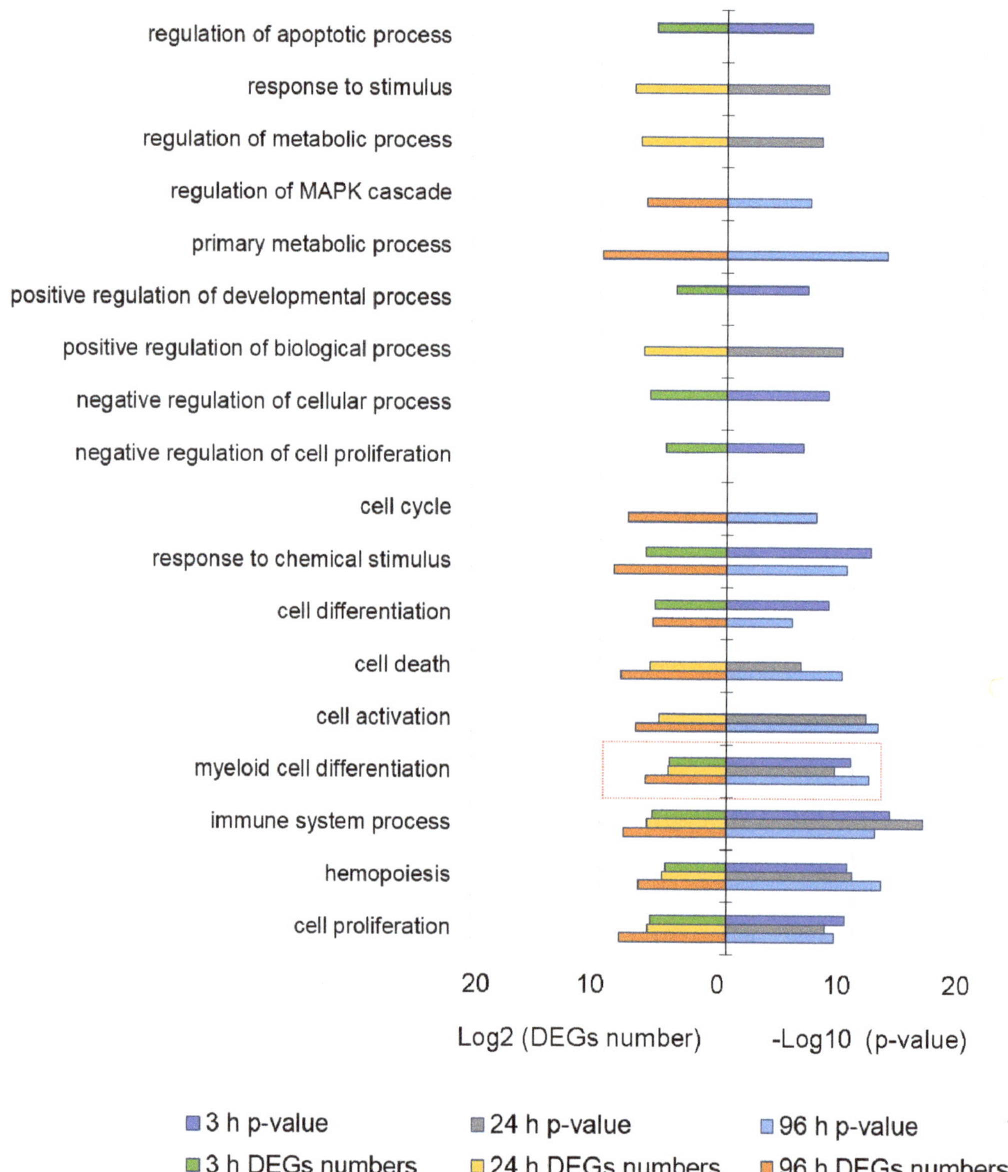

Figure 2. The functional GO analysis of differentially expressed genes (DEGs) of HL-60 cells at 3 h, 24 h, and 96 h after ATRA treatment. The number of DEGs (Log2 transformed) and *p*-value (-Log10 transformed) are provided on the x-axis. The groups from the category of "Biological process" are on the y-axis. The threshold adjusted *p*-value < 10^{-4}. The group of "myeloid cell differentiation" (MCD, GO: 0030099) is marked by red color.

Figure 2 shows the DEGs at all time points were enriched by molecules, which were assigned to the group of "myeloid cell differentiation" (MCD, GO: 0030099). The MCD group was revealed at 3 h after ATRA treatment with 22 DEGs, and then was expanded up to 24 and 81 DEGs at 24 h, and 96 h, respectively.

The results of the interaction analysis by STRING (Figure S2) show that the DEGs of MCD group were enriched in their interaction with the highest confidence (0.9). The KEGG database annotation revealed mapping of the DEGs of MCD group into "Chemokine signaling pathway" at 3 h and 24 h, and into "NOD-like receptor signaling pathway" at 96 h.

While the data for 3 h and 24 h suggest the cytokine signaling as one of the mechanisms of the ATRA-induced granulocytic differentiation, the results for 96 h indicate the manifestation of functions of already mature neutrophils. These observations emphasize that the bioinformatics mapping of molecules with altered expression on known signaling pathways is insufficient for a complete understanding of the regulatory events.

Moreover, the earliest time point (3 h after ATRA treatment) provides transcriptomic data on the granulocytic differentiation onset. The DEGs of the MCD group at 3 h included ASB2, BCL2A1, CCL2, CCL3L1, CCL4, CCR5, CD300A, CD38, CEBPB, FGR, HES1, HNRPLL, IL8, LRG1, LYN, RELB, TNFAIP2, BCL11A, NR2F2, PTGER2, RGS18, and SERPINB2. Among them CEBPB, CCR5, CCL4, FGR, CXCL8 (IL8), and LYN form a putative functional complex according to the STRING interaction analysis (Figure S2a). These data are of great importance for deciphering the very first molecular events of ATRA-induced granulocytic differentiation. Further, the dynamics of transcription factor CEBPB and LYN kinase was assessed by targeted mass-spectrometry approach (selected reaction monitoring (SRM)) at protein level.

The MCD group genes have been used for following upstream regulators search. The lists of the MCD group genes are presented in Supplemental Table S2.

3.2. Proteomic Analysis and Functional Annotation of Differentially Expressed Proteins during ATRA-Induced Differentiation of HL-60 Cells

Proteome dynamics is associated with cell phenotype development and its continuous observation can contribute to understanding of the cell maturation process. Previously, for systems analysis of induced granulocyte differentiation and apoptosis under ATRA/arsenic trioxide treatment starting time points of 6 h at transcriptomic level and 12 h at proteomic level were used [3]. We tried to unveil the molecular onset of differentiation. In our preliminary experiments we did not observed any significant changes in the ATRA induced cell proteome within the first 2 h after ATRA induction (data not shown). We performed proteomic profiling of HL-60 cells at 0, 3 h, 24 h, 48 h, and 96 h after ATRA-treatment.

Using "Composite" search engine in the SPIRE software, we identified 1436, 1470, 1379, 1253, and 1210 proteins with (locFDR) < 0.01 at the 0, 3 h, 24 h, 48 h, and 96 h time points, respectively (Supplemental Tables S3 and S4). Mass-spectrometric data are available via the ProteomeXchange with identifier PXD006768. Based on label free quantitative analysis, 122, 169, 199, and 275 proteins were revealed as differentially expressed proteins (DEPs) (FC $\geq$ 1.5, *p*-value < 0.05, CV < 30%) at 3, 24, 48, and 96 h after ATRA treatment comparing to control (0 h), respectively. Data on label free quantitative analysis and relative expression are presented in Supplemental Table S5. The heatmap of protein expression is presented in Figure S3. The DEPs are listed in Table S5.

The functional analysis of DEPs was performed in the same way as for the DEGs. The results are shown in Figure 3.

Figure 3a shows that the DEPs are enriched with the proteins involved in programmed cell death and its regulation at 3 h and 96 h after ATRA treatment. The five most up-regulated DEPs involved in programmed cell death at 3 h after ATRA-treatment comprise proteasome subunit beta type-2 (PSMB2, P49721), apoptosis-inducing factor 1 (AIFM1, O95831), alpha-actinin-1 (ACTN1, P12814), RNA-binding protein 25 (RBM25, P49756), and apoptosis inhibitor 5 (API5, Q9BZZ5). The top five down-regulated DEPs included 26S proteasome regulatory subunit 8 (PSMC5, P62195), alpha-actinin-2 (ACTN2, P35609), 14-3-3 protein eta (YWHAH, Q04917), CD44 antigen (CD44, P16070), and protein S100-A9 (S100A9, P06702).

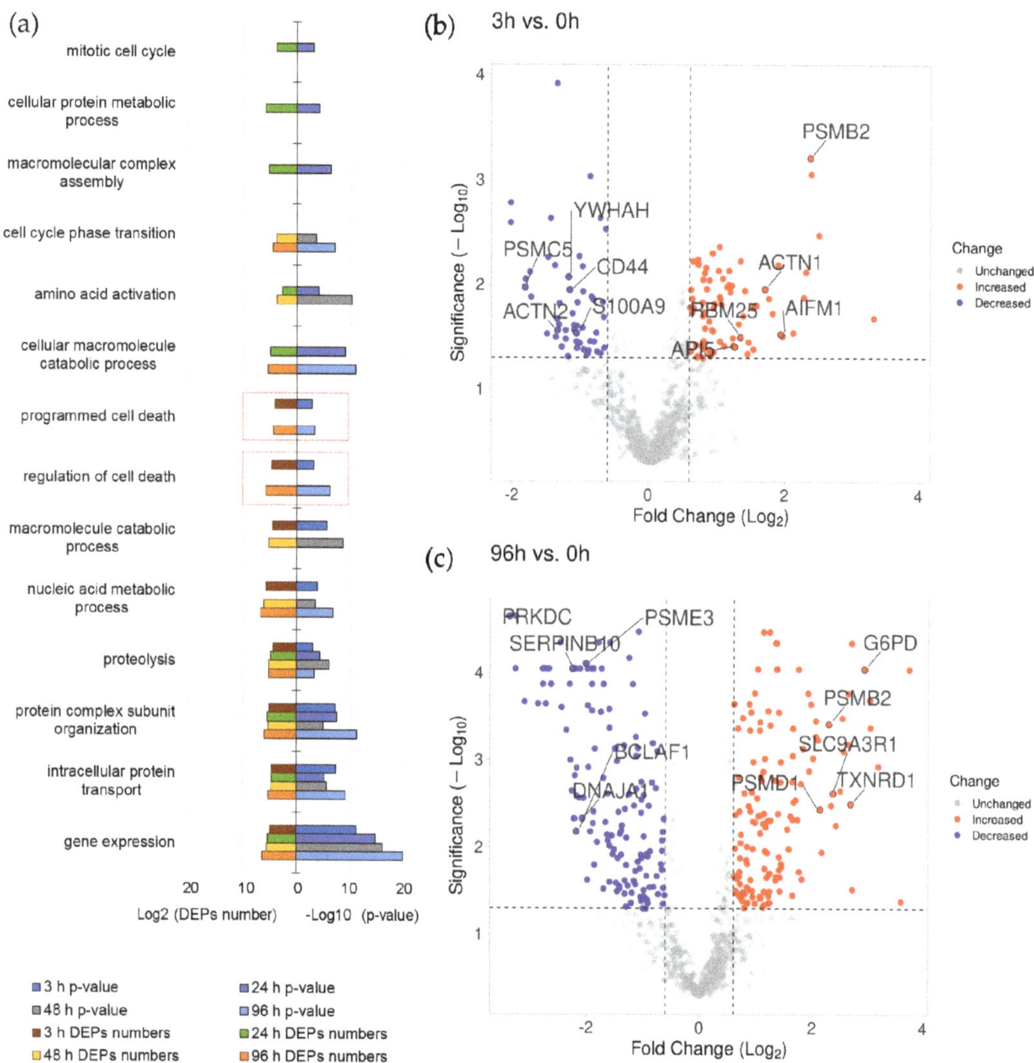

Figure 3. (**a**) The functional GO analysis of differentially expressed proteins (DEPs) of HL-60 cells at 3 h, 24 h, 48 h, and 96 h after ATRA treatment. The number of DEPs (Log2 transformed) and p-value (-Log10 transformed) are provided on the x-axis. The groups from the category of "Biological process" are on the y-axis. The threshold adjusted p-value < 10^{-4}. The groups containing proteins regulating cell death and apoptosis are marked by red. The volcano plots show the differences in proteins abundance at 3 h (**b**) and 96 (**c**) after ATRA treatment; significantly up- and down-regulated proteins are shown as red and blue dots, respectively; names are shown for five most up- and down-regulated proteins that were annotated by GO belonging to groups "programmed cell death" and/or "regulation of cell death".

The five most up-regulated proteins at 96 h after ATRA-treatment included 26S proteasome non-ATPase regulatory subunit (PSMD1, Q99460), proteasome subunit beta type-2 (PSMB2, P49721), glucose-6-phosphate 1-dehydrogenase (G6PD, P11413), thioredoxin reductase 1 (TXNRD1, Q16881), and Na(+)/H(+) exchange regulatory cofactor NHE-RF1 (SLC9A3R1, O14745). Although these DEPs are assigned to the groups regulating cell death, they affect cell fate indirectly through metabolic effects. The 5 most down-regulated

DEPs included DNA-dependent protein kinase catalytic subunit (PRKDC, P78527), Bcl-2-associated transcription factor 1 (BCLAF1, Q9NYF8), DnaJ homolog sub-family A member 1 (DNAJA1, P31689), proteasome activator complex subunit 3 (PSME3, P61289), and serpin B10 (SERPINB10, P48595).

The STRING interaction analysis (Figure S4) revealed that the DEPs of group "programmed cell death" and/or "regulation of cell death" were enriched in their interaction with the highest confidence (0.9) at 3 h and 96 h after ATRA-treatment. Moreover, these proteins were mapped to the "Proteasome" pathway (KEGG database annotation) with high confidence.

3.3. The Workflow of Transcriptome- and Proteome-Based Regulatory Networks Design

The lists of DEGs and DEPs given in Supplemental Tables S2 and S5 have been used as the test sets (Yes-sets). The control sets were formed from the transcripts and proteins with unaltered expression as described in "Materials and Methods". We performed the two-step bioinformatic analysis including:

1. Identification of TFs that can regulate the DEGs (MCD group) and DEPs at different time points after ATRA treatment using TRANSFAC@ database followed by matching putative TFs with the list of all transcripts identified (Supplemental Table S1) to cut-off the molecules that are not expressed in HL-60 cells at the mRNA level;
2. The upstream prediction of key molecules that regulate the TFs determined at the previous step using TRANSFAC@ database followed by visualization of the predicted interaction as a model regulatory networks.

To verify the molecules that are actually expressed in HL-60 cells, we matched the list of all identified and differentially expressed genes (Supplemental Table S1) and/or proteins (Supplemental Table S5) with the elements of model regulatory networks.

3.3.1. The Transcriptome-Based Modeling Pathway

To find TFs responsible for regulation of gene expression we performed a search for the DEGs (MCD group) transcription factors binding sites (TFBS) at each time point studied (see results in Supplemental Table S6). TFs of DEGs determined at the 3/24 h and 24/96 h time were the same in general. So, in the case of time points 3, 24, and 96 h, we have combined all putative TFs in one set in order to perform key regulator search. The upstream analysis of the combined set of TFs, which are involved in regulation of MCD group genes at the 3 h, 24 h, and 96 h, revealed the top five key molecules with the lowest "Rank sum" value. The results are summarized in Table 1.

Table 1. Putative key molecules responsible for regulation of the DEGs related to the myeloid cell differentiation (MCD group) at 3, 24 and 96 h after ATRA treatment.

Time Point.	Key Molecule Name	Reached from TF Set [1]	Reachable Total [2]	Score [3]	FDR [4]	Z-Score [5]	Ranks Sum [6]
3-24-96 h (MCD)	AhR	22	12488	0.34	0.011	2.31	5
	arnt	21	8990	0.34	0.026	2.1	6
	Nrf2	15	9200	0.24	0.033	2.61	6
	(CKII-α)2:(CKII-β)2	22	10803	0.28	0.025	2.45	8
	NF-kappaB1	21	10897	0.29	0.024	2.05	11

[1] "Reached from TF set"—the number of the TFs from the input set (Supplemental Table S6) that is reached from the respective key molecule; [2] "Reachable total"—the total number of molecules that can be reached from the key molecule, independent of the input set; [3] "Score"—the value reflecting how well the respective key molecule is connected with other molecules in the database, and how many molecules from the input set are present in the network triggered by this key molecule, the higher value—the better suitability (threshold value > 0.2); [4] FDR—false discovery rate (from 1000 random input sets); [5] "Z-score"—the value that reflects how specific each key molecule is for the input list, the higher value—the better suitability (threshold value > 1); [6] "Rank sum"—composite value that reflects the impact of Score and Z-score simultaneously, the lower value—the better suitability.

Further, to select the key molecules for visualization, we checked either its expressions were altered at ATRA-induced granulocytic differentiation (of primary importance), and compared FDR statistics. None of the key molecules from Table 1 were significantly changed at the transcript or protein levels. At the same time, AhR and NF-kappaB1 were the most reliable based on FDR value. Moreover, AhR and NF-kappaB1 mutually regulate each other according TRANSFAC@ database. The regulatory network triggered by AhR and NF-kappaB1 is shown in Figure 4.

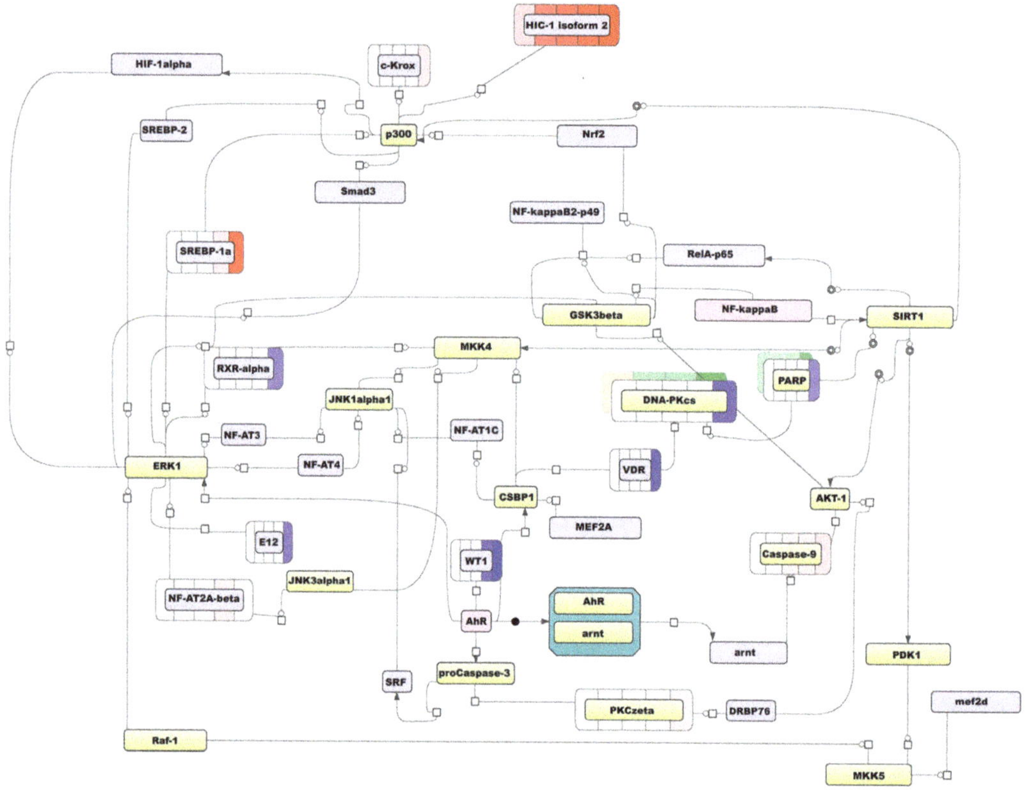

Figure 4. The transcriptome-based model network of regulation of MCD group DEGs during ATRA-induced HL-60 cells differentiation (the time points 3 h, 24 h, and 96 h). Legend: master regulatory molecules are represented by pink ellipses; connecting molecules considered by the graph-analyzing algorithm to find the path from the TF input list to the master molecule are represented by green ellipses; the molecules from the TF input list are represented by lilac ellipses. The colored bars around molecules show changes in the expression level. Transcript expressions are shown in blue (decreased expression) or pink (increased expression) color arrays, color intensity correlates with fold-change (FC), bars are colored if FC $\geq$ 2. From left to right each bar represent experimental time point (the time points at 3 h, 24 h, and 96 h and additional time points at 0.5 h and 1 h). Protein expression is shown in yellow (decreased expression) and green (increased expression) color array, color intensity correlates with fold-change (FC) of relative protein expression, bar is colored if FC $\geq$ 1.5, from left to right each bar represent experimental time point (3 h, 24 h, 48 h, and 96 h).

According to the scheme, the key molecule AhR, apparently, causes down-regulation of proto-oncogene WT1, nuclear receptor RXRα, and transcription factor E12 (TCF3) and up-regulation of PKC zeta. AhR affects GSK3beta that regulates another key molecule, NF-kappaB1. On the other hand, NF-kappaB1 affects SIRT1 deacetylase, which inhibits the transcriptional activity of RelA/p65. NF-kappaB1 also influences GSK3beta kinase, thus performing the feedback and cross-regulation from two key molecules.

The model network also shows, that the NF-kappaB1/SIRT1 tandem down-regulates PARP1 (2-fold mRNA decrease at 96 h), DNA-PKcs (3-fold mRNA decrease at 96 h), and VDR (5-fold mRNA decrease at 96 h). VDR gene have been also indirectly controlled (via CSBP1) by AhR. Furthermore, NF-kappaB1/SIRT1 up-regulates TFs c-Krox, SREBP-1a, NF-AT2A-beta, and HIC1 mRNA expression. Both NF-kappaB1 and AhR trigger the up-regulation of caspase 9. These results indicate the synergistic effect of key molecules. Notably, transcriptome-based MCD-regulating scheme included various protein kinases (ERK, JNKalpha1, MKK4, GSK3beta, CSBP1 (MK14), AKT1, JNK3alpha1, Raf-1, PDK1, MKK5, and PKCzeta). This observation suggests the significant role of MAPK pathway in the regulation of DEGs of MCD group.

3.3.2. The Proteome-Based Modeling Pathway

In the case of proteome data analysis, we have combined TFs which may regulate the expression of genes encoding DEPs (Supplementary Materials, Table S7). The results of the key regulator molecules search for DEPs are presented in Table 2. The Top-5 key molecules with the lowest "Rank sum" value are shown.

Table 2. Putative key molecules that regulate DEPs at 3, 24, 48, and 96 h during ATRA-induced differentiation of HL-60 cells.

Time Point	Key Molecule Name	Reached from TF Set [1]	Reachable Total [2]	Score [3]	FDR [4]	Z-Score [5]	Ranks Sum [6]
Combined 3-24-48-96 h	YY1	22	32835	0.650	0.013	2.779	59
	plk1{p}	22	32762	0.640	0.002	2.803	63
	PARP1	22	32360	0.607	0.006	2.950	63
	faim	22	32361	0.607	0.006	2.950	64
	MKK6	22	33716	0.721	0.004	2.325	88
	NR1B1 (RARA)	22	30223	0.505	0.018	3.007	204

[1] "Reached from TF set"—the number of the TFs from the input set (Supplemental Table S7) that is reached from the respective key molecule; [2] "Reachable total"—the total number of molecules that can be reached from the key molecule, independent of the input set; [3] "Score"—the value reflecting how well the respective key molecule is connected with other molecules in the database, and how many molecules from the input set are present in the network triggered by this key molecule, the higher value—the better suitability (threshold value > 0.2); [4] FDR—false discovery rate (from 1000 random input sets); [5] "Z-score"—the value that reflects how specific each key molecule is for the input list, the higher value—the better suitability (threshold value > 1); [6] "Rank sum"—composite value that reflects the impact of Score and Z-score simultaneously, the lower value—the better suitability.

Further, to select the key molecule for visualization, we checked either its expression was altered during ATRA-induced granulocytic differentiation (of primary importance), and compared their FDR statistics. According to our transcriptomic data, we observed a 2-fold decrease of the PARP1 levels at 96 h. At the same time, PARP1 was identified in a shotgun mass spectrometry experiment. Furthermore, this molecule represents an intermediate node in the SIRT1-mediated signal transduction in the transcriptome-based network triggered by NF-kappaB1 and AhR (see Figure 4). In addition to the five most statistically significant molecular regulators, Table 2 also includes a retinoic acid receptor NR1B1 (RARα) as the key molecule. Although the Rank sum has not included RARα in the top five molecules, it has sufficient Score, Z-Score, and FDR values. Moreover, RARα is the well-known target of retinoic acid, inducing the differentiation of HL-60 cells [29]. The proteome-based scheme of TF regulation based on the selected key molecules, PARP1 and RARα, is shown in Figure 5. This modeling pathway could demonstrate molecular synergy of PARP1 and RARα.

Figure 5 demonstrates that in addition to the TFs with altered expression described previously (VDR, RXRα, and HIC1) the unique TFs were predicted using the proteomic data, including IRF7 and AML3 (RUNX2) (2.6-fold mRNA increased at 96 h), and GATA2 (mRNA reduced by 3.6- and 6.5-fold at 24 h and 96 h, respectively).

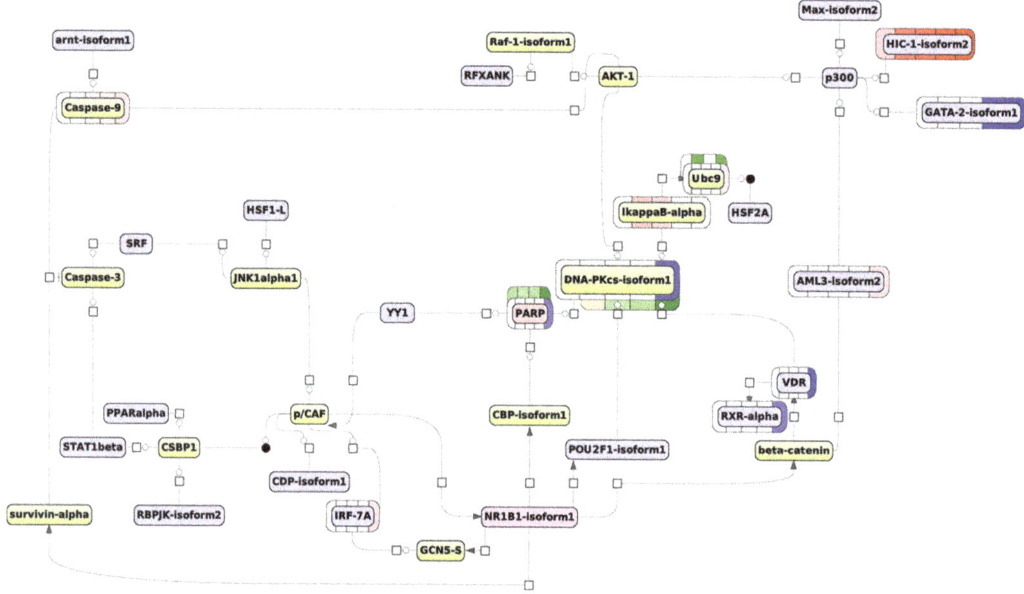

Figure 5. Proteome-based model network of regulation of ATRA-induced HL-60 cell line differentiation (time points 3 h, 24 h, 48 h, and 96 h). Legend: master regulatory molecules are represented by pink ellipses; connecting molecules considered by the graph-analyzing algorithm to find the path from the TF input list to the master molecule are represented by green ellipses; the molecules from the TF input list are represented by lilac ellipses. The colored bars around molecules show changes in the expression. Transcript expression is shown in blue (decreased expression) and pink (increased expression) color array, color intensity correlates with fold-change (FC) of relative mRNA expression, bar is colored if FC $\geq$ 2, from left to right each bar represent experimental time point (the time points at 3 h, 24 h, and 96 h and additional time points at 0.5 h and 1 h). Protein expression is shown in yellow (decreased expression) and green (increased expression) color array, color intensity correlates with fold-change (FC) of relative protein expression, bar is colored if FC $\geq$ 1.5, from left to right each bar represent experimental time point (3 h, 24 h, 48 h, and 96 h).

According to Figure 5, DNA-PKcs also affects IkappaB-alpha (NFKBIA): its expression is 3.2- and 2.9-fold increased at the transcriptome level at the time point 3 h.

Notably, the key molecule RARα (NR1B1 on the scheme) regulates PARP1 through CBP acetylase. In turn, the PARP1-triggered network regulates RARα through the DNK-PKcs/AKT1/CASP9/CASP3/SRF/JNK1α1/pCAF loop. In the case of RAR-dependent transcription, it has been found that PARP1 functions as a co-regulator, which is required to switch the mediator complex in the active state and start the transcription [30].

The same pathway branch (PARP1/DNA-PKcs/VDR) and some TFs (HIC1 and RXRα) belong to both transcriptome and proteome-based model regulatory networks that suggests the importance of these molecules and actual involvement of the pathways in the regulation of ATRA-induced differentiation of HL-60 cells.

3.4. *Verification of Protein Levels of HIC1, PARP1, CEBPB, and LYN During ATRA-Induced Differentiation by SRM Analysis*

To reveal molecules of the transcriptome- and proteome-based pathways, which are actually expressed in HL-60 cells, we have matched the list of all identified and differentially expressed genes (Supplemental Table S1) and proteins (Supplemental Table S5) with molecules in the model regulatory networks. Differentially expressed genes belonging to the transcriptome- and proteome-based modeling networks are shown in Figure 6.

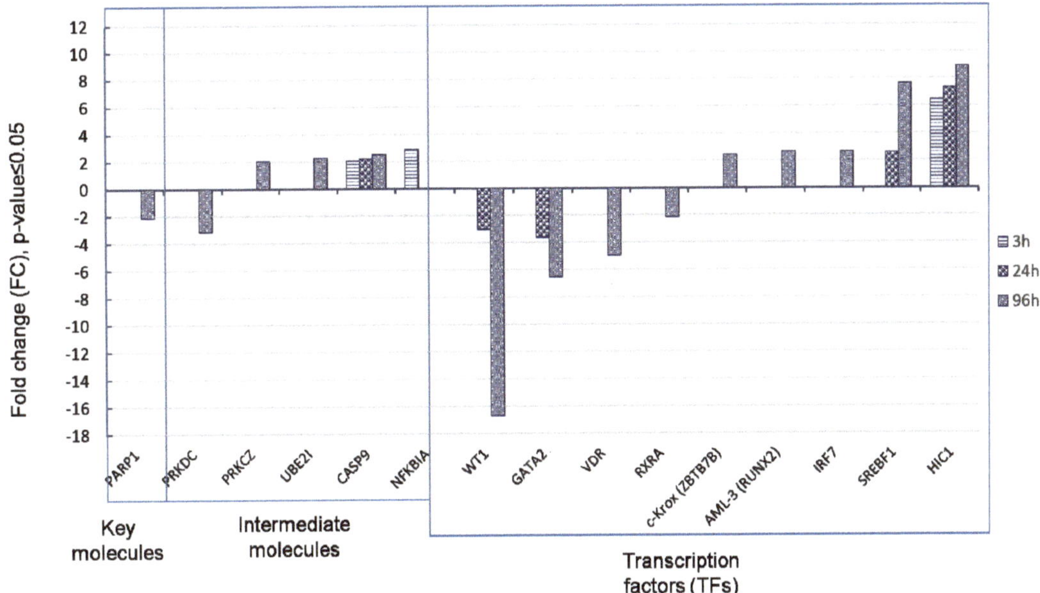

Figure 6. Components of transcriptome- and proteome-based model networks with altered mRNA expression under ATRA treatment. Transcription factors, TFs (predicted by TRANSFAC@ database), intermediate and key molecules (predicted by TRANSPATH@ database) with fold change ≥ 2 (p-value ≤ 0.05) at 3, 24, and 96 h is presented.

Figure 6 shows that 15 molecules, including one key molecule, five intermediate molecules, and nine transcription factors (TFs) of the transcriptome- and proteome-based model networks were characterized by the altered mRNA expression level. Transcriptional repressor HIC1 was strongly up-regulated at all time points studied suggesting its regulatory value. It is noteworthy that CASP9 and NFKBIA were up-regulated at 3 h after ATRA treatment. Transcription factors VDR and RXRA, which are intimately related to induced differentiation, were down-regulated (as well as key molecule PARP1).

Among predicted regulatory molecules we selected transcription factor HIC1 and key molecule PARP1 for measuring abundance in HL-60 cells at different time points by SRM. Next, we have compared transcriptomic and proteomic profiles during ATRA-induced differentiation. We also evaluated levels of transcription factor CEBPB and LYN kinase with altered expression at the earliest time point (3 h) by SRM. Results are shown in Figures 7 and 8.

Figure 7a,d demonstrate the trace of SRM transitions for native (above) and SIS standard (below) peptides LEEAAPPSDPFR of HIC1 protein, and TLGDFAAEYAK of PARP1 protein, respectively. The Figure 7b,c show transcriptomic and proteomic profiles of HIC1 expression. Transcription repressor HIC1 was up-regulated at 3 h and its mRNA abundance gradually increased almost 9 times to 96 h. HIC1 protein has not been identified in shotgun mass-spectrometry experiment. Using SRM technique with stable isotope labeled peptide standard (LEEAAPPSDPFR) the HIC1 abundance was detected at 24 h, 48 h, and 96 h. At these time-points its concentration was 0.63 ± 0.21 fmol/µg, 0.85 ± 0.14 fmol/µg, and 1.2 ± 0.15 fmol/µg of total protein, respectively. The HIC1 protein level was increased approximately 2-fold (FC = 1.9, p-value ≤ 0.05) from 24 h to 96 h after ATRA treatment.

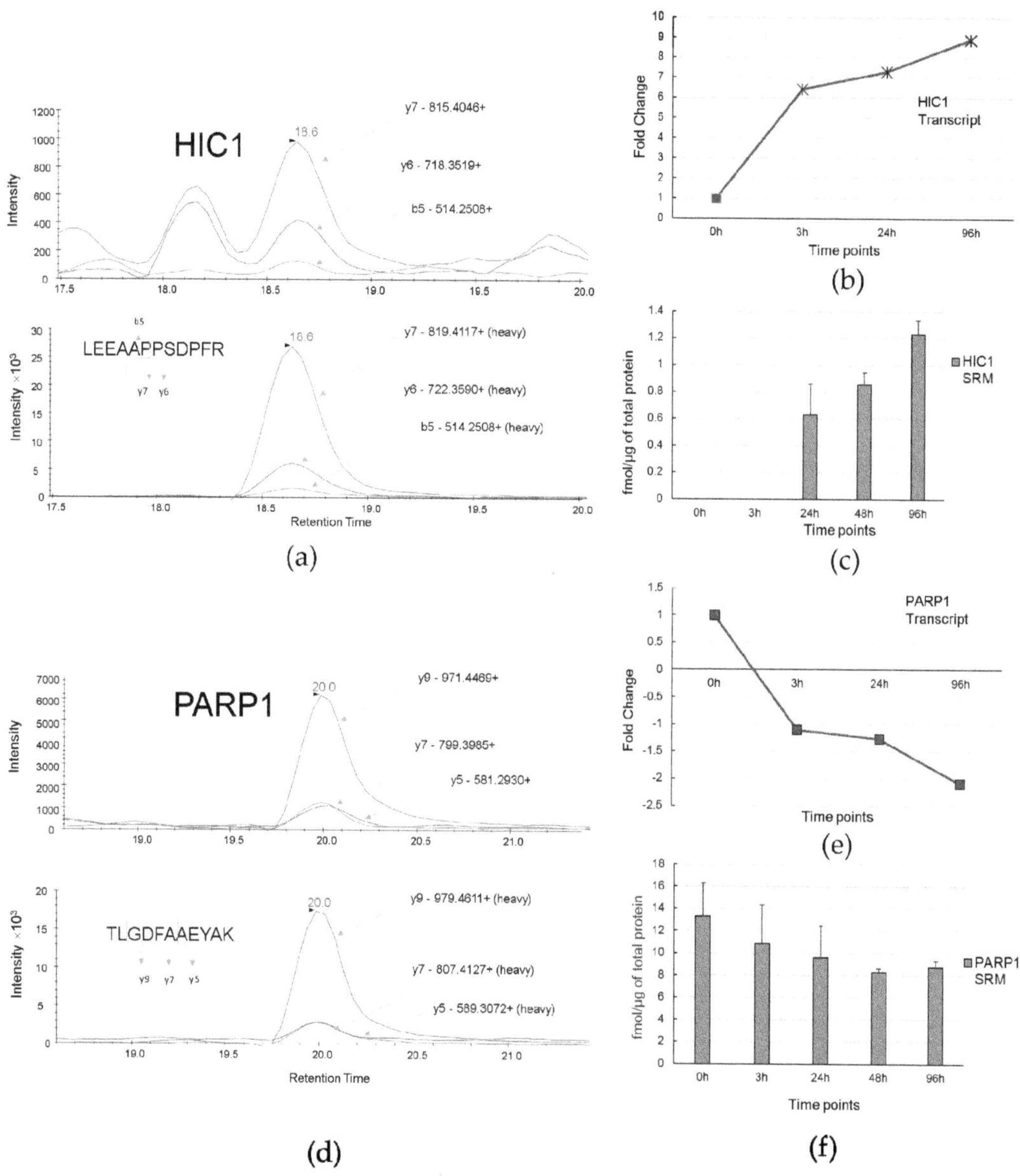

Figure 7. HIC1 and PARP1 expressions at ATRA-induced granulocytic differentiation. (**a**) Trace of SRM transitions for native and stable isotope labeled peptide standard LEEAAPPSDPFR of HIC1. (**b**) Profile of transcript expression HIC1 during HL60 differentiation (fold change ≥ 2, p-value ≤ 0.05 at 3 h, 24 h, and 96 h). (**c**) Protein expression level of HIC1 obtained by SRM (three biological replicates) at 3 h, 24 h, 48 h, 96 h. (**d**) Trace of SRM transitions for native and standard isotopically-labeled peptide TLGDFAAEYAK of PARP1. (**e**) Profile of transcript expression PARP1 during HL60 differentiation (fold change ≥ 2, p-value ≤ 0.05 at 96 h). (**f**) Protein expression level of PARP1 obtained by SRM (three biological replicates) at 3 h, 24 h, 48 h, 96 h.

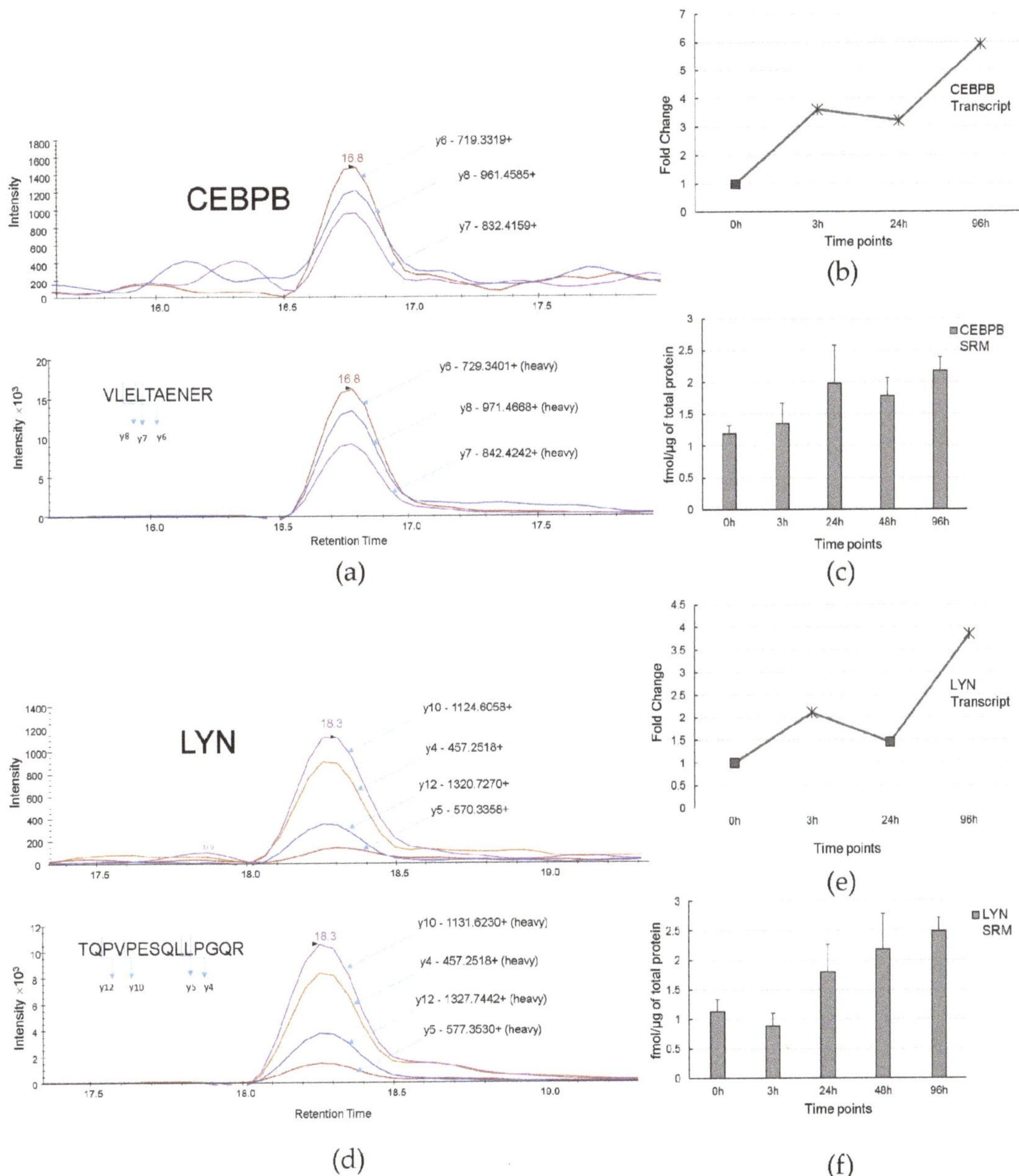

Figure 8. CEBPB and LYN expression during ATRA-induced granulocytic differentiation. (**a**) Trace of SRM transitions for native and stable isotope labeled peptide standard VLELTAENER of CEBPB. (**b**) Profile of CEBPB transcript expression during HL60 differentiation (fold change ≥ 2, p-value ≤ 0.05 at 3 h, 24 h, and 96 h) (**c**) Protein expression level of CEBPB obtained by SRM (three biological replicates) at 3 h, 24 h, 48 h, 96 h. (**d**) Trace of SRM transitions for native and standard isotopically-labeled peptide TQPVPESQLLPGQR of LYN. (**e**) Profile of transcript expression LYN during HL60 differentiation (fold change ≥ 2, p-value ≤ 0.05 at 3 h and 96 h). (**f**) Protein expression level of LYN obtained by SRM (three biological replicates) at 3 h, 24 h, 48 h, 96 h.

The Figure 7e,f show transcriptomic and proteomic profiles of PARP1 expression. PARP1 was selected as a key molecule for the proteome-based model network. At the transcriptome level we revealed a 2-fold decrease in PARP1 mRNA expression at 96 h. The SRM measurements for the TLGDFAAEYAK peptide of PARP1 were 13.28 ± 2.98, 10.83 ± 3.46 fmol/μg, 9.57 ± 2.88 fmol/μg, 8.28 ± 0.35 fmol/μg, and 8.77 ± 0.54 fmol/μg of total protein at 0, 3 h, 24 h, 48 h, and 96 h after ATRA-treatment, respectively. The PARP1 protein level was 1.5-fold (p-value ≤ 0.05) down-regulated by 96 h after ATRA treatment.

Figure 8a,d demonstrate the trace of SRM transitions for native (above) and SIS standard (below) peptides VLELTAENER of CEBPB protein, and TQPVPESQLLPGQR of LYN protein, respectively.

Figure 8b demonstrates that CEBPB was up-regulated starting from 3 h (FC = 3.6, p-value ≤ 0.05) up to 96 h (FC = 5.95, p-value ≤ 0.05) at transcriptome level. Using SRM, we measured CEBPB in amount of 1.2 ± 0.12 fmol/μg, 1.36 ± 0.31 fmol/μg, 1.98 ± 0.59 fmol/μg, 1.78 ± 0.28 fmol/μg, and 2.17 ± 0.21 fmol/μg at 0, 3 h, 24 h, 48 h, and 96 h after ATRA-treatment, respectively (Figure 8c).

Figure 8e,f show transcriptomic and proteomic profiles of expression of LYN kinase. Transcriptomic data demonstrates significant LYN up-regulation at 3 and 96 h. The unique peptide (TQPVPESQLLPGQR, 21-34aa), which has been used for SRM analysis, is the LYN isoform B-specific and is mapped to the region that distinguishes isoform A from isoform B. High-resolution annotated MS2 spectrum of LYN isoform B-specific peptide TQPVPESQLLPGQR is shown in Figure S5. Protein LYN expression was detected in amount of 1.12 ± 0.2 fmol/μg, 0.8 ± 0.21 fmol/μg, 1.8 ± 0.46 fmol/μg, 2.18 ± 0.6 fmol/μg, and 2.49 ± 0.23 fmol/μg of total protein at 0, 3 h, 24 h, 48 h, and 96 h after ATRA-treatment, respectively.

We observed coordinate increase or decrease at the transcript and protein level for HIC1, CEBPB, LYN, and PARP1; this confirms involvement of corresponding genes in the ATRA induced HL60 differentiation. The targeted mass-spectrometric data have been uploaded into PASSEL repository (dataset PASS01678).

4. Discussion

Omics techniques provide a massive amount of data on the molecular state of the biological object studied. Nevertheless, in high-throughput transcriptome and proteome profiling, we always register only certain molecular consequences of regulatory events that occurred in the past (e.g., induction of the expression of the corresponding gene). Especially, proteomic research of differentiation onset is complicated by the fact that observed changes in protein levels take time. Thus, up-stream regulator search provides bioinformatics reconstruction of the molecular events up to one or several trigger points. Consistent with this, our whole-genome transcriptome results indicated activation of myeloid differentiation, whereas proteomic data demonstrated the involvement of the apoptosis pathways under ATRA treatment. However, knowing the expression differences alone does not allow us to reveal the effector that leads a biological system towards the particular molecular state. Applying up-stream regulator search and visualizing its result, we provide the putative "molecular scenarios" of how a dozen regulatory molecules decided the fate of hundreds of proteins and transcripts.

After ATRA treatment leukemic cells, of which the phenotype is generally driven by genetic abnormalities, acquire features of mature granulocytes. As in the case of many others malignancy, HL-60 cells harbor genetic aberrations including the most frequent mutations: extensive deletion of the *p53* gene, amplification of *MYC* oncogene, and monoallelic deletion of granulocyte–macrophage colony stimulating factor (*GM-CSF*) [11,12]. Considering this, we suggest that our model regulatory networks represent a putative way to overcome the effect of these mutations.

Proto-oncogene *MYC* plays a crucial role in the regulation of cell proliferation, differentiation, and apoptosis [31,32]. From 16- to 32-fold *MYC* gene amplification in the HL-60 genome has been reported [33]. Although the decreased expression of *MYC* is not sufficient for triggering differentiation of HL-60 cells, it is accompanied by the inhibition of cell growth [34]. In our study, we observed a 7-fold decrease of MYC mRNA expression during granulocytic differentiation. Notably, TF MAX that binds MYC protein for activation of target genes [35] is the part of our proteome-based model network. Thus, the modeling scheme presented in Figure 4 could represent a way to overcome the deleterious effect of MYC gene amplification.

Normally, the *p53* gene is a crucial component of the molecular response to different kinds of cell stress including DNA damage. Namely, p53 is involved in mismatch repair, DNA double-strand break repair, and nucleotide excision repair that could accompany uncontrolled proliferation [36]. Poly(ADP-ribose) polymerase 1 (PARP1), the key molecule of the proteome-based model network, has intricate interplay with p53 in regulation of cell death and survival. PARP1 affects p53 transcriptional activity, and promotes its oncosupressive function [37]. In turn, the p53 expression level is prominently increased after DNA damage in PARP1-defiecint cells that leads to apoptosis [38]. Moreover, in the case of the multidrug-resistant leukemia cell line HL-60[R] the PARP1 mRNA expression level was up-regulated [39]. At the same time, a branch of components PARP1/DNA-PKcs/VDR, which is presented both the in transcriptome- and proteome-based model pathways (Figures 4 and 5), regulates DNA repair [40,41]. Thus, the proteome-based model network could represent a molecular bypass to overcome consequences of *p53* deletion. It may be assumed that inhibition of PARP1 in p53-deficient HL-60 cells could have the similar antiproliferative effect as on BRCA1-deficient cancer cells of solid tumors [42]. This assumption is in agreement with the fact that primary blasts from patients with acute myeloid leukemia are sensitive to PARP-inhibitor Olaparib [43].

In our study SRM measurements show a trend of the diminution of PARP1 protein abundance, while the mRNA level was significantly down-regulated, 2-fold, to 96 h after ATRA-treatment. Considering the moderate modulation of abundance it is conceivable that PARP1 is regulated by post-translation modification. Figures 4 and 5 demonstrate that PARP1 could be acetylated by CREB-binding protein (CBP) or deacetylate by SIRT1. Both PARP1 and SIRT1 compete for the common NAD+ substrate and modulate each other's activity by mutual modification [44]. PARP1 inhibition by SIRT1 could contribute to the increase in the DNA damage level and cell death in the absence of *p53* expression. SIRT1 stimulation by pharmacological agents could promote PARP1 inhibition. On other hand, SIRT1 can activate apoptosis by direct deacetylation of the RelA-p65 subunit that inhibits the transcription of NF-kappaB and increases cell sensitivity to TNF-alpha-induced apoptosis [40]. TNF-alpha is known to cause p53-independent apoptosis, which promotes the monocytic differentiation of HL-60 cells [45].

At the same time, we observed prominent up-regulation of transcriptional repressor HIC1 that suppresses SIRT1 gene expression. SIRT1 deacetylates and inactivates both p53 and PARP1; HIC1 affects cell cycle, apoptosis, and DNA repair. According to our transcriptome-based model network (Figure 4), HIC1 was triggered by NF-kappaB via SIRT1 and p300. In Figure 5, a proteome-based model network represents HIC1 regulated by cascade triggered by PARP1 through DNA-PKs, AKT, and p300. This suggests a feedback loop involved in maintaining moderate inhibition of SIRT1 via HIC1 that sustains PARP1 activity, resulting in delayed apoptosis and allowing cells to differentiate into neutrophils. Apparently, accumulation of critical amount of HIC1 causes SIRT1 suppression, and further PARP1 down-regulation occurs due to apoptosis-driven cleavage. It seems that the cell machinery involved in the response to the DNA damage plays a key role in induced granulocytic differentiation, and its component could be sensitive to target treatment.

The transcriptome analysis provides biological data on ATRA-induced granulocytic differentiation at the whole genome-scale. However, not all transcripts detected could be traced at the protein level. In turn, despite the proteomic data being limited by the sensitivity of mass-spectrometry, the protein expression underlies the cell phenotype manifestation. As expected, different inputs to up-stream regulator search resulted in different key molecules in transcriptome- and proteome-based modeling pathways. Still, the schemas show common predicted transcription factors (SRF, ARNT, RXRA, VDR, and

HIC1), intermediate molecules (Caspase9, histone acetyltransferase p300, protein kinases ERK1, Raf-1, AKT1, CSBP1 (MK14), JNKaplha1, and AKT), and even whole branches of molecular events (axis PARP1-DNA-PKcs-VDR). The gene transcription and protein synthesis are separated in time, and the above observations suggest different key regulation, but we also observe the general molecular consequences, such as the involvement of the DNA repair system and the MAPK kinase cascade.

Interesting but conflicting results were obtained for LYN kinase. The previous studied demonstrated that constitutively activated LYN was involved in AML pathogenesis and treatment of cells by LYN siRNA resulted in the antiproliferative effect [46,47]. In our study we observed LYN up-regulation at mRNA level under ATRA treatment. SRM technique allows to distinguish different isoforms of the same protein. We used the isoform specific peptide standard to detect LYN isoform B and found it to be up-regulated at the proteome level. Previously it was reported that phosphorylation activity of Lyn isoform B was lower than that of Lyn isoform A [48]. Moreover, the ratio of Lyn isoform A and Lyn isoform B splice forms may represent a biomarker of neoplasm aggressiveness as was shown in the case of breast cancer [49].

Absolute quantification by SRM with SIS peptides demonstrates the almost equimolar abundance of TF CEBPB and Src kinase LYN. Considering their possible interaction (STRING analysis of DEGs, Figure S2), absolute abundances of CEBPB and LYN suggest protein stoichiometry in the putative complex involved in the earliest step of ATRA-induced granulocytic differentiation.

The myeloid-associated TFs (RARa, RXR, VDR, CEBPB, and GATA2) of model schemes confirm the biological relevance of bioinformatics modeling. Notably, a transcriptome-based MCD-regulating scheme included various protein kinases (ERK, JNKalpha1, MKK4, GSK3beta, CSBP1 (MK14), AKT1, JNK3alpha1, Raf-1, PDK1, MKK5, and PKCzeta), that is in accordance with MAPK-based mechanisms for ATRA-induced granulocytic differentiation [14]. Moreover, the current inter-platform study shows the involvement of such less associated with AML TFs as NF-ATs, SMAD3, WT1, and c-Krox, as well as ubiquitous molecules (p300, P/CAF, UBC9), which are involved in posttranslational modifications (acetylation, sumoylation, ubiqutunilation etc.). All the above observations suggest the existence of alternative, RAR/RXR transcription-independent, induced differentiation pathways. However, this assumption should be experimentally proven.

5. Conclusions

Applying transcriptomic, proteomic analysis, and bioinformatics prediction we have suggested a hypothesis on molecular mechanism of ATRA-induced granulocytic differentiation. We aimed to trace dynamics at different molecular levels in a time-course manner. The novelty of the approach used in our study is that molecules with altered expression from omics experiments have not been just mapped to known signaling pathways. Instead, an upstream regulator search aimed to obtain the hierarchical model of ATRA-induced granulocytic differentiation that reconstructs the molecular events affecting differentially expressed mRNA and proteins. Only the TFBS in the promotor region of genes with altered expression and highly validated data on protein–protein interaction were taken into account in upstream regulator search. The resulting modeling schemas are visualizations of the most probable variant of a biological signal transmission, which leads to a change in the expression levels of transcripts and proteins, observed experimentally. The validation of bioinformatics prediction by functional molecular research is an important item, and a subject of our further work. The TF HIC1 and the key molecule PARP1 are contemplated as the most promising targets for validation of the modeling pathways.

The approach combining transcriptomic, proteomic analysis, and computational analysis described here is applicable to various cells models including primary blast cells from patients under different treatment regimens. Thus this platform could be useful for the goals of precision medicine such as monitoring response to treatment especially in case of drug resistance. Our results suggest that the multi-disciplinary platform combining transcriptomics, proteomics, and bioinformatics is a promising approach to reveal regulatory molecules that are hardly detected by convenient omics methods or laborious to derive from convoluted proteomic or transcriptomic data.

Supplementary Materials: The following are available online at https://www.mdpi.com/article/10.3390/biom11060907/s1, Figure S1. Evaluation of HL-60 cell line by the CD38 and CD11b expression level measured by flow cytometry. Figure S2. STRING interaction analysis of differentially expressed genes (DEGs) of Myeloid cell differentiation (MCD) group at 3, 24, and 96 h after ATRA treatment. Figure S3. Heatmap of protein expression during HL-60 cell line differentiation. Figure S4. STRING interaction analysis of differentially expressed proteins (DEPs) assigned to group "programmed cell death" and/or "regulation of cell death" at 3 and 96 h after ATRA treatment. Figure S5. High-resolution annotated MS2 spectrum of LYN isoform B-specific peptide TQPVPESQLLPGQR. Figure S6. Calibration curves plotting of experimentally determined concentrations versus theoretical concentrations of the target analyte using isotopically labeled and label-free synthetic standard peptide A: LEEAAPPSDPFR (HIC1), B: TLGDFAAEYAK (PARP1), C: VLELTAENER (CEBPB), and D: TQPVPESQLLPGQR (LYN). Table S1. All transcripts detected and differentially expressed genes (DEGs) with fold-change equal or above 2 (p-value < 0.05) at 3 h, 24 h, and 96 h after ATRA treatment. Table S2. The transcriptomic test sets (Yes-sets) for pathway modeling: DEGs related to the myeloid cell differentiation (MCD) (GO: 0030099) at the 3, 24, and 96 h time points. Table S3. Data on spectral counting. Number of unique peptides, percent of coverage, number of spectra, q-values and local FDR are shown for each peptide in each time point (0, 3, 24, 48, and 96 h). Table S4. Data on spectral counting. Number of unique proteins, percent of coverage, number of spectra, q-values and local FDR are shown for each protein in each time point (0, 3, 24, 48, and 96 h). Table S5. Summary of relative expression analysis and proteomic test sets (Yes-sets) for pathway modeling. Table S6. Transcription factors (TFs) possibly regulating DEGs expression during ATRA-induced HL-60 cell differentiation at time points 3 h, 24 h, and 96 h. Table S7. Transcription factors (TFs) possibly regulating DEPs expression during ATRA-induced HL-60 cell differentiation at time points 3 h, 24 h, 48 h, and 96 h. Table S8. Transition for SRM method (QqQ TSQ Vantage (Thermo Scientific, USA).

Author Contributions: Conceptualization, V.Z.; methodology, V.Z. and K.Y.; software, O.T.; investigation, I.V., S.N., T.F., O.T., L.K., and A.L.; data curation, S.N. and O.T.; writing—original draft preparation, S.N. and O.T.; writing—review and editing, S.N., O.T., and V.Z.; visualization, S.N. and O.T.; supervision, V.Z.; project administration, V.Z. All authors have read and agreed to the published version of the manuscript.

Funding: This research was funded by Russian Scientific Foundation, grant number 21-74-20122.

Institutional Review Board Statement: Not applicable.

Informed Consent Statement: Not applicable.

Data Availability Statement: Mass-spectrometric data are available via ProteomeXchange with identifier PXD006768.

Acknowledgments: The authors are grateful for opportunity to use the flow cytometry and mass spectrometry equipment of "Human Proteome" Core Facility (Institute of Biomedical Chemistry, Moscow).

Conflicts of Interest: The authors declare no conflict of interest.

References

1. Kambhampati, S.; Verma, A.; Li, Y.; Parmar, S.; Sassano, A.; Platanias, L. Signalling Pathways Activated by All- *trans* -Retinoic Acid in Acute Promyelocytic Leukemia Cells. *Leuk. Lymphoma* **2004**, *45*, 2175–2185. [CrossRef]
2. Jimenez, J.J.; Chale, R.S.; Abad, A.C.; Schally, A.V. Acute promyelocytic leukemia (APL): A review of the literature. *Oncotarget* **2020**, *11*, 992–1003. [CrossRef] [PubMed]
3. Zheng, P.Z.; Wang, K.K.; Zhang, Q.Y.; Huang, Q.H.; Du, Y.Z.; Zhang, Q.H.; Xiao, D.K.; Shen, S.H.; Imbeaud, S.; Eveno, E.; et al. Systems analysis of transcriptome and proteome in retinoic acid/arsenic trioxide-induced cell differentiation apoptosis of promyelocytic leukemia. *Proc. Natl. Acad. Sci. USA* **2005**. [CrossRef] [PubMed]
4. Mengeling, B.J.; Phan, T.Q.; Goodson, M.L.; Privalsky, M.L. Aberrant corepressor interactions implicated in PML-RARα and PLZF-RARα leukemogenesis reflect an altered recruitment and release of specific NCoR and SMRT splice variants. *J. Biol. Chem.* **2011**. [CrossRef]

5. Lin, R.J.; Sternsdorf, T.; Tini, M.; Evans, R.M. Transcriptional regulation in acute promyelocytic leukemia. *Oncogene* **2001**, *20*, 7204–7215. [CrossRef]
6. Isakson, P.; Bjørås, M.; Bøe, S.O.; Simonsen, A. Autophagy contributes to therapy-induced degradation of the PML/RARA oncoprotein. *Blood* **2010**. [CrossRef] [PubMed]
7. Coombs, C.C.; Tavakkoli, M.; Tallman, M.S. Acute promyelocytic Leukemia: Where did we start, where are we now, and the future. *Blood Cancer J.* **2015**, *5*, e304. [CrossRef]
8. Dutta, S.; Pregartner, G.; Rücker, F.G.; Heitzer, E.; Zebisch, A.; Bullinger, L.; Berghold, A.; Döhner, K.; Sill, H. Functional classification of tp53 mutations in acute myeloid leukemia. *Cancers* **2020**, *12*, 637. [CrossRef]
9. Johnson, D.E.; Redner, R.L. An ATRActive future for differentiation therapy in AML. *Blood Rev.* **2015**. [CrossRef]
10. Dalton, W.T.; Ahearn, M.J.; McCredie, K.B.; Freireich, E.J.; Stass, S.A.; Trujillo, J.M. HL-60 cell line was derived from a patient with FAB-M2 and not FAB-M3. *Blood* **1988**, *71*, 242–247. [CrossRef]
11. Birnie, G.D. The HL60 cell line: A model system for studying human myeloid cell differentiation. *Br. J. Cancer. Suppl.* **1988**, *9*, 41–45. [PubMed]
12. Wolf, D.; Rotter, V. Major deletions in the gene encoding the p53 tumor antigen cause lack of p53 expression in HL-60 cells. *Proc. Natl. Acad. Sci. USA* **1985**. [CrossRef]
13. Shima, H.; Nakayasu, M.; Aonuma, S.; Sugimura, T.; Nagao, M. Loss of the MYC gene amplified in human HL-60 cells after treatment with inhibitors of poly(ADP-ribose) polymerase or with dimethyl sulfoxide. *Proc. Natl. Acad. Sci. USA* **1989**. [CrossRef]
14. Tasseff, R.; Jensen, H.A.; Congleton, J.; Dai, D.; Rogers, K.V.; Sagar, A.; Bunaciu, R.P.; Yen, A.; Varner, J.D. An Effective Model of the Retinoic Acid Induced HL-60 Differentiation Program. *Sci. Rep.* **2017**. [CrossRef]
15. Lee, K.-H.; Chang, M.-Y.; Ahn, J.-I.; Yu, D.-H.; Jung, S.-S.; Choi, J.-H.; Noh, Y.-H.; Lee, Y.-S.; Ahn, M.-J. Differential gene expression in retinoic acid-induced differentiation of acute promyelocytic leukemia cells, NB4 and HL-60 cells. *Biochem. Biophys. Res. Commun.* **2002**, *296*, 1125–1133. [CrossRef]
16. Mollinedo, F.; López-Pérez, R.; Gajate, C. Differential gene expression patterns coupled to commitment and acquisition of phenotypic hallmarks during neutrophil differentiation of human leukaemia HL-60 cells. *Gene* **2008**, *419*, 16–26. [CrossRef]
17. Wang, J.; Fong, C.-C.; Tzang, C.-H.; Xiao, P.; Han, R.; Yang, M. Gene expression analysis of human promyelocytic leukemia HL-60 cell differentiation and cytotoxicity induced by natural and synthetic retinoids. *Life Sci.* **2009**, *84*, 576–583. [CrossRef]
18. Harris, M.N.; Ozpolat, B.; Abdi, F.; Gu, S.; Legler, A.; Mawuenyega, K.G.; Tirado-Gomez, M.; Lopez-Berestein, G.; Chen, X. Comparative proteomic analysis of all-trans-retinoic acid treatment reveals systematic posttranscriptional control mechanisms in acute promyelocytic leukemia. *Blood* **2004**, *104*, 1314–1323. [CrossRef]
19. Hofmann, A.; Gerrits, B.; Schmidt, A.; Bock, T.; Bausch-Fluck, D.; Aebersold, R.; Wollscheid, B. Proteomic cell surface phenotyping of differentiating acute myeloid leukemia cells. *Blood* **2010**, *116*, e26–e34. [CrossRef] [PubMed]
20. Kel, A.; Voss, N.; Valeev, T.; Stegmaier, P.; Kel-Margoulis, O.; Wingender, E. ExPlain: Finding upstream drug targets in disease gene regulatory networks. *SAR QSAR Environ. Res.* **2008**, *19*, 481–494. [CrossRef] [PubMed]
21. Fleck, R.A.; Athwal, H.; Bygraves, J.A.; Hockley, D.J.; Feavers, I.M.; Stacey, G.N. Optimization of NB-4 and HL-60 differentiation for use in opsonophagocytosis assays. *Vitr. Cell. Dev. Biol. Anim.* **2003**. [CrossRef]
22. Zgoda, V.G.; Kopylov, A.T.; Tikhonova, O.V.; Moisa, A.A.; Pyndyk, N.V.; Farafonova, T.E.; Novikova, S.E.; Lisitsa, A.V.; Ponomarenko, E.A.; Poverennaya, E.V.; et al. Chromosome 18 transcriptome profiling and targeted proteome mapping in depleted plasma, liver tissue and HepG2 cells. *J. Proteome Res.* **2013**, *12*. [CrossRef] [PubMed]
23. Kolker, E.; Higdon, R.; Morgan, P.; Sedensky, M.; Welch, D.; Bauman, A.; Stewart, E.; Haynes, W.; Broomall, W.; Kolker, N. SPIRE: Systematic protein investigative research environment. *J. Proteomics* **2011**, *75*, 122–126. [CrossRef]
24. Goedhart, J.; Luijsterburg, M.S. VolcaNoseR is a web app for creating, exploring, labeling and sharing volcano plots. *Sci. Rep.* **2020**. [CrossRef]
25. Matys, V.; Kel-Margoulis, O.V.; Fricke, E.; Liebich, I.; Land, S.; Barre-Dirrie, A.; Reuter, I.; Chekmenev, D.; Krull, M.; Hornischer, K.; et al. TRANSFAC(R) and its module TRANSCompel(R): Transcriptional gene regulation in eukaryotes. *Nucleic Acids Res.* **2006**, *34*, D108–D110. [CrossRef] [PubMed]
26. Hood, C.A.; Fuentes, G.; Patel, H.; Page, K.; Menakuru, M.; Park, J.H. Fast conventional Fmoc solid-phase peptide synthesis with HCTU. *J. Pept. Sci.* **2008**, *14*, 97–101. [CrossRef]
27. Fekkes, D. State-of-the-art of high-performance liquid chromatographic analysis of amino acids in physiological samples. *J. Chromatogr. B Biomed. Appl.* **1996**, *682*, 3–22. [CrossRef]
28. Congleton, J.; Jiang, H.; Malavasi, F.; Lin, H.; Yen, A. ATRA-induced HL-60 myeloid leukemia cell differentiation depends on the CD38 cytosolic tail needed for membrane localization, but CD38 enzymatic activity is unnecessary. *Exp. Cell Res.* **2011**, *317*, 910–919. [CrossRef]
29. De The, H.; Marchio, A.; Tiollais, P.; Dejean, A. Differential expression and ligand regulation of the retinoic acid receptor alpha and beta genes. *EMBO J.* **1989**, *8*, 429–433. [CrossRef] [PubMed]
30. Pavri, R.; Lewis, B.; Kim, T.-K.; Dilworth, F.J.; Erdjument-Bromage, H.; Tempst, P.; de Murcia, G.; Evans, R.; Chambon, P.; Reinberg, D. PARP-1 determines specificity in a retinoid signaling pathway via direct modulation of mediator. *Mol. Cell* **2005**, *18*, 83–96. [CrossRef]
31. Adhikary, S.; Eilers, M. Transcriptional regulation and transformation by Myc proteins. *Nat. Rev. Mol. Cell Biol.* **2005**, *6*, 635–645. [CrossRef]

32. Hoffman, B.; Liebermann, D.A. Apoptotic signaling by c-MYC. *Oncogene* **2008**, *27*, 6462–6472. [CrossRef]
33. Collins, S.J. The HL-60 promyelocytic leukemia cell line: Proliferation, differentiation, and cellular oncogene expression. *Blood* **1987**, *70*, 1233–1244. [CrossRef] [PubMed]
34. Ben-Baruch, N.; Reinhold, W.C.; Alexandrova, N.; Ichinose, I.; Blake, M.; Trepel, J.B.; Zajac-Kaye, M. c-myc Down-regulation in suramin-treated HL60 cells precedes growth inhibition but does not trigger differentiation. *Mol. Pharmacol.* **1994**, *46*, 73–78.
35. Amati, B.; Land, H. Myc-Max-Mad: A transcription factor network controlling cell cycle progression, differentiation and death. *Curr. Opin. Genet. Dev.* **1994**. [CrossRef]
36. Williams, A.B.; Schumacher, B. p53 in the DNA-damage-repair process. *Cold Spring Harb. Perspect. Med.* **2016**, *6*. [CrossRef]
37. Lee, M.H.; Na, H.; Kim, E.J.; Lee, H.W.; Lee, M.O. Poly(ADP-ribosyl)ation of p53 induces gene-specific transcriptional repression of MTA1. *Oncogene* **2012**, *31*, 5099–5107. [CrossRef] [PubMed]
38. Beneke, R.; Geisen, C.; Zevnik, B.; Bauch, T.; Müller, W.U.; Küpper, J.H.; Möröy, T. DNA excision repair and DNA damage-induced apoptosis are linked to Poly(ADP-ribosyl)ation but have different requirements for p53. *Mol. Cell. Biol.* **2000**, *20*, 6695–6703. [CrossRef] [PubMed]
39. Liu, S.-M.; Chen, W.; Wang, J. Distinguishing between cancer cell differentiation and resistance induced by all-trans retinoic acid using transcriptional profiles and functional pathway analysis. *Sci. Rep.* **2014**, *4*, 5577. [CrossRef]
40. Ruscetti, T.; Lehnert, B.E.; Halbrook, J.; Le Trong, H.; Hoekstra, M.F.; Chen, D.J.; Peterson, S.R. Stimulation of the DNA-dependent protein kinase by poly(ADP-ribose) polymerase. *J. Biol. Chem.* **1998**, *273*, 14461–14467. [CrossRef]
41. Ting, H.J.; Yasmin-Karim, S.; Yan, S.J.; Hsu, J.W.; Lin, T.H.; Zeng, W.; Messing, J.; Sheu, T.J.; Bao, B.Y.; Li, W.X.; et al. A positive feedback signaling loop between ATM and the vitamin D receptor is critical for cancer chemoprevention by vitamin D. *Cancer Res.* **2012**, *72*, 958–968. [CrossRef]
42. Turk, A.; Wisinski, K.B. PARP Inhibition in BRCA-Mutant Breast Cancer. *Cancer* **2018**, *124*, 2498–2506. [CrossRef]
43. Faraoni, I.; Compagnone, M.; Lavorgna, S.; Angelini, D.F.; Cencioni, M.T.; Piras, E.; Panetta, P.; Ottone, T.; Dolci, S.; Venditti, A.; et al. BRCA1, PARP1 and γH2AX in acute myeloid leukemia: Role as biomarkers of response to the PARP inhibitor olaparib. *Biochim. Biophys. Acta Mol. Basis Dis.* **2015**, *1852*, 462–472. [CrossRef]
44. Cantó, C.; Sauve, A.A.; Bai, P. Crosstalk between poly(ADP-ribose) polymerase and sirtuin enzymes. *Mol. Aspects Med.* **2013**, *34*, 1168–1201. [CrossRef] [PubMed]
45. Yoshida, A.; Ueda, T.; Nakamura, T. The mechanism of apoptosis induced by anticancer agents in human leukemia HL-60 cells. *Rinsho. Ketsueki.* **1996**, *37*, 552–557.
46. Dos Santos, C.; Demur, C.; Bardet, V.; Prade-Houdellier, N.; Payrastre, B.; Récher, C. A critical role for Lyn in acute myeloid leukemia. *Blood* **2008**, *111*, 2269–2279. [CrossRef]
47. Iriyama, N.; Yuan, B.; Hatta, Y.; Takagi, N.; Takei, M. Lyn, a tyrosine kinase closely linked to the differentiation status of primary acute myeloid leukemia blasts, associates with negative regulation of all-trans retinoic acid (ATRA) and dihydroxyvitamin D3 (VD3)-induced HL-60 cells differentiation. *Cancer Cell Int.* **2016**, *16*, 37. [CrossRef] [PubMed]
48. Alvarez-Errico, D.; Yamashita, Y.; Suzuki, R.; Odom, S.; Furumoto, Y.; Yamashita, T.; Rivera, J. Functional analysis of Lyn kinase A and B isoforms reveals redundant and distinct roles in Fc epsilon RI-dependent mast cell activation. *J. Immunol.* **2010**, *184*, 5000–5008. [CrossRef]
49. Tornillo, G.; Knowlson, C.; Kendrick, H.; Buckley, N.E.; Grigoriadis, A.; Smalley Correspondence, M.J. Dual Mechanisms of LYN Kinase Dysregulation Drive Aggressive Behavior in Breast Cancer Cells. *CellReports* **2018**, *25*, 3674–3692.e10. [CrossRef]

Article

Identification and Validation of VEGFR2 Kinase as a Target of Voacangine by a Systematic Combination of DARTS and MSI

Yonghyo Kim [1,2,3], Yutaka Sugihara [2,3], Tae Young Kim [1], Sung Min Cho [1], Jin Young Kim [4], Ju Yeon Lee [4], Jong Shin Yoo [4], Doona Song [5], Gyoonhee Han [5], Melinda Rezeli [2], Charlotte Welinder [2,3], Roger Appelqvist [2], György Marko-Varga [1,2,6] and Ho Jeong Kwon [1,7,*]

1 Chemical Genomics Global Research Lab., Department of Biotechnology, College of Life Science and Biotechnology, Yonsei University, Seoul 03722, Korea; yonghyo.kim@med.lu.se (Y.K.); kty1273@yonsei.ac.kr (T.Y.K.); sungmincho@yonsei.ac.kr (S.M.C.); gyorgy.marko-varga@bme.lth.se (G.M.-V.)
2 Clinical Protein Science & Imaging, Department of Biomedical Engineering, Lund University, BMC D13, SE-221 84 Lund, Sweden; yutaka.sugihara@med.lu.se (Y.S.); melinda.rezeli@bme.lth.se (M.R.); charlotte.welinder@med.lu.se (C.W.); roger.appelqvist@bme.lth.se (R.A.)
3 Division of Oncology and Pathology, Department of Clinical Sciences Lund, Lund University, Barngatan 4, SE-221 85 Lund, Sweden
4 Biomedical Omics Group, Korea Basic Science Institute, Ochang, Chungbuk 28119, Korea; jinyoung@kbsi.re.kr (J.Y.K.); jylee@kbsi.re.kr (J.Y.L.); jongshin@kbsi.re.kr (J.S.Y.)
5 Department of Biotechnology, College of Life Science and Biotechnology, Yonsei University, Seoul 03722, Korea; doona.s@yonsei.ac.kr (D.S.); gyoonhee@yonsei.ac.kr (G.H.)
6 Department of Surgery, Tokyo Medical University, 6-7-1 Nishishinjiku, Shinjuku-ku, Tokyo 160-0023, Japan
7 Department of Internal Medicine, College of Medicine, Yonsei University, Seoul 03722, Korea
* Correspondence: kwonhj@yonsei.ac.kr; Tel.: +82-2-2123-5883

Received: 1 February 2020; Accepted: 25 March 2020; Published: 27 March 2020

Abstract: Although natural products are an important source of drugs and drug leads, identification and validation of their target proteins have proven difficult. Here, we report the development of a systematic strategy for target identification and validation employing drug affinity responsive target stability (DARTS) and mass spectrometry imaging (MSI) without modifying or labeling natural compounds. Through a validation step using curcumin, which targets aminopeptidase N (APN), we successfully standardized the systematic strategy. Using label-free voacangine, an antiangiogenic alkaloid molecule as the model natural compound, DARTS analysis revealed vascular endothelial growth factor receptor 2 (VEGFR2) as a target protein. Voacangine inhibits VEGFR2 kinase activity and its downstream signaling by binding to the kinase domain of VEGFR2, as was revealed by docking simulation. Through cell culture assays, voacangine was found to inhibit the growth of glioblastoma cells expressing high levels of VEGFR2. Specific localization of voacangine to tumor compartments in a glioblastoma xenograft mouse was revealed by MSI analysis. The overlap of histological images with the MSI signals for voacangine was intense in the tumor regions and showed colocalization of voacangine and VEGFR2 in the tumor tissues by immunofluorescence analysis of VEGFR2. The strategy employing DARTS and MSI to identify and validate the targets of a natural compound as demonstrated for voacangine in this study is expected to streamline the general approach of drug discovery and validation using other biomolecules including natural products.

Keywords: target identification; target validation; label-free method for drugs; anti-angiogenesis; mechanism of action; receptor tyrosine kinases; curcumin; natural products

---

1. Introduction

Identifying the protein targets of therapeutic natural products and deciphering the specific mechanisms of action at the molecular level are crucial steps in the development of natural products as drugs to treat human diseases [1,2]. Without the validation of targets and the cellular actions of natural products, these compounds may cause unexpected events, including adverse and toxic effects in patients. Thus, methods for the identification of protein targets and the understanding of molecular mechanisms of action of these natural products are pivotal for treating human diseases.

Given the importance and necessity of deciphering these parameters, there have been several technological attempts to successfully identify the protein targets for natural products [3,4]. Widely-applied approaches include affinity-based matrices that label or tag small molecules, such as affinity pull-downs and phage display methods [5–7]. Nevertheless, these methods have limitations, such as changes in structural properties may occur upon labeling with chemical probes, or upon tagging functional groups for immobilization; biological activities of natural products may change as a result of the alterations in the chemical structure; these processes incur a high cost, and are time and labor-consuming; and difficulty in modifying small molecules due to availability of only virtual three-dimensional structures [8,9]. To overcome these limitations, new strategies have been suggested for target identification and validation using label-free methods [6] with natural products. Specifically, the targets are validated by utilizing changes in thermodynamic properties and structural stability when a natural product directly interacts with a cognate protein target [5,10]. These methods apply thermal [11], proteolytic [12,13] or oxidative stress [14] to analyze changes in the structural stability of protein targets. One of these methods using proteolytic stress for target validation is the DARTS method. DARTS is based on the principle of increased stability of a protein target upon interaction with a natural product, which makes the complex less susceptible to proteolytic effects. The conformational changes induced upon the interaction between the natural product and the target protein thermodynamically stabilize the protein structure [12,13]. Moreover, through unbiased DARTS approaches in combination with mass spectrometry (MS) analysis, quantitative MS based-proteomics is utilized to identify multiple target proteins in drug-treated versus control samples [15].

Although these label-free methods have several advantages that overcome many of the difficulties associated with labeling methods, researchers are yet to identify and validate directly in vivo interactions between natural products and their target proteins at the tissue or cellular level. Recently, matrix-assisted laser desorption ionization-mass spectrometry imaging (MALDI-MSI) has emerged as a new technology to analyze the distribution of a natural product in tissues by directly measuring the molecular mass of said molecule from the tissue sections. MALDI-MSI has also been adapted to investigate the interactions between natural products and protein targets ex vivo. MALDI-MSI is an analytical mass spectrometry technology that identifies ion peaks from a natural product at 25–100 μm resolution [16,17]. This platform can be utilized to either detect all ion masses within a tissue microenvironment or to perform selected-ion monitoring (SIM) of a single, specific ion mass. Furthermore, with the automated computational procedures [18,19], the data generated by MALDI-MSI can exhibit the spatial localization of all the detected natural products as a single integrated image. Accordingly, MALDI-MSI is a powerful tool for validating the interactions between a natural product and its protein targets ex vivo without any labeling probes or chemical immobilization [20,21]. With respect to absorption, distribution, metabolism, and excretion (ADME), MALDI-MSI could prove to be an effective approach to provide valuable information about the in vivo effects of label-free compounds in patients [22,23].

Accordingly, for overcoming the conventional limitations of target identification and analyzing the localization of interaction between natural products and target proteins at the tissue level, we combined the aforementioned methods into a systematic procedure and applied the same for target validation. To validate DARTS-MSI as a successful systematic strategy for target identification of natural products, we applied the same for a target-validated compound, curcumin, as a positive control natural compound. In the previous studies, curcumin was shown to directly and irreversibly bind to aminopeptidase N (APN), which plays a key role in tumor angiogenesis and proliferation, inhibiting its activity and hence angiogenesis [24].

According to a previous report, we identified a small natural molecule with antiangiogenic activity [25]. This molecule is called voacangine and is extracted from *Voacanga africana*, *Trachelospermum jasminoides*, or *Tabernaemontana catharinensis*. Preliminary experiments suggested that voacangine potentially inhibits angiogenesis. This effect was observed in tube formation assays in endothelial cells (ECs) and vascularization of the chick chorioallantoic membrane. Additionally, we also reported that voacangine significantly inhibited VEGF-induced chemoinvasion activity on HUVECs in a dose-dependent manner. However, its mechanistic pathways and molecular targets are still uncovered and not fully understood. Therefore, we focused on the investigation for the mode of action in voacangine as the model natural compound by applying the aforementioned systematic approach and various molecular experiments.

In the current study, we investigated the mode of action of voacangine via label-free DARTS and successfully identified VEGFR2 as a target protein responsible for the observed antiangiogenic properties of voacangine in ECs. The direct interaction between voacangine and VEGFR2 was validated in vivo in animal models and also by analyzing the localization of voacangine by MSI in xenograft tumor tissue sections. In addition, sunitinib, a marketed drug inhibiting tyrosine kinases by targeting not only VEGFR2 but also other RTKs (EGFR, PDGFR, and FGFR), was selected as a reference compound for comparing the potency on angiogenesis and tumor suppression with voacangine [26–29]. The strategy of employing DARTS and MSI to identify and validate the downstream targets of a natural compound as demonstrated for voacangine in this study can streamline the general process of drug discovery and validation of protein targets for other biomolecules including natural products.

2. Materials and Methods

2.1. Materials and General Methods

Curcumin (purity, ≥98%) was obtained from Sigma-Aldrich (St Louis, MO, USA). Voacangine (12-methoxyibogamine-18-carboxylic acid methylester) (purity, ≥98%) was purchased from THC Pharm (Frankfurt, Germany) [25], and the stock solution made in 100% dimethyl sulfoxide (DMSO) was stored at −20 °C, and diluted with the culture medium before the in vitro experiments. The working solution was freshly prepared in basal medium and the control group was treated with the same volume of DMSO as a vehicle control. Sunitinib (purity, ≥98%) was obtained from Sigma-Aldrich. Endothelial growth medium-2 (EGM-2) was purchased from Lonza (Walkersville, MD, USA). Dulbecco"=s Modified Eagle Medium (DMEM), Roswell Park Memorial Institute (RPMI) 1640, and fetal bovine serum (FBS) were purchased from Invitrogen (Grand Island, NY, USA). Vascular endothelial growth factor (VEGF), Tumor Necrosis Factor-α (TNF-α), epidermal growth factor (EGF), basic fibroblast growth factor (bFGF), and platelet-derived growth factor-BB (PDGF-BB) were purchased from KOMA biotech (Seoul, Korea). The Transwell chamber system for chemoinvasion assay and Matrigel (growth factors reduced) were obtained from Corning Costar (Corning, NY, USA) and BD Biosciences (Bedford, MA, USA), respectively. Pronase and protease inhibitor cocktail tablets were obtained from Roche (Mannheim, Germany). Phosphatase inhibitor solution, Triton X-100, and dithiothreitol (DTT) were purchased from Sigma-Aldrich. Sodium chloride (NaCl), Tris, and glycine were obtained from Samchun Chemical Co., Ltd. (Seoul, Korea). Trifluoroacetic acid (TFA), high-performance liquid chromatography (HPLC)-grade methanol (≥99.8%) and the matrix compound, α-cyano-4-hydroxycinnamic acid (CHCA) were purchased from Sigma-Aldrich, and liquid chromatography–mass spectrometry (LC–MS) hypergrade acetonitrile (ACN) was obtained from Merck (Darmstadt, Germany). Primary antibodies of phospho-VEGFR2, VEGFR2, fibroblast growth factor receptor 1 (FGFR1), platelet-derived growth factor receptor α (PDGFRα), platelet-derived growth factor receptor β (PDGFR β), phospho- extracellular signal-regulated protein kinases 1 and 2 (ERK1/2), ERK1/2, phospho-protein kinase B (Akt), Akt, APN, and β-actin were purchased from Cell Signaling Technology (Beverly, MA, USA). Anti-β-III-tubulin was purchased from Millipore (Temecula, CA, USA). Anti-epidermal growth factor receptor 1 (EGFR1) and anti-voltage-dependent anion-selective channel 1 (VDAC1) were purchased from Abcam (Cambridge, UK). Anti-fibroblast growth factor receptor 5 (FGFR5) was purchased from Thermo Fischer Scientific (Waltham, MA, USA). Anti-cluster of differentiation 31 (CD31) was purchased from Novus (Littleton, CO, USA). Secondary antibodies of anti-rabbit immunoglobulin G (IgG) and anti-mouse IgG were purchased from Cell Signaling Technology. U87 glioblastoma cells (U87MG), human umbilical vein endothelial cells (HUVECs), and human hepatoma cell (HepG2) were purchased from Korea Cell Line Bank, Seoul, Korea.

2.2. *In Vivo Mouse Tumor Xenograft Assays*

Mice were housed in the pathogen-free facility of the Laboratory Animal Research Center in Yonsei University, Seoul, Korea. The mice were handled following the Institutional Animal Care and Use Committee (IACUC) (permission number: IACUC-A-201407-254-01, IACUC-A-201503-213-01, IACUC-A-201602-149-02, and IACUC-201603-422-01) and International Guidelines for the Ethical Use of Animals. U87MG cells (5×10^6 cells) suspended in 200 μL phosphate-buffered saline (PBS)/Matrigel (1:1) were subcutaneously implanted into the dorsal flank of athymic nude mice (4-week-old female BALB/c nude mice, Orient Bio, Seoul, Korea). Once the tumors became palpable (50–100 mm^3, ~2 weeks), mice were randomly selected and separated into four groups (6 mice per group), and intraperitoneally treated with vehicle, curcumin (60 mg/kg), and voacangine (10 mg/kg) daily. Sunitinib was administered orally (40 mg/kg) daily. Vehicle and drug solutions were prepared in saline:ethanol:Tween-80 (97.8:2:0.2). Tween-80 was used to enhance drug solubility. Tumor volume and mouse body weight were measured daily using the following formula: $\pi/6 \times$ length $\times$ width $\times$ height. Four hours after the last treatment (on day 12), mice were sacrificed, and tissue samples (tumors, livers, and kidneys) were obtained. The tissues were surgically removed and slowly frozen by placing tumors for 2 min on a plastic boat floating in a bath of isopentane that was supercooled with dry ice (−70 °C) [30]. All animal study protocols were performed following the Guidelines for Animal Experiments and were approved by the Department of Institutional Animal Care and Use Committee, Yonsei University, Seoul, Korea.

2.3. *Growth Factor-Induced Chemoinvasion Assays*

To determine the invasiveness of HUVECs in vitro, a Transwell chamber system with polycarbonate filter inserts containing 8.0 μm pores was used. The lower side of the filter was coated with 10 μL of gelatin (Sigma-Aldrich) (1 mg/mL), and the upper side was coated with 10 μL of Matrigel (growth factors reduced, 3 mg/mL in high-grade pure water). Voacangine was added to the lower chambers in the presence of the growth factors (VEGF, TNF-α, bFGF, PDGF-BB, and EGF, 30 ng/mL each), and HUVECs (FBS starvation for 17 h, 6×10^5 cells/well) were placed in the upper chambers. The chambers were incubated at 37 °C for 16 h. The invasiveness of cells fixed with 70% methanol and stained with hematoxylin and eosin (H&E) was measured by counting the total number of cells on the lower side of the filter, using an Olympus IX70 microscope at 100× *g* magnification.

2.4. *Drug Affinity Responsive Target Stability (DARTS) Assay*

DARTS assay was performed as previously described [13,15]. Briefly, HUVECs were lysed using 0.5% Triton X-100 lysis buffer (50 mM Tris-HCl pH 7.5, 200 mM NaCl, 0.5% Triton X-100, 10% glycerin, 1 mM DTT) containing protease and phosphatase inhibitors. The supernatant from the cell lysates containing 2–3 mg/mL total protein was incubated with voacangine at the indicated concentrations at room temperature (RT) for 1 h, followed by proteolysis with pronase (1 μg/mL per sample) for 2, 5, and 10 min at RT. For curcumin treatment, the supernatant from the membrane fraction of HepG2 cell lysates containing 1.5 mg/mL total protein was incubated with curcumin at the indicated concentrations at RT for 3 h, followed by proteolysis with 10 μg/mL pronase per sample for 2, 5, and 10 min at RT. The final concentration of DMSO was 1% in all samples. To quench proteolysis, 6× sodium dodecyl sulfate (SDS) sample loading buffer (1 M Tris-hydrochloric acid (HCl) pH 6.8, SDS 10%, glycerol 60%, bromophenol blue 0.012%, and 0.6 M DTT) was added to each sample in a 1:3 ratio, thoroughly mixed, and boiled at 100 °C for 5 min. Samples were analyzed by immunoblotting with primary antibodies (APN, VDAC1, β-actin, VEGFR2, FGFR1, PDGFRα, and PDGFRβ) according to the manufacturer's instructions.

2.5. *In Silico Docking Simulation*

All molecular docking analyses were performed with Discovery Studio 2016 software adopting the CHARMM (Chemistry at Harvard Macromolecular Mechanics) force field. The crystal structure of human VEGFR2 (Protein Data Bank code, 4AGD) was obtained from the Research Collaboratory for Structural Bioinformatics (RCSB) protein data bank. The protein structures of VEGFR2 were optimized by the Powell algorithm to minimized energy. To dock the ligands, the Ligandfit docking method was used. The parameters of Ligandfit were validated using the ligand from the VEGFR2 crystal structure. Voacangine was docked to the binding site of the protein and 10 poses were generated. The most predictive binding mode were determined based on various scoring functions (Ligscore1_Dreiding, Ligscore2_Dreiding, PLP1, PLP2, PMF, and DOCK_SCORE), and the binding energies were determined by calculating the binding energy of the most predictive binding mode.

2.6. *Immunoblotting Analysis*

Cell lysates were separated by 10% SDS–polyacrylamide gel electrophoresis (PAGE), and the proteins were transferred to polyvinylidene difluoride (PVDF) membranes using standard electroblotting procedures. Blots were blocked and immunolabeled overnight at 4 °C with primary antibodies. For immunoblotting of tumor samples, sections (10 μm thickness) from tumor tissues were collected and lysed in radioimmunoprecipitation assay (RIPA) buffer. Lysates from tumor samples were separated by 10% SDS-PAGE, and the proteins were transferred on PVDF membranes using standard electroblotting procedures. Blots were blocked and immunolabeled overnight at 4 °C with primary antibodies (phospho VEGFR2, VEGFR2 [31,32], phospho ERK, ERK, phospho Akt, Akt, and β3-tubulin) according to manufacturer's instructions. Then membranes were washed with TBST (Tris-buffered saline with 0.05% Tween-20,) 3 times for 10 min each, and the secondary antibody was added and incubated for 1 h at RT. Immunolabeling was detected using an enhanced chemiluminescence (ECL) kit according to the manufacturer's instructions.

2.7. *Cell Growth Condition and Cell Proliferation Assays*

Cell lines were grown according to the recommendations and protocols of the supplier. All cells were maintained at 37 °C in a humidified 5% CO_2 incubator. All cells were seeded in 96-well plates at a density of 2000 cells/well. Voacangine was added to the cells to determine their effect on cell proliferation. Cells were grown for 72 h and growth was analyzed using the 3-(4,5-dimehylthiazol-2-yl)-2,5-diphenyl tetrazolium bromide (MTT) colorimetric assay.

2.8. *Quantification of Microvessel Density*

To measure the expression levels of the vascular marker CD31 in tumor sections, frozen sections were incubated with a primary anti-CD31 antibody. Frozen xenograft tumor sections (10 μm thickness) were incubated in primary antibody in 1% bovine serum albumin (BSA) overnight blocking buffer. After rinsing the primary antibody, the tumors were labeled with anti-rabbit Alexa-594 labeled secondary antibody (Invitrogen, 1:1000) for 1 h at RT and then counterstained with 4′,6-Diamidino-2-phenylindole (DAPI). Microvessel density was measured by counting the number of positive structures in three random fields. The images were obtained using a confocal laser scanning microscope LSM 700 (Carl Zeiss, Jena, Germany) from the whole tumor tissue at 400× *g* magnification.

2.9. *Compound Detection and Analysis of Drug Distribution Using MSI*

A MALDI LTQ Orbitrap XL mass spectrometer (Thermo Fisher Scientific, Waltham, MA, USA) was utilized for compound characterization, drug detection, and tissue imaging. For the matrix, 7.5 mg/mL α-CHCA was dissolved in 50% ACN and 50% Milli-Q water (high-grade pure water) containing 0.2% TFA.

For tissue drug imaging, the freshly frozen tissues were cut into 10 μm sections using a cryotome (Thermo Fisher Scientific, Waltham, MA, USA) and placed on glass microscope slides (Superfrost ultra plus). After drying the tissue for 1 h at RT, 0.5 mL of the matrix solution was deposited stepwise onto the tissue by an airbrush. To control spraying conditions, the position of the airbrush was constantly maintained. Mass spectra were obtained using the Orbitrap mass analyzer (Thermo Fisher Scientific, Waltham, MA, USA) at 60,000 resolution (at *m*/*z* 400). Tissue sections were sampled in the 150–800 Da mass range in positive-ion mode with a 50 μm raster size. The nitrogen laser was operated at 10.0 μJ with activated automatic gain control. For MS/MS, the curcumin peak observed at *m*/*z* 369.14, and voacangine peak observed at *m*/*z* 369.21 were isolated with a 1.0 Da window, and fragmented at 40% normalized collision energy with a 30 ms activation time, 0.250 activation Q and the fragment ions were scanned at a normal scan rate in the linear ion-trap analyzer. The minimum signal required for MS/MS spectra generation was 500 counts. Spectra were analyzed with Xcalibur v 2.1.0. software. Visualization of the compound and fragment ions was performed with ImageQuest software (Version 1.0.1., Thermo Fisher Scientific, Waltham, MA, USA).

2.10. *Quantitation of the Precursor Compound*

For tissue quantitation, calibration curves of the drug and compounds were established in control tissue sections of the mice. Voacangine was diluted in 50% ACN containing 0.2% TFA. For each concentration, aliquots of 0.5 μL were applied to the tissue surface within the concentration range of 10 nM–1 mM. Spraying and detection conditions were identical to those used for the tissue sample analysis. The calibration curve was then used to estimate the tissue drug concentrations in in vivo-treated tumor sections.

2.11. *Histochemical Analysis of Protein Target in Tissues and Compound Colocalization*

To compare the immunofluorescence staining of the target protein and voacangine localization in tissues, frozen sections were sequentially cut from each tumor. Voacangine distribution was determined in the sections using MALDI-MSI and H&E staining. Sequential sections were labeled with anti-APN (1:100), anti-EGFR1 (1:50), anti-FGFR1 (1:50), anti-FGFR5 (1:50), and anti-VEGFR2 (1:50) [31]. The primary antibody incubation was followed by incubation with a fluorescent-tagged secondary antibody of anti-rabbit Alexa-488 (Invitrogen, 1:500). Nuclei were stained with DAPI (Invitrogen). The images were obtained using a confocal laser scanning microscope (LSM 700, Carl Zeiss) at a 200× *g* and 400× *g* magnification by title scanning the whole tumor tissue. Overlapping regions in the tumor tissue between compounds (curcumin and voacangine)-MSI and immunofluorescence staining of receptor tyrosine kinases (RTKs) were quantitated by Image J, Adobe Photoshop, and Qupath software [33] (0.2.0-m1) by counting the pixels of each merged region at identical image sizes.

2.12. *Statistical Analysis*

All data fitting and statistical analysis in different experimental groups are expressed as the mean ± standard deviation (S.D.) using GraphPad Prism and Microsoft Excel. The data shown in the study were obtained from at least three independent experiments. Statistical analyses were performed using an unpaired, two-tailed Student's *t*-test. *P*-values less than 0.05 were considered statistically significant (* indicates $p < 0.05$, ** indicates $p < 0.01$, *** indicates $p < 0.001$).

3. Results and Discussions

3.1. *Validation of the Systematic Combination of DARTS and MSI for Natural Product-Target Protein Interaction*

Firstly, we performed a DARTS assay to validate the interaction between curcumin and APN (Figure 1a). The stability of APN significantly decreased after 2 min of treatment with pronase, but APN pretreated with curcumin before pronase treatment retained its stability. Secondly, we identified curcumin by mass spectra using MALDI-MS and MS/MS for the detection of curcumin at the tissue level (Figure S1a,b). Based on results obtained from mass spectra, quantitation of curcumin in tissue sections was conducted using MALDI-MSI. The precursor ion (curcumin, *m*/*z* 369.14) and fragment ions 1, 2, and 3 (*m*/*z* 176.08, *m*/*z* 245.08, and *m*/*z* 285.17, respectively) in the curcumin-treated tumor tissue sections were readily detected using MALDI-MSI (Figure 1b). In contrast, the precursor ion signals and fragment ions were not detected in control tissues (vehicle solution-treated tumors) (Figure S1c). Additionally, the intensity of curcumin was weaker in the liver and kidney tissue sections than in the tumor tissues of treated animals (Figure S1d). A merged image visualizing the transparent MSI signal of curcumin and the immunofluorescence image of APN was obtained (Figure 1c, red color). Notably, the highest concentrations of curcumin were observed in the tumor regions that expressed the highest concentrations of APN (yellow color in the high APN image). Through the quantitation of the merged pixel count, we observed that curcumin MSI showed high colocalization (72.65%) in the regions with the highest APN expression (Table 1).

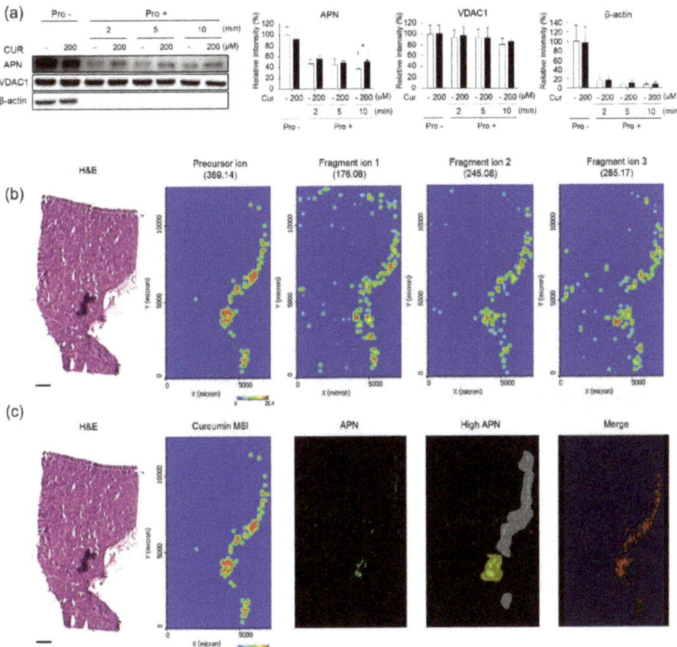

Figure 1. Validation of a systematic combination of drug affinity responsive target stability and mass spectrometry imaging (DARTS-MSI) for curcumin-aminopeptidase N (APN) interaction. (**a**) Analysis of direct binding of curcumin with APN in human umbilical vein endothelial cells (HUVECs) using the DARTS assay and immunoblotting. HUVEC lysates were incubated with curcumin and digested with pronase (0.1 µg/mL) at each incubation time. All images are the representative of three independent experiments. Each value represents the mean ± S.D. from three independent experiments. * $p < 0.05$ versus control. Cur: curcumin-treated, APN: aminopeptidase N. Pro -:Pronase non-treated; Pro +:Pronase treated; -:Non-treated (**b**) MALDI-MSI images of curcumin (precursor ion, fragment ion 1, fragment ion 2, and fragment ion 3) on curcumin-treated tumor tissue. The results shown are representative of three independent experiments. Scale bar, 1 mm. (**c**) Comparison of curcumin MSI (precursor ion) and immunofluorescence staining of the target protein, APN, in curcumin-treated tumor tissue. The transparent MSI image of curcumin is overlaid on the immunofluorescence staining image for APN and is visualized in the merged region (red). Scale bar, 1 mm.

Table 1. Comparison with the highest concentration of curcumin and the highest APN expressed regions in the tumor regions. Each value represents the mean from three independent experiments.

Quantitation	Pixels	Merged Regions/Cur-MSI (%)
MALDI-MSI (curcumin)	44,408	100
Merged regions (high curcumin and high APN)	32,264	72.65

3.2. Identification of VEGFR2 as a Potent Target for Voacangine

As the first step to determine the molecular mechanisms underlying the antiangiogenic effect and the signaling pathways involved in the process, the effect of voacangine on growth factor-induced chemoinvasion of ECs was investigated. HUVECs were treated with 10 and 20 μM voacangine in the presence of growth factors such as VEGF, TNF-α, bFGF, PDGF-BB, and EGF, and voacangine exhibited specific and potent suppression of VEGF-induced EC chemoinvasion (Figure 2a). In addition, a human phospho RTKs assay was also performed to further validate the inhibitory effect of voacangine on VEGF-mediated signaling activity (Figure 2b and Figure S2). The effect of voacangine on the phosphorylation of various RTKs was investigated in the HUVEC lysates.

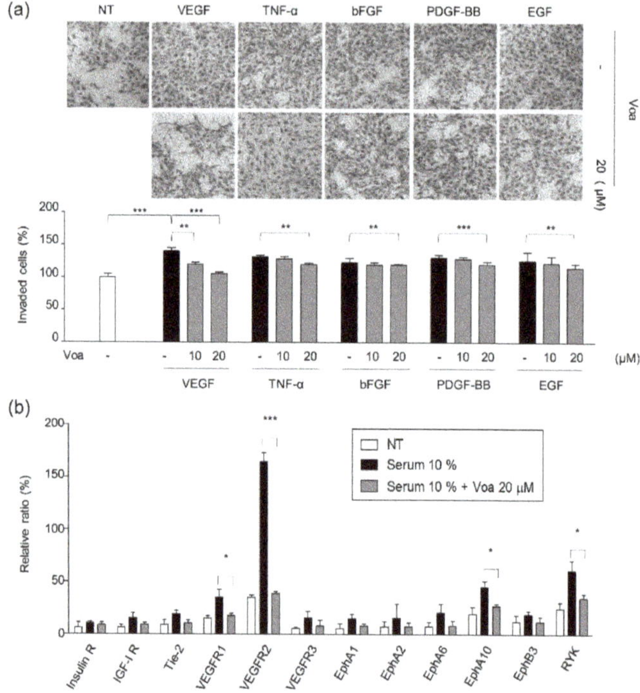

Figure 2. Voacangine specifically inhibited vascular endothelial growth factor (VEGF)-induced angiogenesis. (**a**) Effect of voacangine on chemoinvasion induced by various growth factors (VEGF, TNF-α, bFGF, PDGF-BB, and EGF). *** $p < 0.001$, ** $p < 0.01$ versus control of representative growth factors. -: Non-treated (**b**) Quantitation of the results from human p-RTKs array assay in HUVECs. The images are the representative of three independent experiments. Each value represents the mean ± S.D. from three independent experiments. *** $p < 0.001$, * $p < 0.05$ versus control. NT: non-treated control, Voa: voacangine-treated.

3.3. Validation of In Vitro Direct Interaction between Voacangine and VEGFR2 by DARTS

Next, DARTS was performed to explore whether voacangine directly binds to VEGFR2. As shown in Figure 3a, the stability of VEGFR2 significantly decreased after a 2 min pronase digestion. In contrast, VEGFR2 pretreated with voacangine before pronase treatment retained stability even after 5 and 10 min.

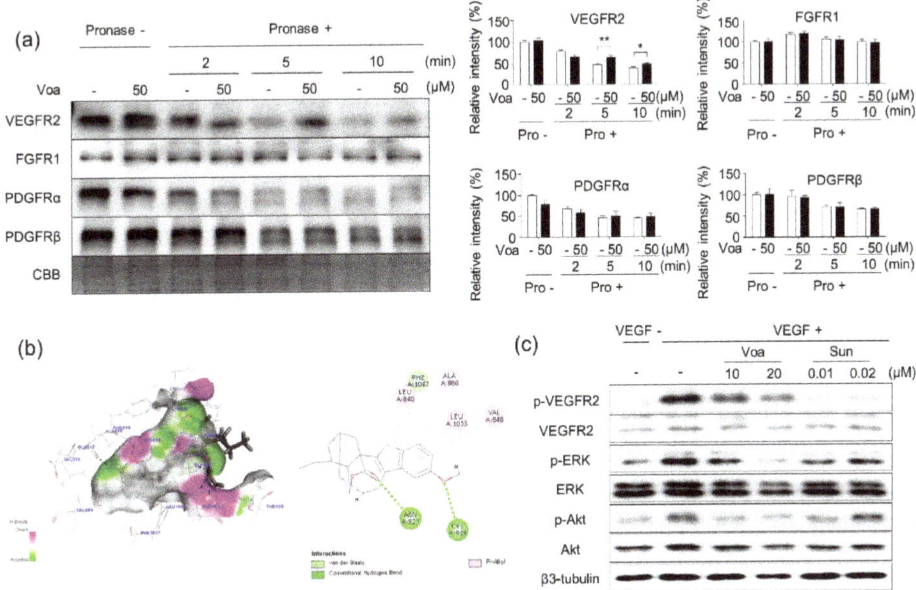

Figure 3. Voacangine specifically and directly binds to vascular endothelial growth factor receptor 2 (VEGFR2) (**a**) DARTS results from direct binding of voacangine to various receptor tyrosine kinases (RTKs) in HUVECs. HUVEC lysates were incubated with voacangine and digested with pronase (1 μg/mL) at each incubation time. Each value represents the mean ± S.D. from three independent experiments. ** $p < 0.01$ versus control, * $p < 0.05$ versus control. NT: non-treated control, Voa: voacangine-treated, Pro -:Pronase non-treated; Pro +:Pronase treated; -:Non-treated (**b**) In silico docking analysis using a 2D-diagram for validating the interaction between voacangine and VEGFR2 (juxtamembrane and kinase domains, RCSB Protein Data Bank number: 4AGD). Left panel, green (voacangine) is superimposed with VEGFR2 (grey). Right panel, binding motifs are illustrated with various interactions of voacangine with the ATP-binding pocket of VEGFR2. (**c**) Effect of voacangine on VEGF-induced VEGFR2 signaling. Protein levels were determined by immunoblotting using specific antibodies. The results shown are representative of three independent experiments. Voa: voacangine-treated; Sun: sunitinib-treated.

To evaluate the binding of voacangine with VEGFR2, a docking model of human VEGFR2 based on its crystal structure (juxtamembrane and kinase domains, PDB number: 4AGD) was examined. In the virtual docking model of VEGFR2, two oxygen atoms of voacangine were found to reside in the hydrophobic pocket of VEGFR2 and form hydrogen bonds with Asn 923 and Cys 919, resulting in a high affinity and direct interaction with VEGFR2. The indole moiety of core directly interacted with the active residues in the VEGFR2 kinase domain (Leu 840, Val 848, Ala 866, and Leu 1035) via hydrophobic interactions (Pi-Alkyl interaction). The in-silico docking data suggested that voacangine directly interacts with VEGFR2 (Figure 3b).

In pathological states, such as cancer development and tumor progression and other conditions exhibiting abnormal angiogenic phenotypes, growth factors such as VEGF are secreted from preexisting blood vessels to promote excessive cell growth. The secretion of VEGF by tumor cells ultimately leads to a remarkable promotion of angiogenesis [34,35]. Furthermore, VEGF-induced VEGFR2 signaling activates ERK and Akt by downstream phosphorylation of VEGFR2 kinase(s) and promotes angiogenesis by regulating the expression of the target gene, VEGF [36,37]. Accordingly, the effect of voacangine on VEGF-induced VEGFR2 activation and subsequent downstream signaling were investigated in HUVECs. Treatment with voacangine and sunitinib significantly suppressed the VEGF-induced phosphorylation of VEGFR2 and the downstream activation of ERK and Akt in a dose-dependent manner (Figure 3c).

3.4. Voacangine Inhibits Xenograft Tumor Growth and Angiogenesis In Vivo

Validation of voacangine as a new VEGFR2-targeting antiangiogenic and antitumor compound was achieved by investigating its effect on tumor growth in the U87MG cell glioblastoma xenograft mouse model. U87MG glioblastoma cells form aggressive angiogenic solid tumors that exhibit high levels of VEGF and VEGFR2 [38]. As a reference, data were compared with those for sunitinib (SU11248, Sutent), a known VEGFR2-targeting anticancer drug. Both voacangine and sunitinib significantly inhibited tumor growth (Figure 4a) in 6 to 12 days without causing overt toxicity, as no significant weight loss was observed in the mice (Figure S4). As shown in Figure 4b, the expression levels of the blood vessel marker, CD31, were significantly lower in tumor-bearing mice treated with either voacangine (47.3%) or sunitinib (50%) than that of control.

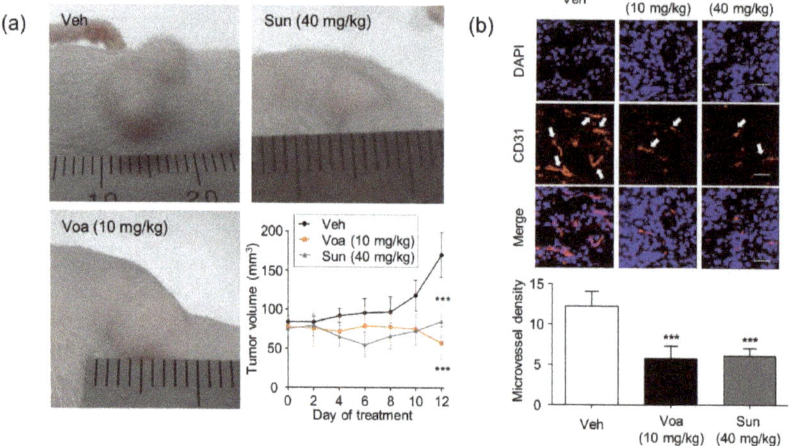

Figure 4. Voacangine inhibits xenograft tumor growth and angiogenesis in vivo. (**a**) Representative images of U87MG tumor xenograft on 12 days. Athymic nude mice bearing glioblastoma tumors consisting of U87MG glioblastoma cells were treated with vehicle, voacangine (10 mg/kg, intraperitoneal treatment), or sunitinib (40 mg/kg, oral treatment). *** $p < 0.001$ versus vehicle treatment. (**b**) Effect of voacangine or sunitinib treatment on the expression levels of the vascular marker, CD31 in tumor tissues. All images shown are representative of three independent experiments. White arrows indicate CD31 expression. Original magnification of fluorescence images for CD31 staining: 400× *g*. Scale bar, 50 μm. Microvessel density was measured by counting the number of CD31-positive structures in three random fields. Each value represents the mean ± S.D. from three independent experiments. *** $p < 0.001$ versus vehicle treatment. Veh, vehicle-treated; Voa, voacangine-treated; Sun, sunitinib-treated.

High-resolution histological inspection revealed a cellular presentation within these tumors that showed a high degree of heterogeneity in cell size and shape. The sunitinib dosage (40 mg/kg, oral treatment) administered in these studies was selected from the dosage administered in the previous mice experiments [29,39,40]. Voacangine at a dosage of 10 mg/kg was administered at 48-h intervals for 14 days in the disease mouse model. A similar tendency of stabilized tumor growth was observed in mice treated with both, voacangine and sunitinib, with a fast growth onset that began on day 5, and then considerably expanded by day 8. These xenograft studies were performed with 6 animals per group, which showed a high degree of tumor growth consistency.

Next, the identification of voacangine in vivo by MSI was conducted. The precursor ion (voacangine, *m/z* 369.2162) (Figure S5a) fragment ions 1 and 2 (*m/z* 309.17 and *m/z* 337.17, respectively) (Figure S5b) in voacangine-treated tumor tissue sections were readily-detected using MALDI-MSI (Figure 5a). In contrast, the precursor compound signal and fragment ions for voacangine were not detected in the vehicle solution-treated tumor tissues (Figure S7). Additionally, in the liver and kidney tissue sections of treated animals, the voacangine signal was detected with weaker intensity than that in the tumor tissue (Figure 5b).

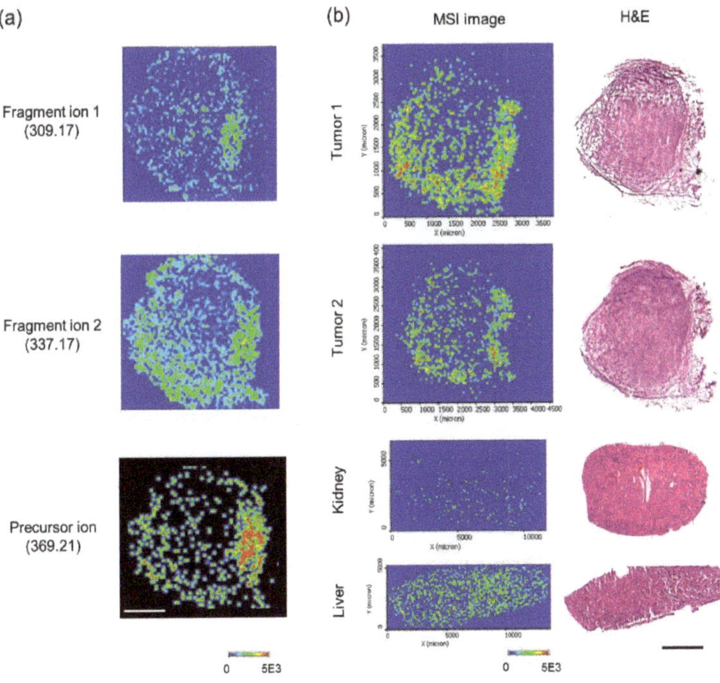

Figure 5. Identification of voacangine in tissue sections using mass spectrometry imaging (MSI). (**a**) MALDI-MSI signal for voacangine (precursor ion, fragment ion 1, and fragment ion 2) from voacangine-treated tumor tissue. The results shown are representative of three independent experiments. Scale bar, 1 mm. (**b**) Comparison of voacangine MSI (the precursor ion) between other voacangine-treated tumors and organ tissues (kidney and liver).

3.5. Validation of Target Interaction by Colocalization of Voacangine with VEGFR2 and Other RTKs

As shown in Figure S8, each tumor-bearing mouse showed high VEGFR2 expression, indicating that the tumor-bearing tissues also expressed high levels of VEGFR2 which were significantly reduced by voacangine treatment. From these observations, an overlay image was visualized with a transparent MSI signal of voacangine and the immunofluorescence image of VEGFR2 in a merged region (Figure 6a, red color). Next, immunofluorescence was performed to analyze the interaction between voacangine and various other RTKs (EGFR1, FGFR1, FGFR5, and VEGFR2) in voacangine-treated tumor tissues (Figure 6b) by quantifying the pixel counts from the merged regions (Table 2). The overlaid transparent

MSI image was quantitated with the immunofluorescence images of various RTKs. The merged regions are highlighted in red.

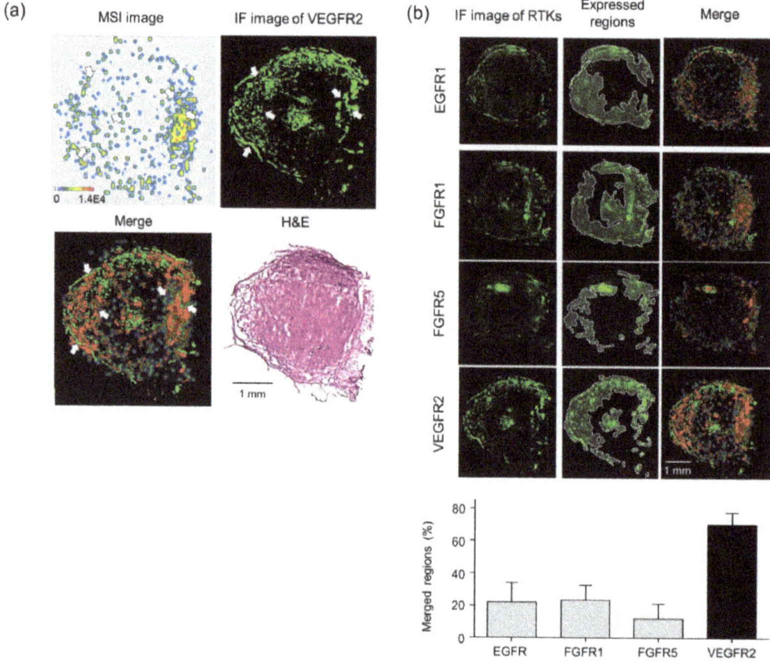

Figure 6. Validation of target interaction by colocalization of voacangine with VEGFR2 and other RTKs. (**a**) Comparison of voacangine MSI (precursor ion) and immunofluorescence staining of the target candidate, VEGFR2, in voacangine-treated tumor tissue. The overlaid image with transparent MSI signal on the immunofluorescence staining image for VEGFR2 is visualized in the merged region (red). Scale bar, 1 mm. (**b**) Comparisons and quantitation of merged regions for voacangine distribution and RTK receptors (EGFR1, FGFR1, FGFR5, and VEGFR2) in tumor tissues. Regions of expression for each RTK are indicated with white dashed lines on the immunofluorescence images. The merged regions are visualized in red. The results shown are representative of three independent experiments. Each value represents the mean ± S.D. from three independent experiments. Scale bar, 1 mm.

Table 2. Comparison with other RTKs in voacangine-treated tissues. Each value represents the mean from three independent experiments. IF: immunofluorescence-stained regions. Voa, voacangine.

Quantitation		Pixels	IF/Voa-MSI (%)
MALDI-MSI (voacangine)		71,406	100
Merged regions (red)	EGFR1	15,572	21
	FGFR1	16,745	23
	FGFR5	8616	12
	VEGFR2	50,039	70

3.6. Discussions

Natural products have been widely used as pharmacological or nutraceutical agents for effectively treating various human diseases due to their significant biological activities with diverse chemical structures. Given the advantages of natural products, there have been many obstacles to address the exact mode of action and in vivo effects on possible target proteins. Most often these problems arise due to difficulties in chemical modification of these natural products and alteration in the chemical and biological properties during these processes.

In this study, we attempted to overcome these hurdles by developing a new systematic procedure to identify targets of label-free natural products in vitro and in vivo with DARTS and MALDI-MSI, respectively. In Figure 1 and Table 1, curcumin pretreatment significantly protected APN digestion from pronase even at 10 min (p = 0.0185); but the digestion of other proteins, VDAC1 and β-actin remained unaffected, suggesting that curcumin specifically binds to APN, as reported by the earlier studies [24]. Further, these results strongly suggested that there is a stronger curcumin-binding and localization in the tumor tissue with elevated levels of APN expression. From these observations with curcumin, we suggest that the combination of DARTS and MALDI-MSI could be used as a systematic procedure for validation of in vivo target interaction with label-free natural products.

After validating the interaction of curcumin with APN as a proof of concept case, we utilized this approach for voacangine, a natural antiangiogenic compound, without any known targets. As shown in Figure 2a, HUVECs were treated with 10 and 20 μM voacangine in the presence of growth factors such as VEGF, TNF-α, bFGF, PDGF-BB, and EGF, and voacangine exhibited specific and potent suppression of VEGF-induced EC chemoinvasion. These results demonstrated that voacangine did not show inhibitory activity against the other growth factors but specifically inhibit VEGF-mediated signaling. In VEGF-induced conditions, voacangine specifically suppressed the phosphorylation of VEGFR2. This was further validated by the p-RTK array assay, wherein the phosphorylation profiles of various RTKs were analyzed. Furthermore, from the 12 RTKs activated by serum out of the 45 RTKs, voacangine specifically inhibited the VEGF-induced phosphorylation of VEGFR2 (Figure 2b, Figure S2). Accordingly, the following experiments on the activity of voacangine focused on VEGFR2.

Using DARTS technology, VEGFR2 was identified as a cellular target protein of voacangine. Notably, pretreatment with voacangine resulted in limited VEGFR2 digestion upon pronase treatment; however, the digestion of other RTKs, FGFR1, PDGFRα, and PDGFRβ remained unaffected, suggesting that voacangine specifically binds to VEGFR2, but not other RTKs (Figure 3a). The in vitro inhibitory activity of voacangine significantly inhibited the phosphorylation of VEGFR2 and downstream signaling proteins, such as ERK and Akt, in HUVECs (Figure 3c). These results demonstrated that the antiangiogenic mechanisms of voacangine action affect the VEGFR2 mediated signaling pathway. These also have been well-known that VEGF-mediated signaling plays a key role in tumor angiogenesis [34,35,41], and secreted VEGF binds to VEGFR2 that is expressed on the vascular endothelium. Subsequently, an angiogenic response is evoked that leads to the activation of VEGFR2 signaling [41]. VEGF-stimulated VEGFR2 induces the phosphorylation of downstream signaling kinases, including ERK and Akt which promote migration, proliferation, invasion, adhesion, and tube formation in ECs [36,37]. Therefore, targeting VEGFR2 is considered a promising strategy and an important therapeutic approach for treating diseases associated with angiogenesis, such as cancer [42].

Further evidence revealed that the effects of voacangine significantly correlated with the levels of VEGFR2 expression in different cell lines. The expression of VEGFR2 in HUVEC, U87MG, and Panc-1 cells was higher than that in the remaining cell lines (Figure S3a). As shown in Figure S3b, voacangine significantly inhibited cell proliferation in HUVEC, U87MG, and Panc-1 cells (Table S1). From the examination of the effects of voacangine on several cell lines with different VEGFR2 expression levels demonstrated that voacangine exerted its biological effects by specifically inhibiting proliferation in cells with high levels of VEGFR2 expression. Further, these results suggested that voacangine could be used as a specific inhibitor targeting VEGFR2-overexpressing cells (Figure S3).

In the U87MG xenograft tumor mouse model, where VEGFR2 is highly expressed, voacangine significantly suppressed tumor growth and microvessel density in vivo without significant toxicity (Figure S4). The outcome of voacangine treatment suggested that it reduced the tumor growth to a basal level (as determined in the control groups), similar to treatment with sunitinib, which returned the tumors to baseline levels during the 14 days (Figure 4). Notably, this observation suggests that voacangine may cause dual inhibition of tumor angiogenesis and tumor growth, thus resulting in a potent antitumor activity. These results also demonstrated that voacangine might be a promising candidate for effective treatment of aggressive glioblastomas which are resistant to various chemotherapies [43]. Additionally, these data are in accord with the effects of the known VEGFR inhibitor, sunitinib. Based on these results, voacangine inhibited angiogenesis and tumor proliferation in vivo by directly targeting VEGFR2-overexpressing cancer cells, similar to sunitinib.

To further validate VEGFR2 as the target receptor of voacangine, the interaction was analyzed in vivo. We established a novel systematic combination to compare the localization of small molecules and their protein targets using MALDI-MSI and immunofluorescent staining. From the MALDI-MS analyses, voacangine was identified at m/z 369.21, as well as its two major fragment ions were identified (Figure S5). On the surface of the tissue sections, a single droplet of voacangine could be detected by MALDI-MSI (Figure S6). The concentration of the voacangine and the precursor ion signal intensity linearly correlated from 0.01–100 μM. Voacangine precursor ions and its fragment ions were detected and colocalized with the tumor tissue from voacangine-treated mice. Furthermore, voacangine precursor ions and its fragment ions were not detected in control tissues, confirming that voacangine and the fragment ions observed by MALDI-MSI did not originate from the matrix or tissue. A comparative analysis of MALDI-MSI of voacangine ions and immunofluorescence images of other RTKs (EGFR1, FGFR1, FGFR5, PDGFRα, and PDGFβ), revealed that the highest concentrations of voacangine were observed in the tumor regions that expressed the highest concentrations of VEGFR2. This suggested a stronger voacangine-binding and localization in tumor tissues with elevated VEGFR2 expression (Figure 6).

In a previous study, high concentrations of sunitinib colocalized with high expression levels of VEGFR2 [31]. Akin to the marketed drug sunitinib, these results demonstrated that voacangine directly interacts with VEGFR2 and therefore, can potentially be considered as a promising natural compound for suppressing angiogenesis and tumor growth by targeting VEGFR2. Among VEGFR2-targeting drugs, sunitinib is the most widely used drug for treating cancer patients [26]. Sunitinib is a multi-targeting drug and a receptor tyrosine kinase inhibitor that inhibits signaling via key angiogenic receptors, including VEGFRs, PDGFRs, and FGFRs [27]. In this study, the newly identified antiangiogenic small molecule, voacangine, was shown to interact specifically with VEGFR2. Voacangine treatment resulted in a decrease in VEGFR2 kinase activity in vitro (Figure 2b and Figure S2) and reduced its expression levels in vivo (Figure S8).

Recently, many reports have demonstrated that sunitinib directly targets VEGFR2 to inhibit cancer progression in patients [44,45]. It is specifically administered as a first-line treatment to patients with advanced renal cell carcinoma (RCC) and imatinib-resistant gastrointestinal stromal tumors [46,47]. Despite the significant benefits of sunitinib treatment related to progression-free survival and disease stabilization in patients, almost all patients acquire resistance to sunitinib and relapse [48,49]. Approximately, 70% of patients show an initial response, while the remaining 30% show primary (intrinsic) resistance. Furthermore, 70% of patients acquired extrinsic resistance within 6–15 months. To treat patients with sunitinib resistance, the development of new VEGFR2 inhibitors with distinct structures and pharmacological activities is imperative for improved cancer therapy. As a possible drug candidate targeting VEGFR2 kinase, voacangine significantly inhibited in vitro and in vivo angiogenesis by directly and specifically interacting with VEGFR2. Hence, further development of voacangine as a new scaffold compound targeting the VEGFR2 kinase could provide a new option to treat cancer patients with resistance to sunitinib. Additionally, identifying and developing drug replacements from natural products such as voacangine will potentially reduce unpredicted and adverse side effects, and provide a promising strategy to improve the efficiency of small molecules in preclinical and clinical stages.

4. Conclusions

Here, we demonstrate an effective and novel systematic combination method, consisting of a label-free method for target identification of natural products and their in vivo validation with information on "on-target" effects and bioimaging data consisting of molecular interactions in tissue samples (Figure 7). This combinatorial technique is effective not only for voacangine but could also be effectively used for many tricky natural products and could boost target identification and hence, drug development. This study provides a new systematic approach to overcome many of the problems associated with currently available methods used for in vitro and in vivo target identification and validation. Our study represents a new means to identify and validate protein targets of natural compounds as "cold compounds" and eases the exploration of the mode of action of these natural products in vitro and in vivo without any chemical modifications. These results also provide new insights into the evaluation of drug actions in tissues and the colocalization of drugs and their respective targets in vivo.

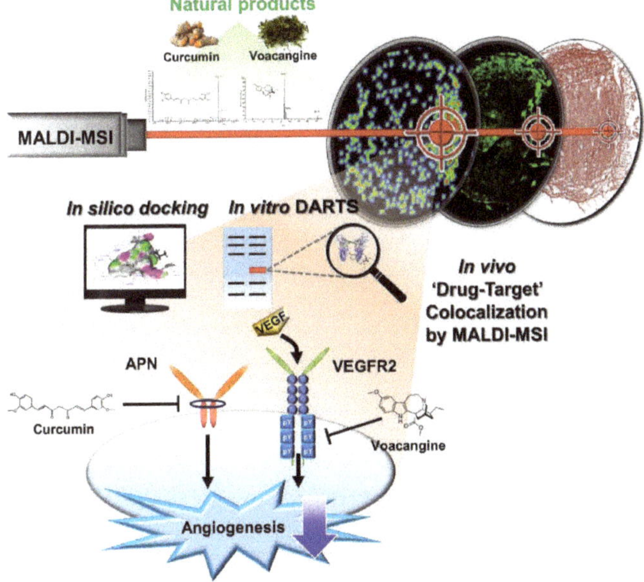

Figure 7. Summary of the study. The systematic approach using the combination of DARTS-MSI for in vitro and in vivo target identification and validation of natural products.

Supplementary Materials: The following are available online at http://www.mdpi.com/2218-273X/10/4/508/s1, Figure S1: Characterization of curcumin by MALDI-MS, MS/MS, and MSI, Figure S2: Voacangine specifically inhibits VEGFR2 signaling, Figure S3: Effects of voacangine on tumor growth and angiogenesis-related to VEGFR2 expression levels, Figure S4: Body weight of in vivo xenograft model, Figure S5: Characterization of curcumin by MALDI-MS and MS/MS, Figure S6: Quantitation of a droplet of voacangine on tumor tissue sections, Figure S7: Comparison of voacangine MSI between vehicle-treated tumor tissue, Figure S8: Effect of voacangine on the expression of VEGFR2, Table S1: IC50 values of voacangine in study cell lines.

Author Contributions: Y.K. and H.J.K. participated in the conception and experimental designs. Y.K. and S.M.C. performed the cell and molecular biology assays, DARTS assay, tumor xenograft mouse model assay, and MALDI-MSI experiments, and analyzed the data; Y.S., T.Y.K., and M.R. carried out the MALDI-MSI and reviewed the manuscript; J.Y.K., J.Y.L., and J.S.Y. analyzed data of DARTS assay and mass spectrometry; D.S. and G.H. performed the in silico docking analysis; C.W. and R.A. contributed in the data analysis and reviewed the manuscript; Y.K., G.M.-V., and H.J.K. wrote the paper. All authors edited and approved the final manuscript.

Funding: This work was partly supported by grants from the National Research Foundation of Korea, funded by the government of Korea (MSIP; 2012M3A9D1054520, 2015K1A1A2028365, 2015M3A9B6027818,

2016K2A9A1A03904900, 2018M3A9C4076477) and the Brain Korea 21Plus Project in the Korea and ICONS (Institute of Convergence Science), Yonsei University; as well as the Berta Kamprad Foundation, Lund, Sweden.

Acknowledgments: We thanks that Jae-Ho Cheong (Yonsei Medical School, Korea) kindly provided the gastric cancer cells (YCC16) for cell experiments.

Conflicts of Interest: The authors declare no conflict of interest.

References

1. Schreiber, S.L. Target-oriented and diversity-oriented organic synthesis in drug discovery. *Science* **2000**, *287*, 1964–1969. [CrossRef] [PubMed]
2. Schenone, M.; Dančík, V.; Wagner, B.K.; Clemons, P.A. Target identification and mechanism of action in chemical biology and drug discovery. *Nat. Chem. Biol.* **2013**, *9*, 232–240. [CrossRef] [PubMed]
3. Gibbs, J.B. Mechanism-based target identification and drug discovery in cancer research. *Science* **2000**, *287*, 1969–1973. [CrossRef]
4. Drews, J. Drug discovery: A historical perspective. *Science* **2000**, *287*, 1960–1964. [CrossRef] [PubMed]
5. Pan, S.; Zhang, H.; Wang, C.; Yao, S.C.L.; Yao, S.Q. Target identification of natural products and bioactive compounds using affinity-based probes. *Nat. Prod. Rep.* **2016**, *33*, 612–620. [CrossRef]
6. Ziegler, S.; Pries, V.; Hedberg, C.; Waldmann, H. Target identification for small bioactive molecules: Finding the needle in the haystack. *Angew. Chemie-Int. Ed.* **2013**, *52*, 2744–2792. [CrossRef] [PubMed]
7. Rix, U.; Superti-Furga, G. Target profiling of small molecules by chemical proteomics. *Nat. Chem. Biol.* **2009**, *5*, 616–624. [CrossRef]
8. Chang, J.; Kim, Y.; Kwon, H.J. Advances in identification and validation of protein targets of natural products without chemical modification. *Nat. Prod. Rep.* **2016**, *33*, 719–730. [CrossRef]
9. Annis, D.A.; Nickbarg, E.; Yang, X.; Ziebell, M.R.; Whitehurst, C.E. Affinity selection-mass spectrometry screening techniques for small molecule drug discovery. *Curr. Opin. Chem. Biol.* **2007**, *11*, 518–526. [CrossRef]
10. Schirle, M.; Bantscheff, M.; Kuster, B. Mass spectrometry-based proteomics in preclinical drug discovery. *Chem. Biol.* **2012**, *19*, 72–84. [CrossRef]
11. Molina, D.M.; Jafari, R.; Ignatushchenko, M.; Seki, T.; Larsson, E.A.; Dan, C.; Sreekumar, L.; Cao, Y.; Nordlund, P. Monitoring drug target engagement in cells and tissues using the cellular thermal shift assay. *Science* **2013**, *341*, 84–87. [CrossRef] [PubMed]
12. Lomenick, B.; Jung, G.; Wohlschlegel, J.A.; Huang, J. Target Identification using drug affinity responsive target stability (DARTS). *Curr. Protoc. Chem. Biol.* **2011**, *3*, 163–180. [CrossRef] [PubMed]
13. Lomenick, B.; Hao, R.; Jonai, N.; Chin, R.M.; Aghajan, M.; Warburton, S.; Wang, J.; Wu, R.P.; Gomez, F.; Loo, J.A.; et al. Target identification using drug affinity responsive target stability (DARTS). *Proc. Natl. Acad. Sci. USA* **2009**, *106*, 21984–21989. [CrossRef]
14. West, G.M.; Tang, L.; Fitzgerald, M.C. Thermodynamic analysis of protein stability and ligand binding using a chemical modification- and mass spectrometry-based strategy. *Anal. Chem.* **2008**, *80*, 4175–4185. [CrossRef] [PubMed]
15. Pai, M.Y.; Lomenick, B.; Hwang, H.; Schiestl, R.; McBride, W.; Loo, J.A.; Huang, J. Drug affinity responsive target stability (DARTS) for small-molecule target identification. *Methods Mol. Biol.* **2015**, *1263*, 287–298. [CrossRef] [PubMed]
16. Andersson, M.; Groseclose, M.R.; Deutch, A.Y.; Caprioli, R.M. Imaging mass spectrometry of proteins and peptides: 3D volume reconstruction. *Nat. Methods* **2008**, *5*, 101–108. [CrossRef] [PubMed]
17. Stoeckli, M.; Chaurand, P.; Hallahan, D.E.; Caprioli, R.M. Imaging mass spectrometry: A new technology for the analysis of protein expression in mammalian tissues. *Nat. Med.* **2001**, *7*, 493–496. [CrossRef]
18. Végvári, Á. Drug localizations in tissue by mass spectrometry imaging. *Biomark. Med.* **2015**, *9*, 869–876. [CrossRef]
19. Marko-Varga, G.; Fehniger, T.E.; Rezeli, M.; Döme, B.; Laurell, T.; Végvári, Á. Drug localization in different lung cancer phenotypes by MALDI mass spectrometry imaging. *J. Proteom.* **2011**, *74*, 982–992. [CrossRef]
20. Seeley, E.H.; Caprioli, R.M. MALDI imaging mass spectrometry of human tissue: Method challenges and clinical perspectives. *Trends Biotechnol.* **2011**, *29*, 136–143. [CrossRef]
21. Cornett, D.S.; Reyzer, M.L.; Chaurand, P.; Caprioli, R.M. MALDI imaging mass spectrometry: Molecular snapshots of biochemical systems. *Nat. Methods* **2007**, *4*, 828–833. [CrossRef] [PubMed]

22. Kim, Y.H.; Fujimura, Y.; Hagihara, T.; Sasaki, M.; Yukihira, D.; Nagao, T.; Miura, D.; Yamaguchi, S.; Saito, K.; Tanaka, H.; et al. In situ label-free imaging for visualizing the biotransformation of a bioactive polyphenol. *Sci. Rep.* **2013**, *3*. [CrossRef] [PubMed]
23. Schulz, S.; Becker, M.; Groseclose, M.R.; Schadt, S.; Hopf, C. Advanced MALDI mass spectrometry imaging in pharmaceutical research and drug development. *Curr. Opin. Biotechnol.* **2019**, *55*, 51–59. [CrossRef] [PubMed]
24. Shim, J.S.; Kim, J.H.; Cho, H.Y.; Yum, Y.N.; Kim, S.H.; Park, H.J.; Shim, B.S.; Choi, S.H.; Kwon, H.J. Irreversible inhibition of CD13/aminopeptidase N by the antiangiogenic agent curcumin. *Chem. Biol.* **2003**, *10*, 695–704. [CrossRef]
25. Kim, Y.; Jung, H.J.; Kwon, H.J. A natural small molecule voacangine inhibits angiogenesis both in vitro and in vivo. *Biochem. Biophys. Res. Commun.* **2012**, *417*, 330–334. [CrossRef] [PubMed]
26. Mendel, D.B.; Douglas Laird, A.; Xin, X.; Louie, S.G.; Christensen, J.G.; Li, G.; Schreck, R.E.; Abrams, T.J.; Ngai, T.J.; Lee, L.B.; et al. In vivo antitumor activity of SU11248, a novel tyrosine kinase inhibitor targeting vascular endothelial growth factor and platelet-derived growth factor receptors: Determination of a pharmacokinetic/pharmacodynamic relationship. *Clin. Cancer Res.* **2003**, *9*, 327–337.
27. Sun, L.; Liang, C.; Shirazian, S.; Zhou, Y.; Miller, T.; Cui, J.; Fukuda, J.Y.; Chu, J.Y.; Nematalla, A.; Wang, X.; et al. Discovery of 5-[5-fluoro-2-oxo-1,2-dihydroindol-(3Z)-ylidenemethyl]-2, 4-dimethyl-1H-pyrrole-3-carboxylic acid (2-diethylaminoethyl)amide, a novel tyrosine kinase inhibitor targeting vascular endothelial and platelet-derived growth factor receptor tyrosin. *J. Med. Chem.* **2003**, *46*, 1116–1119. [CrossRef]
28. Patyna, S.; Laird, A.D.; Mendel, D.B.; O'Farrell, A.M.; Liang, C.; Guan, H.; Vojkovsky, T.; Vasile, S.; Wang, X.; Chen, J.; et al. SU14813: A novel multiple receptor tyrosine kinase inhibitor with potent antiangiogenic and antitumor activity. *Mol. Cancer Ther.* **2006**, *5*, 1774–1782. [CrossRef]
29. Adelaiye, R.; Ciamporcero, E.; Miles, K.M.; Sotomayor, P.; Bard, J.; Tsompana, M.; Conroy, D.; Shen, L.; Ramakrishnan, S.; Ku, S.Y.; et al. Sunitinib dose escalation overcomes transient resistance in clear cell renal cell carcinoma and is associated with epigenetic modifications. *Mol. Cancer Ther.* **2015**, *14*, 513–522. [CrossRef]
30. Döme, B.; Paku, S.; Somlai, B.; Tmr, J. Vascularization of cutaneous melanoma involves vessel co-option and has clinical significance. *J. Pathol.* **2002**, *197*, 355–362. [CrossRef]
31. Torok, S.; Vegvari, A.; Rezeli, M.; Fehniger, T.E.; Tovari, J.; Paku, S.; Laszlo, V.; Hegedus, B.; Rozsas, A.; Dome, B.; et al. Localization of sunitinib, its metabolites and its target receptors in tumour-bearing mice: A MALDI-MS imaging study. *Br. J. Pharmacol.* **2015**, *172*, 1148–1163. [CrossRef] [PubMed]
32. Olsson, A.K.; Dimberg, A.; Kreuger, J.; Claesson-Welsh, L. VEGF receptor signalling-In control of vascular function. *Nat. Rev. Mol. Cell Biol.* **2006**, *7*, 359–371. [CrossRef] [PubMed]
33. Bankhead, P.; Loughrey, M.B.; Fernández, J.A.; Dombrowski, Y.; McArt, D.G.; Dunne, P.D.; McQuaid, S.; Gray, R.T.; Murray, L.J.; Coleman, H.G.; et al. QuPath: Open source software for digital pathology image analysis. *Sci. Rep.* **2017**, *7*. [CrossRef] [PubMed]
34. Folkman, J. Angiogenesis-dependent diseases. *Semin. Oncol.* **2001**, *28*, 536–542. [CrossRef]
35. Flier, J.S.; Underhill, L.H.; Folkman, J. Clinical applications of research on angiogenesis. *N. Engl. J. Med.* **1995**, *333*, 1757–1763. [CrossRef]
36. Ivy, S.P.; Wick, J.Y.; Kaufman, B.M. An overview of small-molecule inhibitors of VEGFR signaling. *Nat. Rev. Clin. Oncol.* **2009**, *6*, 569–579. [CrossRef]
37. Sarkar, S.; Mazumdar, A.; Dash, R.; Sarkar, D.; Fisher, P.B.; Mandal, M. ZD6474, a dual tyrosine kinase inhibitor of EGFR and VEGFR-2, inhibits MAPK/ERK and AKT/PI3-K and induces apoptosis in breast cancer cells. *Cancer Biol. Ther.* **2010**, *9*, 592–603. [CrossRef]
38. Plate, K.H.; Breier, G.; Weich, H.A.; Risau, W. Vascular endothelial growth factor is a potential tumour angiogenesis factor in human gliomas in vivo. *Nature* **1992**, *359*, 845–848. [CrossRef]
39. Emmenegger, U.; Shaked, Y.; Man, S.; Bocci, G.; Spasojevic, I.; Francia, G.; Kouri, A.; Coke, R.; Cruz-Munoz, W.; Ludeman, S.M.; et al. Pharmacodynamic and pharmacokinetic study of chronic low-dose metronomic cyclophosphamide therapy in mice. *Mol. Cancer Ther.* **2007**, *6*, 2280–2289. [CrossRef]
40. Zhang, L.; Smith, K.M.; Chong, A.L.; Stempak, D.; Yeger, H.; Marrano, P.; Thorner, P.S.; Irwin, M.S.; Kaplan, D.R.; Baruchel, S. In vivo antitumor and antimetastatic activity of Sunitinib in preclinical Neuroblastoma mouse model. *Neoplasia* **2009**, *11*, 426–435. [CrossRef]

41. Holmes, K.; Roberts, O.L.; Thomas, A.M.; Cross, M.J. Vascular endothelial growth factor receptor-2: Structure, function, intracellular signalling and therapeutic inhibition. *Cell. Signal.* **2007**, *19*, 2003–2012. [CrossRef] [PubMed]
42. Yi, T.; Cho, S.G.; Yi, Z.; Pang, X.; Rodriguez, M.; Wang, Y.; Sethi, G.; Aggarwal, B.B.; Liu, M. Thymoquinone inhibits tumor angiogenesis and tumor growth through suppressing AKT and extracellular signal-regulated kinase signaling pathways. *Mol. Cancer Ther.* **2008**, *7*, 1789–1796. [CrossRef] [PubMed]
43. Fischer, I.; Gagner, J.-P.; Law, M.; Newcomb, E.W.; Zagzag, D. Angiogenesis in Gliomas: Biology and Molecular Pathophysiology. *Brain Pathol.* **2006**, *15*, 297–310. [CrossRef] [PubMed]
44. Ko, J.S.; Zea, A.H.; Rini, B.I.; Ireland, J.L.; Elson, P.; Cohen, P.; Golshayan, A.; Rayman, P.A.; Wood, L.; Garcia, J.; et al. Sunitinib mediates reversal of myeloid-derived suppressor cell accumulation in renal cell carcinoma patients. *Clin. Cancer Res.* **2009**, *15*, 2148–2157. [CrossRef] [PubMed]
45. Jain, R.K. Normalization of tumor vasculature: An emerging concept in antiangiogenic therapy. *Science* **2005**, *307*, 58–62. [CrossRef] [PubMed]
46. Rock, E.P.; Goodman, V.; Jiang, J.X.; Mahjoob, K.; Verbois, S.L.; Morse, D.; Dagher, R.; Justice, R.; Pazdur, R. Food and Drug Administration Drug Approval Summary: Sunitinib Malate for the Treatment of Gastrointestinal Stromal Tumor and Advanced Renal Cell Carcinoma. *Oncologist* **2007**, *12*, 107–113. [CrossRef]
47. Choueiri, T.K. Clinical treatment decisions for advanced renal cell cancer. *JNCCN J. Natl. Compr. Cancer Netw.* **2013**, *11*, 694–697. [CrossRef]
48. Porta, C.; Sabbatini, R.; Procopio, G.; Paglino, C.; Galligioni, E.; Ortega, C. Primary resistance to tyrosine kinase inhibitors in patients with advanced renal cell carcinoma: State-of-the-science. *Expert Rev. Anticancer Ther.* **2012**, *12*, 1571–1577. [CrossRef]
49. Heng, D.Y.; MacKenzie, M.J.; Vaishampayan, U.N.; Bjarnason, G.A.; Knox, J.J.; Tan, M.H.; Wood, L.; Wang, Y.; Kollmannsberger, C.; North, S.; et al. Primary anti-vascular endothelial growth factor (VEGF)-refractory metastatic renal cell carcinoma: Clinical characteristics, risk factors, and subsequent therapy. *Ann. Oncol.* **2012**, *23*, 1549–1555. [CrossRef]

Review

Interrogating Host Antiviral Environments Driven by Nuclear DNA Sensing: A Multiomic Perspective

Timothy R. Howard and Ileana M. Cristea *

Department of Molecular Biology, Princeton University, Washington Road, Princeton, NJ 08544, USA; th12@princeton.edu

* Correspondence: icristea@princeton.edu; Tel.: +1-609-258-9417

Received: 12 November 2020; Accepted: 23 November 2020; Published: 24 November 2020

Abstract: Nuclear DNA sensors are critical components of the mammalian innate immune system, recognizing the presence of pathogens and initiating immune signaling. These proteins act in the nuclei of infected cells by binding to foreign DNA, such as the viral genomes of nuclear-replicating DNA viruses herpes simplex virus type 1 (HSV-1) and human cytomegalovirus (HCMV). Upon binding to pathogenic DNA, the nuclear DNA sensors were shown to initiate antiviral cytokines, as well as to suppress viral gene expression. These host defense responses involve complex signaling processes that, through protein–protein interactions (PPIs) and post-translational modifications (PTMs), drive extensive remodeling of the cellular transcriptome, proteome, and secretome to generate an antiviral environment. As such, a holistic understanding of these changes is required to understand the mechanisms through which nuclear DNA sensors act. The advent of omics techniques has revolutionized the speed and scale at which biological research is conducted and has been used to make great strides in uncovering the molecular underpinnings of DNA sensing. Here, we review the contribution of proteomics approaches to characterizing nuclear DNA sensors via the discovery of functional PPIs and PTMs, as well as proteome and secretome changes that define a host antiviral environment. We also highlight the value of and future need for integrative multiomic efforts to gain a systems-level understanding of DNA sensors and their influence on epigenetic and transcriptomic alterations during infection.

Keywords: DNA sensing; IFI16; cGAS; innate immunity; protein interactions; virus–host interactions; post-translational modifications; mass spectrometry; proteomics; transcriptomics

---

1. Introduction

Eukaryotic cells are relentlessly assailed by a myriad of pathogens, thereby needing to constantly evolve and expand their mechanisms for pathogen detection and host defense. During infection, pathogens bring foreign sugars, lipids, proteins, and nucleic acids into host cells. These foreign molecules can act as pathogen-associated molecular patterns (PAMPs), and the ability of the cell to detect them is critical for the initiation of host defense mechanisms and the inhibition of virus production and spread. Thus, cells utilize specialized proteins known as pattern-recognition receptors (PRRs) to detect PAMPs [1]. A common PAMP detected by host cells is the pathogenic double-stranded DNA (dsDNA) from bacteria, DNA viruses, and some RNA viruses (i.e., retroviruses) [2]. PRRs for dsDNA, known as DNA sensors, bind to the pathogenic DNA and initiate defense programs that include innate immune signaling, inflammatory responses, and apoptosis. It was long believed that DNA sensors can only function outside of the nucleus, in order to avoid recognition of self-DNA and spurious activation of immune responses. However, the majority of the known human dsDNA viruses replicate within the nucleus, thereby depositing their viral genomes in the nuclei of infected cells. Examples of nuclear-replicating DNA viruses are herpesviruses, such as herpes simplex virus type 1 (HSV-1), human cytomegalovirus (HCMV), and Kaposi's sarcoma-associated herpesvirus (KSHV). Herpesviruses are ancient viruses that arose hundreds of millions of years ago, having ample time to co-diverge with their hosts [3–5]. The co-evolution and co-adaptation of viruses with hosts are evidenced by the diversification of PRRs and their ligand-recognition abilities [6]. Indeed, research during the past decade has demonstrated the existence of PRRs that function in nuclear sensing of pathogenic DNA [7,8].

To date, four proteins have been shown to have the ability to perform nuclear DNA sensing—in chronological order of discovery of nuclear function: interferon-inducible protein 16 (IFI16 [9–11]), interferon-inducible protein X (IFIX [12]), cyclic GMP-AMP synthase (cGAS [13–16]), and heterogeneous nuclear ribonucleoprotein A2/B1 (hnRNPA2B1 [17]). The structures of these four proteins and their currently understood mechanisms for induction of antiviral responses are illustrated in Figure 1. Each nuclear DNA sensor was shown to help to induce *ifnβ* expression, which in turn activates numerous critical antiviral signaling pathways in adjacent cells that aim to slow the spread of infection. *Ifnβ* expression is thought to rely primarily on a signaling axis involving the endoplasmic reticulum membrane protein stimulator of interferon genes (STING), although STING-independent signaling has also been proposed [18]. Activation of STING leads to the phosphorylation of TANK binding kinase 1 (TBK1), which in turn phosphorylates the interferon regulatory factor 3 (IRF3). IRF3 then dimerizes, shuttles into the nucleus, and binds to the interferon-stimulated response element upstream of *ifnβ* to transcriptionally activate the expression of antiviral cytokines [19–22].

IFI16 was discovered as a sensor ten years ago [9], becoming the first known nuclear DNA sensor. Both IFI16 and IFIX belong to the PYHIN family of proteins [12]. These DNA sensors consist of an N-terminal pyrin domain (PYD) [23] and either one (IFIX) or two (IFI16) C-terminal HIN-200 domains [24,25] (Figure 1A). The HIN-200 domains facilitate sequence-independent binding of the sensor to the viral DNA [25], while the PYD mediates homotypic oligomerization [26,27]. IFI16 was shown to bind incoming viral dsDNA at the nuclear periphery, immediately following the docking of the virus capsid at the nuclear pore, and the PYD was found to be necessary for the IFI16 recruitment to the nuclear periphery [15]. The IFI16 oligomerization upon binding to viral DNA and recruitment of other host factors is thought to build an antiviral scaffold capable of both activating immune signaling [9,10,26,28,29] and suppressing viral transcription [29–32] (Figure 1B). A subset of IFI16 was shown to be able to shuttle between the nucleus and the cytoplasm to function in DNA sensing in a localization-dependent manner [9,10]. However, during the early stages of infection with nuclear-replicating viruses, IFI16 does not appear to move to the cytoplasm, remaining predominantly nuclear. Thus, a still unanswered question is how IFI16 communicates with STING or whether a STING-independent mechanism also contributes to *ifnβ* induction.

IFIX was also shown to bind dsDNA in a sequence-independent manner and to help induce antiviral cytokine expression upon herpesvirus infection [12]. Furthermore, similar to IFI16, this PYHIN protein displayed pronounced ability to undergo nuclear oligomerization via its PYD [26] and was shown to also function in suppressing viral gene expression [33]. However, very few studies have so far focused on IFIX during infection, and the mechanisms involved in IFIX-mediated antiviral responses remain poorly understood.

The mechanism by which cytoplasmic cGAS induces STING activation is well defined. cGAS contains an NTase core domain (Figure 1A) that catalyzes the formation of 2′3′-cyclic GMP-AMP (cGAMP) (Figure 1B). After binding to dsDNA, cGAS dimerizes and initiates cGAMP production. This small molecule then binds to STING, causing a conformational change and dimerization that leads to TBK1 phosphorylation. The additional presence of cGAS in the nucleus has been initially the subject of debate, although it was shown to form a functional nuclear interaction with IFI16 [14]. However, in recent years, it has become accepted that cGAS indeed has nuclear localization in different cell types, and studies have characterized mechanisms that prevent its autoreactivity [34] or that underlie its nuclear function in inhibiting DNA damage repair [16,35].

Finally, the most recently discovered nuclear DNA sensor, the heterogeneous nuclear ribonucleoproteins A2/B1 (hnRNPA2B1), has classically been understood to play a role in transporting mRNA into the cytoplasm [36,37]. In 2019, it was found that, during HSV-1 infection, hnRNPA2B1 both facilitates the export of IFI16, cGAS, and STING mRNA molecules to the cytoplasm and binds viral DNA within the nucleus, shuttles to the cytoplasm, and activates STING–TBK1–IRF3 signaling [17].

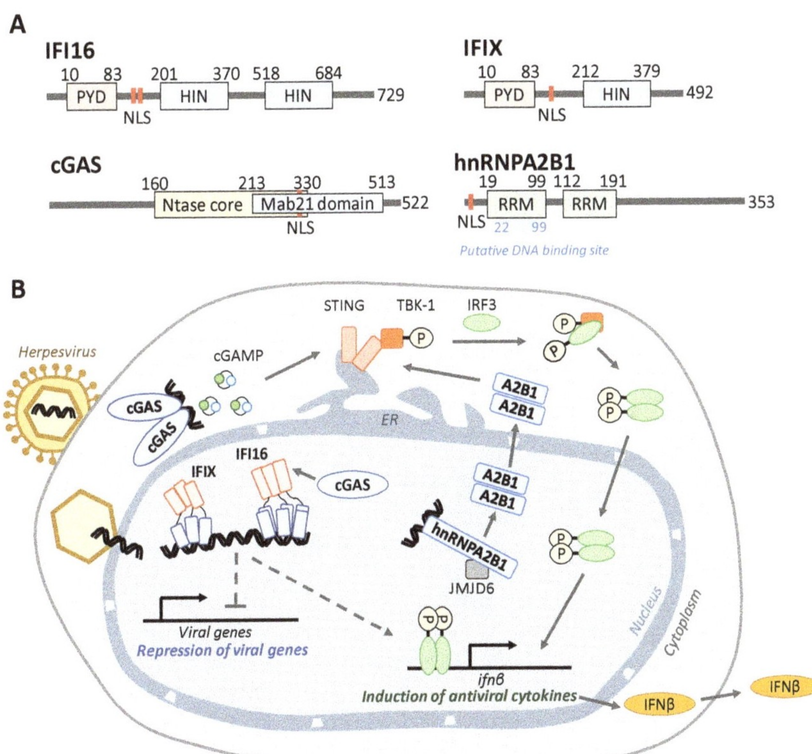

Figure 1. Nuclear DNA sensors bind to viral DNA and activate antiviral cytokine signaling. (**A**) Domain maps for each nuclear DNA sensor. IFI16 and IFIX belong to the PYHIN family of proteins and each contain an N-terminal pyrin domain that mediates protein interactions and one or two HIN-200 domains that bind dsDNA in a sequence-independent manner. cGAS consists of overlapping Ntase core (cGAMP production) and Mab21 (DNA binding) domains. hnRNPA2B1 possesses two RNA recognition motifs, the first of which has been proposed to also contain the DNA binding site. Each protein contains a nuclear localization signal (red bars). (**B**) Model for the intrinsic and innate immune activity of IFI16, IFIX, cGAS, and hnRNPA2B1. During infection, IFI16 and IFIX bind viral DNA entering the nucleus through a nuclear pore complex. After binding to viral DNA via their HIN domains (blue), these proteins each form homo-oligomers mediated by the PYD in order to build antiviral signaling scaffolds necessary for the repression of viral transcription and induction of IFNß. cGAS was shown to stabilize nuclear IFI16 levels during HSV-1 infection to promote immune signaling. In the cytoplasm, cGAS binds to foreign DNA and produces cGAMP, which in turn activates the STING–TBK1–IRF3 signaling axis to induce IFNß. hnRNPA2B1 binds viral DNA and is then demethylated by JMJD6. This is required for hnRNPA2B1 dimerization and subsequent translocation into the cytosol, where it activates the STING–TBK1–IRF3 axis. In each case, IFNß protein is secreted from the cell in order to communicate with and initiate antiviral programs in neighboring cells.

The importance of these nuclear DNA sensors is highlighted by the various strategies acquired by viruses during their co-evolution with their hosts and adaptation to human cells to inhibit these DNA sensors and their antiviral functions. For example, HSV-1 promotes the degradation of IFI16 by targeting this pyrin domain. Several studies have showed this degradation to be primarily driven by the viral E3 ubiquitin ligase, ICP0 [12,15,28], while other studies suggested the contribution of other factors [38]. IFIX was also found to be degraded during HSV-1 infection, and this, yet to be discovered, inhibitory mechanism was shown not to be dependent on the ICP0 E3 ubiquitin ligase activity [33]. HSV-1 further utilizes the tegument protein pUL37 to suppress the cGAS-mediated catalysis of cGAMP through deamidation of a single arginine residue in the cGAS activation loop [39]. HCMV also acquired a mechanism to inhibit the function of nuclear sensors by preventing PYD oligomerization of IFI16 and IFIX [26]. This virus immune evasion strategy uses the major tegument protein of HCMV, pUL83, to clamp the PYD, block oligomerization, and inhibit subsequent immune signaling [26].

The mechanisms described above paint a picture of intricate signaling pathways that underlie the cellular intrinsic and innate immune systems that nuclear DNA sensors feed into and the opposing virus immune evasion strategies. On the host defense side, pathogenic DNA is bound by nuclear DNA sensors which then fulfill two roles: (1) activate immune programming and (2) suppress viral gene expression. These processes rely on interactions between biomolecules, are regulated by these interactions and post-translational modifications (PTMs) and affect the expression of hundreds of cellular and viral transcripts and proteins. Therefore, understanding nuclear DNA sensing requires a holistic approach in which all these factors are considered.

Knowledge of DNA sensor mechanisms is also relevant for understanding human diseases and the development of therapies. Dysregulation of DNA sensors contributes to several autoimmune disorders. For example, patients with systemic lupus erythematosus, Sjögren Syndrome, and systemic sclerosis exhibit significantly elevated levels of anti-IFI16 antibodies [40–42], which can result from aberrant overexpression and mislocalization of IFI16 [43]. Further, autoreactivity of cGAS contributes to Aicardi–Goutières syndrome (AGS) [44,45], and small molecule inhibition of cGAS activity alleviates constitutive interferon expression in an AGS mouse model [46]. Therefore understanding mechanisms regulating DNA sensors can provide important insights into driving factors of autoimmune disorders. Targeting DNA sensors or their activated pathways is also relevant in the development of both antiviral treatments and vaccines. For example, the STING–TBK1–IFNα/β signaling axis mediates the adjuvant effects required for successful immunogenicity with plasmid DNA vaccines [21,47]. Thus, we must consider how DNA sensors upstream of interferon induction react during the administration of DNA vaccines. So far, only the cytosolic PYHIN protein absent in melanoma 2 (AIM2), which directs the maturation of proinflammatory cytokines IL-18 and IL-1β, has been demonstrated to act as a sensor for DNA vaccines [48]. Interestingly, immune responses elicited by DNA vaccines in vivo seem to be cGAS- and IRF3-independent [49]. Further investigations can help elucidate the relative contributions of these DNA sensors to aiding immune memory upon DNA vaccine administration.

Omic methods have significantly contributed to the emergence of the research field of nuclear DNA sensing, helping to build the current level of understanding of the underlying molecular mechanisms. Mass spectrometry (MS)-based proteomic approaches have allowed the discovery of functional regulatory hubs for nuclear DNA sensors, including protein interactions and PTMs, as well as the monitoring of DNA sensor activation (e.g., cGAMP production). Whole-cell proteome analyses and secretome investigations have informed of global cellular changes that take place during the host activation of immune signaling cascades. Transcriptome studies have started to uncover the contribution of some of these DNA sensors to repression of viral gene expression. Here, we review findings stemming from the application of proteomics and other omic methods to characterizing the function and regulation of nuclear DNA sensors and explore the future promise of multiomic approaches in understanding human immune responses to nuclear-replicating viral pathogens.

2. DNA Sensor Identification and Characterization through the Lens of Proteomics

The use of proteomics directly led to the discovery of all known nuclear DNA sensors. As research into DNA sensing has intensified over the past decade, proteomics studies have been crucial for examining the functions and regulations of nuclear DNA sensors (Figure 2). These investigations have focused on proteome changes, protein–protein interactions (PPIs), and PTMs connected to nuclear DNA sensors in order to uncover the mechanisms of DNA sensing in response to viral infections. Here, we discuss the main MS-based approaches used for discovering DNA sensor interactions and PTMs that contribute to either promoting or inhibiting their host defense functions during viral infections (Table 1).

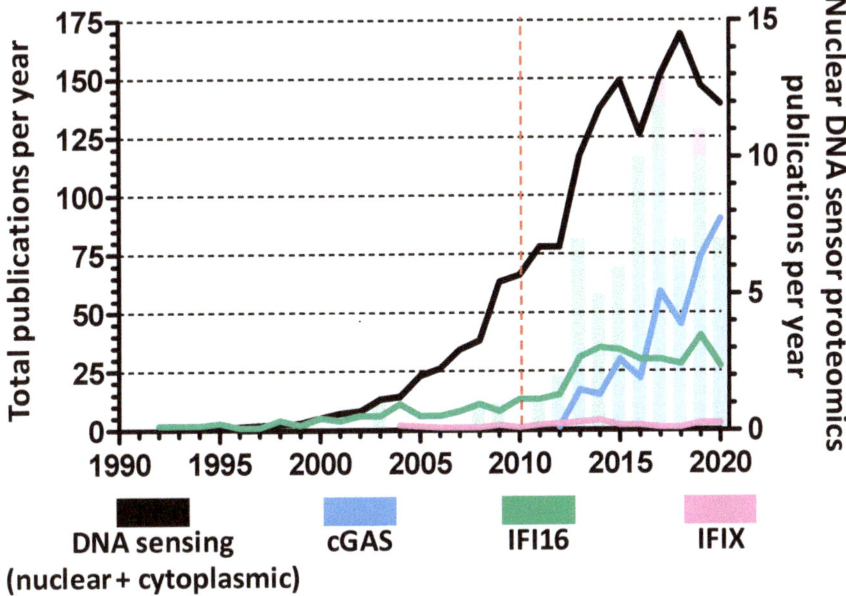

Figure 2. Yearly research articles investigating nuclear DNA sensors. Research papers focused on each nuclear DNA sensor, obtained from PubMed search when considering published research articles each year since 1990. The sum of each year's articles for each protein is represented by line graphs (left *Y* axis) while articles specifically utilizing proteomics approaches to investigate proteome changes, protein–protein interactions, post-translational modifications, etc., are shown as stacked bars (right *Y* axis). Of note, the black line represents the number of articles concerning all kinds of DNA sensing, including non-nuclear sensors such as the cytoplasmic AIM2 and endosomal TLR9. The red dashed line marks the discovery of IFI16 as the first nuclear DNA sensor.

Table 1. Omics techniques used for the discovery and characterization of nuclear DNA sensors and related host antiviral processes.

Strategy	Advantages	Disadvantages	Purpose	Application	References
AP-MS isolating DNA	Unbiased detection of proteins bound to DNA or to DNA sensor; high sensitivity; enrichment of proteins of interest; ability to detect multiple PTM types	Could miss transient interactions; does not discriminate between direct and indirect interactions; nonspecific interactions are possible	Identify DNA sensors	IFI16, hnRNPA2B1	[9,17]
IP-MS isolating DNA sensors			Identify DNA sensors	IFIX	[12]
			Interactome	IFI16, IFIX, cGAS, hnRNPA2B1	[12,15,17,29,33,50,51]
			PTMs	IFI16, cGAS	[10,52,53]
Shotgun MS (whole proteome)	High throughput, unbiased, high sensitivity	Complex datasets; computationally intensive; possible missing values in quantitative proteome measurements	Identify DNA sensors	cGAS	[13]
			Proteome	cGAS	[51]
			Secretome	Herpesvirus infection	[54]
			Metabolome	Herpesvirus infection	[55]
Targeted MS	High accuracy and sensitivity; specific detection; requires low sample amount	Needs prior detection or defining signature detection parameters; needs specialized MS instrumentation	Protein abundance	Immune factor quantification	[51]
			Confirmation of protein interactions	cGAS, IFI16	[14,29]
			PTMs	cGAS	[53]
			Small molecule detection	cGAMP	[13]
DNA microarrays	High throughput; inexpensive; customizable to detect specific sequences and isoforms	High background noise; requires high sample amount; biased approach	Transcriptome	Herpesvirus infection	[56]
RNA sequencing	High throughput; unbiased; requires low sample amount; single base resolution	Requires library preparation; computationally intensive; expensive	Transcriptome	IFI16 (mouse homolog)	[57]

2.1. DNA Sensor Molecular Interactions Drive Host Antiviral and Virus Immune Evasion Mechanisms

Affinity purification-mass spectrometry (AP-MS) has been the cornerstone of identifying and quantifying protein–protein and protein-nucleic acid interactions [58]. In this approach, either a protein of interest or DNA is isolated and the accompanying interacting proteins are analyzed using mass spectrometry. Immunoaffinity purification (IP) is carried out by using an antibody conjugated to a resin, such as magnetic beads, which can be easily separated from the cell lysate and captured via centrifugation or application of a magnet (reviewed in [59]). The antibodies used can be raised against the endogenous protein of interest. However, as the efficiency and specificity of the isolation relies on the quality of the available antibody, antibodies against tags such as FLAG, HA, and GFP are often used to facilitate protein isolation [60]. DNA can be purified from cells through similar methods, usually using biotinylated DNA and streptavidin-coupled beads to isolate DNA–protein complexes [9]. Following complex isolation, the identities and abundances of the accompanying proteins are then characterized using MS.

It has long been understood that viral DNA activates innate immune responses, including *ifn-β* expression [61], but the identities of the DNA sensors and subsequent signaling pathways remained undetermined. AP-MS approaches have been at the core of discovering the identities of DNA sensors. IFI16 was recognized as a DNA sensor in 2010, when Unterholzner et al. performed AP-MS after transfecting THP-1 cells with a biotinylated 70 base-pair vaccinia virus DNA fragment (VACV 70mer) [9]. It is of note that IFI16 is expressed and localized to both the nucleus and cytoplasm in macrophages such as the macrophage-like differentiated THP-1 cells. Further studies demonstrated that IFI16 has DNA sensor activity in the nucleus after different types of infections with nuclear-replication DNA viruses, including HSV-1 [9,10,28], KSHV [11], and HCMV [30], as well as after retrovirus infection, recognizing DNA intermediates of human immunodeficiency virus 1 (HIV-1) [6,62]. The interaction between IFI16 and HSV-1 DNA was also demonstrated in an elegant study that utilized 5-ethynyl-2′deoxycytidine (EdC) labeling of viral genomes coupled with AP-MS to investigate temporal viral genome-protein interactions. Here, IFI16 was found to associate with the viral genome by 2 h post-infection [63]. Recently, IFI16 was identified in an AP-MS study isolating the RNA genome of Chikungunya virus [64]. This is an unexpected finding as IFI16 has no known RNA sensing capability, but it implicates IFI16 in immune sensing pathways beyond dsDNA virus infection.

AP-MS was also integral in the discovery of the most recently identified nuclear DNA sensor, hnRNPA2B1, which was shown to function during HSV-1 infection [17]. In this study, HSV-1 genome biotinylation and AP-MS was integrated with a characterization of the nuclear and cytoplasmic proteomes following cellular fractionation. This allowed the authors to identify hnRNPA2B1 as a protein that both binds to viral DNA and shuttles to the nucleus to activate STING–TBK1–IRF3 signaling.

As nuclear DNA sensors do not directly stimulate interferon expression, interaction with other cellular proteins is crucial for initiating immune signaling pathways. Furthermore, the importance of PPIs in the regulation of immunity is highlighted by the virus–host protein interactions through which viruses inhibit DNA sensors. Thus, IP-MS studies that define the interactomes of DNA sensors have led to a better understanding of both their action and regulation.

The first interactome study of IFI16 during HSV-1 infection used AP-MS to characterize interactions with both endogenous and tagged IFI16 [50]. This study revealed IFI16 interactions with many cellular transcription and chromatin regulators, such as the upstream binding transcription factor (UBTF) and ND10 body components, as well as with the nuclear architecture proteins SUN1 and SUN2. Several viral proteins were also found to associate with IFI16 [50], including the E3 ubiquitin ligase ICP0 that was previously implicated in targeting IFI16 for degradation (Figure 3) [28]. Both UBTF and ND10 bodies (also known as PML nuclear bodies) were shown to function in host defense by repressing HSV-1 transcription [65,66], and ND10 bodies were also found to be targeted for degradation by ICP0 [67].

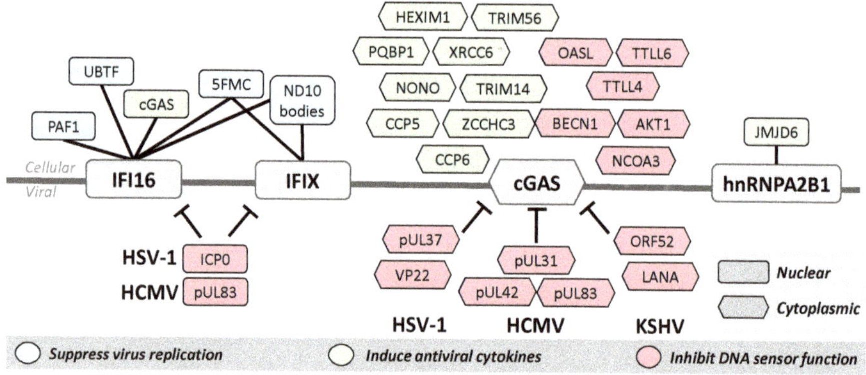

Figure 3. Protein–protein interactions contribute to the activation or inhibition of DNA sensor. Over the course of viral infection and immune signaling, DNA sensors interact with other cellular and viral proteins. Several of these cellular proteins are important for the function of the DNA sensors for both suppressing virus replication by repressing viral transcription and inducing antiviral cytokines. Protein interactions are also used to regulate DNA sensor function. Viruses have evolved distinct mechanisms to facilitate immune evasion and cells must also possess mechanisms to prevent excessive immune signaling. Although localized to both the nucleus and cytoplasm, protein interactions with cGAS are best characterized in the cytoplasm. Nuclear proteins are shown here as rectangles and cytoplasmic interactions as hexagons.

To further clarify how these interactions are facilitated and regulated during HSV-1 infection, the domain-specific interactomes of IFI16 were investigated by performing separate IP-MS experiments for the PYD and HIN domains [15]. This study revealed that the PYD interacts with members of ND10 bodies, cGAS, and the RNA polymerase II-associated factor 1 (PAF1). More recently, IP-MS with oligomerization-deficient IFI16 mutants demonstrated that IFI16 oligomerization is needed for the formation of these interactions with PAF1 and other members of the PAF1 complex during HSV-1 infection [29]. Additional experiments uncovered an antiviral role for PAF1, showing its ability to repress virus gene transcription.

Similar IP-MS interactome studies of PYHIN proteins related to IFI16 led to the discovery and characterization of IFIX as an antiviral nuclear DNA sensor [12]. At the time, very little was known about the cellular role of IFIX, but through IP-MS it was found to interact with many of the same proteins as IFI16, including ND10 body components and other chromatin remodeling and immune signaling proteins. These interactions, in conjunction with its structural similarities to IFI16, suggested that IFIX may also have antiviral properties and function in DNA sensing. Follow-up experiments demonstrated that IFIX binds viral DNA, suppresses HSV-1 replication, and induces interferon expression [12]. Probing the IFIX interactome even further during HSV-1 infection revealed associations with several components of the five friends of methylated chromatin target of Prmt1 (5FMC) complex [33], which functions in epigenetic regulation [68] and was later found to also interact specifically with oligomerized IFI16 [29].

Several important discoveries of cGAS function have been made using AP-MS, and we must also emphasize that the discovery of cGAS as a DNA sensor was initially enabled by the MS characterization of the cellular proteome. Stimulation of STING by cGAMP was discovered in 2013 [69], but the source of the cyclic GMP-AMP synthase activity remained unclear. Thus, cGAS was identified by integrating shotgun proteomics and cellular fractionation in order to pinpoint the protein whose expression pattern matched that of cGAS activity [13]. Since then, targeted IP-MS studies focused on specific interactions of interest uncovered cGAS associations with several cellular proteins that support immune function, including TRIM56 [70], PARP1 [16], and IFI16 [14], among many others

(Figure 3). The interaction between cGAS and IFI16 is particularly interesting because it touches on the question of redundancy for these proteins in the nuclear DNA sensing pathway. It was determined that, during HSV-1 infection, nuclear cGAS interacts with IFI16 for the purpose of stabilizing IFI16 in order to promote immune signaling [14,71]. The knowledge of cGAS interactions was later expanded with an IP-MS study of its interactome, which was further integrated with quantitative profiling of cellular proteome alterations during HSV-1 infection [51]. This interactome revealed the cGAS interaction with the RNA sensor OASL, which was demonstrated to repress cGAS activity as a host negative feedback loop for regulating cytokine induction [51].

Currently, the only study to have utilized AP-MS to study hnRNPA2B1 in the context of DNA sensing is the one in which it was discovered [17]. As indicated above, here, biotinylated HSV-1 genomes were isolated early during infection and the interacting proteins were identified via MS. These data were then cross-referenced with shotgun MS of nuclear/cytoplasmic fractionated cells in order to identify proteins that undergo nucleocytoplasmic translocation during infection. This approach enabled the authors to identify proteins that both bind viral DNA and shuttle to the cytoplasm, potentially for the purpose of activating STING–TBK1–IRF3. IP-MS was then utilized to gain a mechanistic understanding of interferon induction by hnRNPA2B1, showing that it does indeed interact with STING and TBK1 following HSV-1 infection.

The discovery of interactions with nuclear DNA sensors has also led to the characterization of mechanisms by which viruses evade cellular innate immunity. For example, recognizing the ability of the HCMV tegument protein pUL83 to inhibit the nuclear oligomerization of the pyrin domains of IFI16 and IFIX (Figure 3) derived from the identification of their interactions from an IP-MS study [26]. In agreement with its reported ability to target IFI16 for degradation during HSV-1 infection [28], the ICP0 interaction with IFI16 was demonstrated by IP-MS [50]. IP studies followed by targeted assays were valuable for identifying other mechanisms of virus immune evasion, such as the inhibition of cGAMP production by the KSHV virion protein ORF52 [72] and the HSV-1 tegument protein pUL37 (detailed in the PTM section below) [39] (Figure 3).

2.2. Post-Translational Modifications for Finely Tuning DNA Sensor Function

Beyond interactions with other biomolecules, the ability of DNA sensors to detect and respond to pathogenic invasion is closely tied to their regulation by PTMs. Changes to protein structure via phosphorylation, acetylation, ubiquitination, and SUMOylation, among others, enable the rapid regulation of protein function, and the addition or removal of PTMs is a tightly regulated cellular process in response to stress. MS has been well-established as the main method for accurate and unbiased detection of site-specific PTMs in different cellular contexts and has also contributed to the discovery of a multitude of DNA sensor PTMs (Table 2).

Broadly speaking, PTMs are inherent to the ability of a cell to induce immune signaling cascades in response to pathogen infection. The necessity of PTMs for immune signaling is exemplified by the activation of IFNβ expression that hinges upon phosphorylation of both TBK1 and IRF3 in STING-dependent signaling [1]. Further, PTMs of DNA sensors have been shown to directly contribute to immune activation. The hnRNPA2B1 interactome also revealed an interaction with the nuclear protein JMJD6, which facilitates demethylation of hnRNPA2B1 at Arg226. This alteration in hnRNPA2B1 structure is necessary for its dimerization, nucleocytoplasmic translocation, and subsequent interferon induction [17]. Thus, the necessity of Arg226 demethylation for hnRNPA2B1 DNA sensing highlights the importance of protein modification in this immune response.

Table 2. Known post-translational modifications of nuclear DNA sensors.

DNA Sensor	Modification	Residues	Reference in Which First Identified
IFI16	Acetylation	K45, K99, K128, K214, K444, K451, K505, K542, K558	[10]
	Phosphorylation	S95, S106, S153, S168, S174, S724	[10]
		S575	[73]
	SUMOylation	K116, K561	[74]
		K128	[75]
		K683	[76]
cGAS	Acetylation	K7, K50, K384, K392, K394, K414	[52]
		K198, K285, K292, K355, K432, K479	[53]
	Deamidation	N196, N377, Q436, Q439 in mice (N210, N389, Q451, Q454 in human)	[39]
	Glutamylation	E272 (poly), E302 (mono)	[77]
	Phosphorylation	S37, S116, S201, S221, S263	[53]
		S143	[73]
		Y215	[16]
		S305	[78]
	SUMOylation	K217 and K464 in mice (K231 and K479 in human)	[79]
	Ubiquitination	K271 and K464 (poly) in mice	[79]
		K335 (mono)	[70]
hnRNPA2B1	Acetylation	M1	[80]
		K168, K173	[81]
	Demethylation	R226	[17]
	Methylation	R203, R213, R228, R238, R266, R325, R350	[82]
	Phosphorylation	T4, S29, T140, T159, T176, S189, S201, S212, S225, S259, S324, Y331, S341, S344	[83]
		S85, S212, S259, S344	[84]
		S149, S231	[73]
		S236	[85]
		S347	[86]
	SUMOylation	K22, K104, K112, K120, K137, K152, K168, K173	[75]
		K120, K186	[74]

The initial discovery of IFI16 as a viral DNA sensor pointed to its ability to recognize pathogenic DNA in the cytoplasm, and further characterization of this sensor also solidified its nuclear DNA sensing function. However, the mechanisms regulating IFI16 subcellular localization remained unknown. Furthermore, its relative nuclear or cytoplasmic distribution was shown to be cell type dependent, with its localization being predominantly nuclear in lymphoid, epithelial, endothelial, and fibroblast cells, tissues that tend to be among the first infected by an invading virus. In 2012, our group reported that IFI16 contains a bipartite nuclear localization signal (NLS) and, using MS, identified several acetylation sites within the NLS [10]. IFI16 mutation experiments indicated that NLS acetylation at Lys99 and Lys128 inhibits nuclear import and abrogates IFI16 DNA sensing during HSV-1 infection. This discovery was critical for supporting that IFI16 predominantly senses viral DNA within the nucleus during herpesvirus infection. A number of studies have since demonstrated that IFI16 is regulated by different types of PTMs during viral infections, which additionally include phosphorylation and SUMOylation (Table 2) [10,73–76]. PTM-driven mechanisms also underly the ability of the cell to activate DNA sensors by modifying viral immune evasion proteins, thereby crippling their functions. For example, eight phosphorylation sites were discovered on the HCMV tegument protein pUL83 and mutational analyses demonstrated that its binding to the IFI16 PYD can be compromised by Ser364 phosphorylation within the pUL83 pyrin association domain [26].

PTMs of cGAS during DNA sensing have also started to be recognized for their importance in cGAS regulation and function, and MS-based PTM analysis has been crucial for identifying key regulatory hubs. For example, Zhang et al. found that the HSV-1 tegument protein pUL37 antagonizes cGAS during infection [39]. This protein is a known deamidase that acts on the dsRNA sensing protein RIG-I [87] to prevent immune signaling during HSV-1 infection; thus, the authors proposed a similar deamidation event would prevent cGAS signaling. Using tandem MS, they discovered several deamidation sites within the Mab21 enzyme domain and further identified that deamidation of Asn210 indeed impairs the ability of cGAS to produce cGAMP upon binding to dsDNA [39].

Several other important cGAS PTMs have been identified in recent years that function to either suppress or activate cGAS activity during DNA sensing. These PTMs include phosphorylation, glutamylation, ubiquitination, and SUMOylation (Table 2). An IP-MS study of cGAS followed by mutational analysis of the identified modified sites led to the finding that the kinase Akt phosphorylates cGAS Ser305, suppressing cGAMP production and interferon expression [78]. Additionally, glutamylation of cGAS at two distinct sites have been shown to impede cGAS activity [77]. After identifying that the cytosolic carboxypeptidases 5 and 6 (CCP5 and CCP6) contribute to activation of IRF3 during infection with DNA viruses HSV-1 and VACV, Xia et al. used MS to identify cGAS as a substrate of these protein. As CCP5 and CCP6 reverse glutamylation, this then led to the discovery that cGAS activity is suppressed through Glu302 monoglutamylation by tubulin tyrosine ligase-like protein 4 (TTLL4), which prevents cGAMP production, and through Glu272 polyglutamylation by TTLL6, which weakens the cGAS DNA binding ability [77]. More recently, MS analyses led to the discovery that cGAS is also acetylated at several lysine residues, with acetylation at Lys384, Lys394, and Lys414 suppressing cGAS-mediated cGAMP production [52] and apoptosis [53], and Lys198 acetylation promoting cGAS-induced antiviral cytokine expression [53]. Targeted MS/MS quantification of site-specific acetylation during infection demonstrated that the level of Lys198 acetylation decreased during HSV-1 and HCMV infections [53], pointing to the possible presence of a viral immune evasion strategy targeting this residue to control host immune response.

Targeted studies that do not utilize MS have also identified important cGAS PTMs (Table 2). Mutational analysis of cGAS revealed that phosphorylation at Tyr215 inhibits cGAS nuclear translocation upon DNA damage, and a tyrosine kinase knockdown screen showed that B-lymphoid tyrosine kinase controls phosphorylation at this residue [16]. As another example, SUMOylation of murine cGAS by TRIM38 enhanced cGAS DNA sensing by preventing polyubiquitination and subsequent degradation of cGAS [79]. Further investigations of the aforementioned interaction between cGAS and TRIM56 revealed that TRIM56 acts to monoubiquitinate cGAS in order to promote its dimerization and facilitate cytosolic DNA sensing [70].

3. Defining the Cellular Landscape Representative of Immune Activation

In addition to providing specific information regarding the regulation of nuclear DNA sensors, omic studies have also informed of the global alterations occurring in host cells during immune activation. Infections with DNA viruses result in major changes in mRNA expression, protein abundances, interaction networks and PTMs, cellular metabolism, and secretion. During infection, the virus seeks to inhibit host defenses, co-opt cellular machinery, and rewire the cellular metabolome to facilitate production of progeny virions. Meanwhile, the host attempts to reduce energy expenditure while producing and secreting antiviral cytokines that will slow the spread of infection. Transcriptome, proteome, metabolome, and secretome studies have been critical for gaining an understanding of these broad cellular alterations occurring during the progression of virus infections. Temporal transcriptomic and proteomic investigations have been carried out to determine whether a regulation occurs through changes at the transcript or protein level during infection and to correlate expression trends with phenotypes.

Given that viruses appropriate the host cell transcription machinery and RNA processing, a range of transcriptome studies have been performed to monitor temporal cellular and viral transcript levels during different types of infections. For example, DNA microarrays have been used extensively to study the effect infection on transcription by HSV-1 [88,89], HCMV [56,90–93], KSHV [94,95], and the porcine alphaherpesvirus pseudorabies virus [96,97], among others. Similar to proteomic technologies, improvements in sequencing methods have greatly impacted our understanding of host cell response to viral infection. The emergence of RNA sequencing (RNA-seq) as an unbiased method that is both more sensitive and precise than microarrays [98] has benefitted the fields of virology and immunology by more broadly capturing the cellular and viral transcriptional landscape during infection, including the expression of interferon-stimulated genes (ISGs). This technique was used to demonstrate that HSV-1 infection of skin fibroblasts led to the upregulation of 596 genes, downregulation of only 61 genes, and 1032 alternative splicing events [99]. RNA-seq analysis of HCMV infection in human fibroblasts showed that genes involved in the epithelial-to-mesenchymal transition (EMT) are downregulated, while genes that support mesenchymal-to-epithelial transition (MET) are induced, suggesting HCMV prefers an epithelial cellular state for replication [100]. Furthermore, RNA-seq has recently been used to explore transcriptomic differences between endemic Kaposi's sarcoma (EnKS) and epidemic Kaposi's sarcoma (EpKS), which results from KSHV and HIV-1 co-infection in sub-Saharan Africa [101]. This study found that a subset of genes involved in tumorigenesis and immune responses displayed increased dysregulation in EnKS lesions, but the overall gene expression profiles between EnKS and EpKS correlated strongly.

Investigation of cellular transcriptomes through RNA-seq have also revealed important aspects of nuclear DNA sensor regulation outside of the context of virus infection. To provide a few examples, expression of IFI16, among several other innate immunity proteins, was upregulated in macrophages infected with the bacterium *Campylobacter concisus* [102]; tumor-bearing mice with deletion of the IFI16 homolog p204, when compared to WT mice, lacked the ability to induce the upregulation of 382 genes, indicating the extensive involvement of IFI16 in antitumor immunity [57]; and RNA-seq studies of an alcohol-related liver disease model in mice revealed that liver damage from excessive alcohol consumption is mediated by cGAS activation of the STING–TBK1–IRF3 pathway [103].

Similar to transcriptome studies, whole-cell proteome investigations with mass spectrometry have led to a wealth of information about both viral and cellular protein abundances during virus infection, uncovering changes linked to innate immune responses and virus immune evasion strategies. Given the finely tuned temporal regulation of virus replication steps, assessments of the cellular proteomes have been carried out at multiple time points as the infection progresses, as reported for infection with HSV-1 [51,104], HCMV [105,106], and KSHV [107,108]. In conjunction with temporal studies, infection with virus strains that lack the ability to inhibit DNA sensors offered a view of proteome changes during an active host immune response. For example, the *d106* HSV-1 strain contains mutations in four of five immediate-early proteins (ICP4, ICP22, ICP27, and ICP47) but expresses functional ICP0 [109]. Infection with this virus results in increased induction of cytokines and apoptosis when compared to infection with WT HSV-1 [50,110]. By comparing temporal proteome changes during WT and *d106* HSV-1 infections, we discovered the upregulation of several proteins involved in innate immunity and apoptosis, and integration with cGAS IP-MS led to the discovery of OASL-mediated cGAS inhibition [51]. Additional MS studies have been carried out to characterize proteome changes during HSV-1 infection in a range of cell types and to compare alterations induced by different virus strains [51,54,104,111–121]. Spatial proteomics [122] has further provided the ability to characterize changes in proteome organization during infection [123], as well as discover viral proteins that localize to distinct organelles to regulate their functions, as shown for HCMV infection [124]. Recent years have also seen the increased integration of proteome studies with global PTM studies, where the infection-induced host phosphorylation, acetylation, SUMOylation, ubiquitination landscapes, to name just a few, have been started to be characterized [125–128]. Knowledge of global PTM changes have furthered the understanding of signaling cascades during infection and have helped to identify regulatory hubs at the interface between host defense and virus production. Another proteomic perspective of regulatory hubs is provided by the identification of functional protein complexes that are activated or inhibited during an infection process. The use of thermal co-aggregation profiling MS was recently demonstrated to offer a global view of temporal assembly and disassembly of host–host, host–viral, and viral–viral protein interaction events during HCMV infection, including the regulation of complexes involved in host immunity [106]. Altogether, these MS-based proteomic investigations of whole-cell and subcellular proteomes, interactomes, and PTMs provide rich information regarding host cell changes in response to viral infections. The integration of these different datasets promises to reveal a systems-view of the host environment during infection, which can aid in the formulation of specific biological hypotheses, the identification of changes linked to viral pathologies, and the discovery of therapeutic targets. Therefore, efforts have been and continue to be placed in the development of computational platforms that facilitate data integration in a user-friendly manner [129–135]. One platform specifically applied to studying viral infections is the Interaction Visualization in Space and Time Analysis (Inter-ViSTA), a web-accessible platform that enables integration of interactome, proteome, and functional traits to build animated temporal interaction networks [136]. For example, this analysis platform readily illustrated dynamic localization-dependent interactions of the HCMV protein pUL37 that function to either inhibit immune responses early in infection or promote peroxisome metabolic functions that benefit virus assembly late in infection.

Metabolome profiling brings another powerful omic tool to understanding the biology of virus infection and host defense mechanisms. Replication and assembly of virions is an energy-intensive process that requires the virus to trigger the cellular machinery to increase protein and lipid production for building progeny virions, as shown for numerous viruses [137]. Great effort has been put into understanding the mechanisms underlying metabolic reprogramming during a number of viral infections, including with HCMV and HSV-1 [55,138]. Integrating MS-based metabolomics with molecular virology techniques has proved valuable towards this goal; for example, a recent study of HCMV infection found that the viral protein pUL37 is critical for remodeling cellular metabolism by increasing production of very-long-chain fatty acids [139]. Given that pUL37 is an important immune evasion protein, such as by inhibiting cGAS function [39], it is likely that pUL37 bridges proviral metabolism with innate immune regulation during HCMV infection. Future studies geared towards elucidating the relationships between these fundamental infection processes promise to reveal key players in virus replication and spread.

Finally, the secretion of proteins into the extracellular space is crucial for communication with adjacent cells and is the foundation of innate immunity. Interferons secreted by infected cells bind to receptors on neighboring cells to induce immunomodulatory and antiproliferative effects, a phenomenon that has been known for several decades [140]. Upon binding to the interferon receptor and activating the JAK–STAT signaling pathway, dozens of transcripts are upregulated, including additional cytokines [141], altogether leading to inflammatory response and impacting disease pathology. Therefore, examining the secretome of infected cells is a necessary component for understanding these complex intercellular communications [142]. MS-based studies have leveraged proteomics and lipidomics methods to define the composition of secreted biomolecular complexes during infection, including extracellular vesicles known as exosomes [143]. For example, quantitative proteomic analysis of exosomes from HSV-1-infected macrophages demonstrated that specific subsets of cytokines, inflammatory proteins, and transcription factors are secreted rapidly upon infection, thus priming immune response in neighboring cells [144]. Virus-driven secretomes can also impact cellular and tissue physiology, as demonstrated by two recent studies that examined how molecules secreted by herpesvirus infected cells determine local immune and growth responses in neutrophils [145] and cortical brain cells [54], respectively.

4. The Missing Link: Genomics for Understanding the Viral DNA–DNA Sensor Interface

AP-MS isolations of viral DNA during infection have been fundamental for the discovery of nuclear DNA sensors. However, the regulation and complete outcome of the interactions between DNA sensors and viral DNA remain to be fully characterized. In this section, we discuss the conundrum of how DNA sensors bind to pathogenic DNA in a sequence-independent manner, while also being shown to specifically function in repression of viral gene expression.

Though nuclear DNA sensors avoid autoreactivity with host DNA, they do not appear to recognize any specific virus nucleotide sequence motifs or DNA modifications. In fact, for a protein to be classified as a DNA sensor, one requirement is that it should bind to DNA in a sequence-independent manner, thereby having the capacity to recognize multiple DNA pathogens. For example, for the HIN-200 domains of IFI16 and IFIX, their sequence-independent binding to dsDNA is accomplished via weak electrostatic interactions between positively charged amino acids and the negatively charged DNA phosphate backbone [25,146,147]. It was also demonstrated that IFI16 preferentially binds to specific DNA forms, namely cruciform structures, superhelical, and quadruplex DNA, which could maximize contact between the phosphate backbone and the basic amino acids in the HIN-200 oligonucleotide/oligosaccharide binding folds [148,149]. However, there remains no evidence of DNA sequence preference, and it is hypothesized that the activation of immune responses by IFI16 relies on cooperative assembly of IFI16 oligomers, which is limited on host DNA by tight chromatin packing [29,150]. Examinations of crystal structures of cGAS with a dsDNA ligand have similarly shown that the cGAS Mab21 domain binds to the phosphate backbone of B-form DNA without any sequence specificity [151–154]. In contrast with IFI16, it is proposed that cGAS-mediated autoreactivity is inhibited by tight tethering of cGAS to host chromatin through a salt-resistant interaction that is independent of the domains required for cGAS activation [34,35].

Such *in vitro* experiments indicate that DNA binding is sequence independent, but the propensity of DNA sensors to interact with transcriptional regulatory proteins that are sequence specific (e.g., the HSV-1 transcriptional activator ICP4 [155]) could induce preferential accumulation at certain DNA loci. Furthermore, given that IFI16 and IFIX have also been shown to function in host antiviral response by repressing virus transcription [29–33], how does DNA sensor binding affect the chromatin structure at specific binding sites? Are other protein–DNA interactions increased or decreased at these loci, and how does this affect viral transcription and replication?

After entering the nucleus, herpesvirus genomes are subjected to chromatinization by host cell histones [156], and it has been demonstrated that IFI16 promotes the addition of the repressive heterochromatin mark H3K9me3 on viral DNA [31,32,157]. Thus far, these studies investigating where IFI16 and H3K9me3 interact with viral genomes have been conducted using chromatin immunoaffinity purification (ChIP) coupled with PCR or RT-qPCR [31,32,157]. Herpesviruses have large genomes (e.g., HSV-1 is ~152 kilobase pairs and contains ~80 genes), yet this approach is limited by only examining protein–DNA interactions at a few viral genes. Higher throughput techniques can help to more broadly represent interactions between viral DNA and DNA sensors and the subsequent effects on the viral genome chromatin landscape.

To assess where DNA sensors bind to the viral genome, ChIP sequencing (ChIP-seq) is an appropriate technique that has previously been used to study how the HSV-1 genome interacts with ICP4 [155], RNA polymerase II [158], and the transcription factor CCCTC-binding factor (CTCF) [159]. Applying this technique with nuclear DNA sensors would help determine whether DNA sensing is fully a sequence-independent process or whether additional factors within the cell can also cause accumulation of the DNA sensor at specific DNA loci.

Histone PTMs such as H3K9me3 are often used as proxies for determining whether a DNA locus resides in a euchromatin or heterochromatin region of DNA [160]. To investigate how DNA sensors affect the chromatinization of viral genomes, knockout studies can be followed by H3, H3K4me3, and H3K9me3 ChIP-seq. However, these modifications only act as a proxy for the chromatin structure and are not a direct readout of chromatin structure. Additionally, the cost of such experiments must also be considered, as the requirement for multiple conditions per sample considerably increases the amount of sequencing required. Measuring chromatin accessibility is often a better way to examine chromatin structure and can be probed through techniques such as MNase-seq [161], DNase-seq [162], FAIRE-seq [163], and ATAC-seq [164]. Furthermore, integration of protein–DNA interaction mapping data with chromatin accessibility data following DNA sensor knockout can help to identify how DNA sensor binding both globally and locally affects viral DNA structure. Thus, high-throughput sequencing techniques that explore epigenomic changes will be pivotal to continuing to expand our understanding of nuclear DNA sensor mechanisms.

5. Concluding Remarks

The development of omics techniques has helped to greatly expedite biological research. The topic discussed in this paper, the elegantly complex process of nuclear DNA sensing during virus infection has benefited immensely from the ability to examine the identities and PTM states of all proteins within the host cell. The general idea behind DNA sensors is rather simple: bind pathogenic DNA and initiate antiviral signaling pathways. However, the mechanisms by which the nuclear DNA sensors IFI16, IFIX, cGAS, and hnRNPA2B1 activate large-scale transcriptome, proteome, and secretome changes rely on the precise coordination of a multitude protein interactions and PTMs. Here, we have discussed how omics techniques, particularly those implementing mass spectrometry, have led to the discovery and characterization of these nuclear DNA sensors. The future expansion of these investigations to integrative multiomics studies that include epigenomic assays promise to substantially contribute to a more in-depth understanding of the intricacies of DNA sensing, its dysregulation, and connected pathologies.

Author Contributions: Conceptualization, T.R.H. and I.M.C.; writing—original draft preparation, T.R.H. and I.M.C.; writing—review and editing, T.R.H. and I.M.C. All authors have read and agreed to the published version of the manuscript.

Funding: We are grateful for funding from the NIH (R01 GM114141) and Mallinckrodt Scholar Award to IMC, and from the NIH training grant from NIGMS (T32GM007388).

Conflicts of Interest: The authors declare no conflict of interest.

References

1. McNab, F.; Mayer-Barber, K.; Sher, A.; Wack, A.; O'Garra, A. Type I interferons in infectious disease. *Nat. Rev. Immunol.* **2015**, *15*, 87–103. [CrossRef]
2. Paludan, S.R.; Bowie, A.G. Immune Sensing of DNA. *Immunity* **2013**, *38*, 870–880. [CrossRef] [PubMed]
3. Wertheim, J.O.; Smith, M.D.; Smith, D.M.; Scheffler, K.; Kosakovsky Pond, S.L. Evolutionary origins of human herpes simplex viruses 1 and 2. *Mol. Biol. Evol.* **2014**, *31*, 2356–2364. [CrossRef] [PubMed]

4. Renner, D.W.; Szpara, M.L. Impacts of Genome-Wide Analyses on Our Understanding of Human Herpesvirus Diversity and Evolution. *J. Virol.* **2017**, *92*, 92. [CrossRef] [PubMed]
5. Cagliani, R.; Forni, D.; Mozzi, A.; Sironi, M. Evolution and genetic diversity of primate cytomegaloviruses. *Microorganisms* **2020**, *8*, 624. [CrossRef] [PubMed]
6. Hurst, T.P.; Aswad, A.; Karamitros, T.; Katzourakis, A.; Smith, A.L.; Magiorkinis, G. Interferon-inducible protein 16 (IFI16) has a broad-spectrum binding ability against ss DNA targets: An evolutionary hypothesis for antiretroviral checkpoint. *Front. Microbiol.* **2019**, *10*, 1426. [CrossRef] [PubMed]
7. Unterholzner, L.; Bowie, A.G. Innate DNA sensing moves to the nucleus. *Cell Host Microbe* **2011**, *9*, 351–353. [CrossRef] [PubMed]
8. Diner, B.A.; Lum, K.K.; Cristea, I.M. The Emerging Role of Nuclear Viral DNA Sensors. *J. Biol. Chem.* **2015**, *290*, 26412–26421. [CrossRef]
9. Unterholzner, L.; Keating, S.E.; Baran, M.; Horan, K.A.; Jensen, S.B.; Sharma, S.; Sirois, C.M.; Jin, T.; Latz, E.; Xiao, T.S.; et al. IFI16 is an innate immune sensor for intracellular DNA. *Nat. Immunol.* **2010**, *11*, 997–1004. [CrossRef]
10. Li, T.; Diner, B.A.; Chen, J.; Cristea, I.M. Acetylation modulates cellular distribution and DNA sensing ability of interferon-inducible protein IFI16. *Proc. Natl. Acad. Sci. USA* **2012**, *109*, 10558–10563. [CrossRef]
11. Kerur, N.; Veettil, M.V.; Sharma-Walia, N.; Bottero, V.; Sadagopan, S.; Otageri, P.; Chandran, B. IFI16 Acts as a Nuclear Pathogen Sensor to Induce the Inflammasome in Response to Kaposi Sarcoma-Associated Herpesvirus Infection. *Cell Host Microbe* **2011**, *9*, 363–375. [CrossRef] [PubMed]
12. Diner, B.A.; Li, T.; Greco, T.M.; Crow, M.S.; Fuesler, J.A.; Wang, J.; Cristea, I.M. The functional interactome of PYHIN immune regulators reveals IFIX is a sensor of viral DNA. *Mol. Syst. Biol.* **2015**, *11*, 787. [CrossRef] [PubMed]
13. Sun, L.; Wu, J.; Du, F.; Chen, X.; Chen, Z.J. Cyclic GMP-AMP synthase is a cytosolic DNA sensor that activates the type I interferon pathway. *Science* **2013**, *339*, 786–791. [CrossRef] [PubMed]
14. Orzalli, M.H.; Broekema, N.M.; Diner, B.A.; Hancks, D.C.; Elde, N.C.; Cristea, I.M.; Knipe, D.M. cGAS-mediated stabilization of IFI16 promotes innate signaling during herpes simplex virus infection. *Proc. Natl. Acad. Sci. USA* **2015**, *112*, E1773–E1781. [CrossRef] [PubMed]
15. Diner, B.A.; Lum, K.K.; Toettcher, J.E.; Cristea, I.M. Viral DNA Sensors IFI16 and Cyclic GMP-AMP Synthase Possess Distinct Functions in Regulating Viral Gene Expression, Immune Defenses, and Apoptotic Responses during Herpesvirus Infection. *MBio* **2016**, *7*, e01553-16. [CrossRef]
16. Liu, H.; Zhang, H.; Wu, X.; Ma, D.; Wu, J.; Wang, L.; Jiang, Y.; Fei, Y.; Zhu, C.; Tan, R.; et al. Nuclear cGAS suppresses DNA repair and promotes tumorigenesis. *Nature* **2018**, *563*, 131–136. [CrossRef]
17. Wang, L.; Wen, M.; Cao, X. Nuclear hnRNPA2B1 initiates and amplifies the innate immune response to DNA viruses. *Science* **2019**, *365*, eaav0758. [CrossRef]
18. Burleigh, K.; Maltbaek, J.H.; Cambier, S.; Green, R.; Gale, M.; James, R.C.; Stetson, D.B. Human DNA-PK activates a STING-independent DNA sensing pathway. *Sci. Immunol.* **2020**, *5*, eaba4219. [CrossRef]
19. Sato, M.; Tanaka, N.; Hata, N.; Oda, E.; Taniguchi, T. Involvement of the IRF family transcription factor IRF-3 in virus-induced activation of the IFN-beta gene. *FEBS Lett.* **1998**, *425*, 112–116. [CrossRef]
20. Lin, R.; Heylbroeck, C.; Pitha, P.M.; Hiscott, J. Virus-dependent phosphorylation of the IRF-3 transcription factor regulates nuclear translocation, transactivation potential, and proteasome-mediated degradation. *Mol. Cell. Biol.* **1998**, *18*, 2986–2996. [CrossRef]
21. Ishikawa, H.; Ma, Z.; Barber, G.N. STING regulates intracellular DNA-mediated, type I interferon-dependent innate immunity. *Nature* **2009**, *461*, 788–792. [CrossRef] [PubMed]
22. Tanaka, Y.; Chen, Z.J. STING specifies IRF3 phosphorylation by TBK1 in the cytosolic DNA signaling pathway. *Sci. Signal.* **2012**, *5*, ra20. [CrossRef] [PubMed]
23. Bertin, J.; DiStefano, P.S. The PYRIN domain: A novel motif found in apoptosis and inflammation proteins. *Cell Death Differ.* **2000**, *7*, 1273–1274. [CrossRef] [PubMed]
24. Albrecht, M.; Choubey, D.; Lengauer, T. The HIN domain of IFI-200 proteins consists of two OB folds. *Biochem. Biophys. Res. Commun.* **2005**, *327*, 679–687. [CrossRef] [PubMed]
25. Jin, T.; Perry, A.; Jiang, J.; Smith, P.; Curry, J.A.; Unterholzner, L.; Jiang, Z.; Horvath, G.; Rathinam, V.A.; Johnstone, R.W.; et al. Structures of the HIN Domain:DNA Complexes Reveal Ligand Binding and Activation Mechanisms of the AIM2 Inflammasome and IFI16 Receptor. *Immunity* **2012**, *36*, 561–571. [CrossRef]

26. Li, T.; Chen, J.; Cristea, I.M.M. Human Cytomegalovirus Tegument Protein pUL83 Inhibits IFI16-Mediated DNA Sensing for Immune Evasion. *Cell Host Microbe* **2013**, *14*, 591–599. [CrossRef] [PubMed]
27. Stratmann, S.A.; Morrone, S.R.; van Oijen, A.M.; Sohn, J. The innate immune sensor IFI16 recognizes foreign DNA in the nucleus by scanning along the duplex. *Elife* **2015**, *4*, e11721. [CrossRef] [PubMed]
28. Orzalli, M.H.; DeLuca, N.A.; Knipe, D.M. Nuclear IFI16 induction of IRF-3 signaling during herpesviral infection and degradation of IFI16 by the viral ICP0 protein. *Proc. Natl. Acad. Sci. USA* **2012**, *109*, E3008–E3017. [CrossRef]
29. Lum, K.K.; Howard, T.R.; Pan, C.; Cristea, I.M. Charge-mediated pyrin oligomerization nucleates antiviral IFI16 sensing of herpesvirus DNA. *MBio* **2019**, *10*, 10. [CrossRef]
30. Gariano, G.R.; Dell'Oste, V.; Bronzini, M.; Gatti, D.; Luganini, A.; De Andrea, M.; Gribaudo, G.; Gariglio, M.; Landolfo, S. The intracellular DNA sensor IFI16 gene acts as restriction factor for human cytomegalovirus replication. *PLoS Pathog.* **2012**, *8*, e1002498. [CrossRef]
31. Orzalli, M.H.; Conwell, S.E.; Berrios, C.; DeCaprio, J.A.; Knipe, D.M. Nuclear interferon-inducible protein 16 promotes silencing of herpesviral and transfected DNA. *Proc. Natl. Acad. Sci. USA* **2013**, *110*, E4492–E4501. [CrossRef] [PubMed]
32. Johnson, K.E.; Bottero, V.; Flaherty, S.; Dutta, S.; Singh, V.V.; Chandran, B. IFI16 Restricts HSV-1 Replication by Accumulating on the HSV-1 Genome, Repressing HSV-1 Gene Expression, and Directly or Indirectly Modulating Histone Modifications. *PLoS Pathog.* **2014**, *10*, e1004503. [CrossRef] [PubMed]
33. Crow, M.S.; Cristea, I.M. Human antiviral protein IFIX suppresses viral gene expression during herpes simplex virus 1 (HSV-1) infection and is counteracted by virus-induced proteasomal degradation. *Mol. Cell. Proteom.* **2017**, *16*, S200–S214. [CrossRef] [PubMed]
34. Volkman, H.E.; Cambier, S.; Gray, E.E.; Stetson, D.B. Tight nuclear tethering of cGAS is essential for preventing autoreactivity. *Elife* **2019**, *8*, 8. [CrossRef] [PubMed]
35. Michalski, S.; de Oliveira Mann, C.C.; Stafford, C.; Witte, G.; Bartho, J.; Lammens, K.; Hornung, V.; Hopfner, K.P. Structural basis for sequestration and autoinhibition of cGAS by chromatin. *Nature* **2020**, 1–8. [CrossRef]
36. Munro, T.P.; Magee, R.J.; Kidd, G.J.; Carson, J.H.; Barbarese, E.; Smith, L.M.; Smith, R. Mutational analysis of a heterogeneous nuclear ribonucleoprotein A2 response element for RNA trafficking. *J. Biol. Chem.* **1999**, *274*, 34389–34395. [CrossRef]
37. He, Y.; Smith, R. Nuclear functions of heterogeneous nuclear ribonucleoproteins A/B. *Cell. Mol. Life Sci.* **2009**, *66*, 1239–1256. [CrossRef]
38. Cuchet-Lourenco, D.; Anderson, G.; Sloan, E.; Orr, A.; Everett, R.D. The Viral Ubiquitin Ligase ICP0 Is neither Sufficient nor Necessary for Degradation of the Cellular DNA Sensor IFI16 during Herpes Simplex Virus 1 Infection. *J. Virol.* **2013**, *87*, 13422–13432. [CrossRef]
39. Zhang, J.; Zhao, J.; Xu, S.; Li, J.; He, S.; Zeng, Y.; Xie, L.; Xie, N.; Liu, T.; Lee, K.; et al. Species-Specific Deamidation of cGAS by Herpes Simplex Virus UL37 Protein Facilitates Viral Replication. *Cell Host Microbe* **2018**, *24*, 234–248.e5. [CrossRef]
40. Mondini, M.; Vidali, M.; De Andrea, M.; Azzimonti, B.; Airò, P.; D'Ambrosio, R.; Riboldi, P.; Meroni, P.L.; Albano, E.; Shoenfeld, Y.; et al. A novel autoantigen to differentiate limited cutaneous systemic sclerosis from diffuse cutaneous systemic sclerosis: The interferon-inducible gene IFI16. *Arthritis Rheum.* **2006**, *54*, 3939–3944. [CrossRef]
41. Mondini, M.; Vidali, M.; Airò, P.; De Andrea, M.; Riboldi, P.; Meroni, P.L.; Gariglio, M.; Landolfo, S. Role of the interferon-inducible gene IFI16 in the etiopathogenesis of systemic autoimmune disorders. *Ann. N. Y. Acad. Sci.* **2007**, *1110*, 47–56. [CrossRef] [PubMed]
42. Caneparo, V.; Cena, T.; De Andrea, M.; Dell'Oste, V.; Stratta, P.; Quaglia, M.; Tincani, A.; Andreoli, L.; Ceffa, S.; Taraborelli, M.; et al. Anti-IFI16 antibodies and their relation to disease characteristics in systemic lupus erythematosus. *Lupus* **2013**, *22*, 607–613. [CrossRef] [PubMed]
43. Costa, S.; Borgogna, C.; Mondini, M.; De Andrea, M.; Meroni, P.L.; Berti, E.; Gariglio, M.; Landolfo, S. Redistribution of the nuclear protein IFI16 into the cytoplasm of ultraviolet B-exposed keratinocytes as a mechanism of autoantigen processing. *Br. J. Dermatol.* **2011**, *164*, 282–290. [CrossRef] [PubMed]
44. Gao, D.; Li, T.; Li, X.D.; Chen, X.; Li, Q.Z.; Wight-Carter, M.; Chen, Z.J. Activation of cyclic GMP-AMP synthase by self-DNA causes autoimmune diseases. *Proc. Natl. Acad. Sci. USA* **2015**, *112*, E5699–E5705. [CrossRef] [PubMed]

45. Gray, E.E.; Treuting, P.M.; Woodward, J.J.; Stetson, D.B. Cutting Edge: cGAS Is Required for Lethal Autoimmune Disease in the Trex1-Deficient Mouse Model of Aicardi–Goutières Syndrome. *J. Immunol.* **2015**, *195*, 1939–1943. [CrossRef] [PubMed]
46. Vincent, J.; Adura, C.; Gao, P.; Luz, A.; Lama, L.; Asano, Y.; Okamoto, R.; Imaeda, T.; Aida, J.; Rothamel, K.; et al. Small molecule inhibition of cGAS reduces interferon expression in primary macrophages from autoimmune mice. *Nat. Commun.* **2017**, *8*, 750. [CrossRef]
47. Ishii, K.J.; Kawagoe, T.; Koyama, S.; Matsui, K.; Kumar, H.; Kawai, T.; Uematsu, S.; Takeuchi, O.; Takeshita, F.; Coban, C.; et al. TANK-binding kinase-1 delineates innate and adaptive immune responses to DNA vaccines. *Nature* **2008**, *451*, 725–729. [CrossRef]
48. Suschak, J.J.; Wang, S.; Fitzgerald, K.A.; Lu, S. Identification of Aim2 as a Sensor for DNA Vaccines. *J. Immunol.* **2015**, *194*, 630–636. [CrossRef]
49. Suschak, J.J.; Wang, S.; Fitzgerald, K.A.; Lu, S. A cGAS-Independent STING/IRF7 Pathway Mediates the Immunogenicity of DNA Vaccines. *J. Immunol.* **2016**, *196*, 310–316. [CrossRef]
50. Diner, B.A.; Lum, K.K.; Javitt, A.; Cristea, I.M. Interactions of the Antiviral Factor Interferon Gamma-Inducible Protein 16 (IFI16) Mediate Immune Signaling and Herpes Simplex Virus-1 Immunosuppression. *Mol. Cell. Proteom.* **2015**, *14*, 2341–2356. [CrossRef]
51. Lum, K.K.; Song, B.; Federspiel, J.D.; Diner, B.A.; Howard, T.; Cristea, I.M. Interactome and Proteome Dynamics Uncover Immune Modulatory Associations of the Pathogen Sensing Factor cGAS. *Cell Syst.* **2018**, *7*, 627–642.e6. [CrossRef] [PubMed]
52. Dai, J.; Huang, Y.J.; He, X.; Zhao, M.; Wang, X.; Liu, Z.S.; Xue, W.; Cai, H.; Zhan, X.Y.; Huang, S.Y.; et al. Acetylation Blocks cGAS Activity and Inhibits Self-DNA-Induced Autoimmunity. *Cell* **2019**, *176*, 1447–1460.e14. [CrossRef] [PubMed]
53. Song, B.; Greco, T.M.; Lum, K.K.; Taber, C.E.; Cristea, I.M. The DNA Sensor cGAS is Decorated by Acetylation and Phosphorylation Modifications in the Context of Immune Signaling. *Mol. Cell. Proteom.* **2020**, *19*, 1193–1208. [CrossRef] [PubMed]
54. Hensel, N.; Raker, V.; Förthmann, B.; Buch, A.; Sodeik, B.; Pich, A.; Claus, P. The Proteome and Secretome of Cortical Brain Cells Infected With Herpes Simplex Virus. *Front. Neurol.* **2020**, *11*, 11. [CrossRef] [PubMed]
55. Vastag, L.; Koyuncu, E.; Grady, S.L.; Shenk, T.E.; Rabinowitz, J.D. Divergent effects of human cytomegalovirus and herpes simplex virus-1 on cellular metabolism. *PLoS Pathog.* **2011**, *7*, e1002124. [CrossRef]
56. Zhu, H.; Cong, J.P.; Mamtora, G.; Gingeras, T.; Shenk, T. Cellular gene expression altered by human cytomegalovirus: Global monitoring with oligonucleotide arrays. *Proc. Natl. Acad. Sci. USA* **1998**, *95*, 14470–14475. [CrossRef]
57. Jian, J.; Wei, W.; Yin, G.; Hettinghouse, A.; Liu, C.; Shi, Y. RNA-Seq analysis of interferon inducible p204-mediated network in anti-tumor immunity. *Sci. Rep.* **2018**, *8*, 6495. [CrossRef]
58. Greco, T.M.; Kennedy, M.A.; Cristea, I.M. Proteomic Technologies for Deciphering Local and Global Protein Interactions. *Trends Biochem. Sci.* **2020**, *45*, 454–455. [CrossRef]
59. Miteva, Y.V.; Budayeva, H.G.; Cristea, I.M. Proteomics-based methods for discovery, quantification, and validation of protein-protein interactions. *Anal. Chem.* **2013**, *85*, 749–768. [CrossRef]
60. Ten Have, S.; Boulon, S.; Ahmad, Y.; Lamond, A.I. Mass spectrometry-based immuno-precipitation proteomics—The user's guide. *Proteomics* **2011**, *11*, 1153–1159. [CrossRef]
61. Pichlmair, A.; Reis e Sousa, C. Innate Recognition of Viruses. *Immunity* **2007**, *27*, 370–383. [CrossRef] [PubMed]
62. Monroe, K.M.; Yang, Z.; Johnson, J.R.; Geng, X.; Doitsh, G.; Krogan, N.J.; Greene, W.C. IFI16 DNA sensor is required for death of lymphoid CD4 T cells abortively infected with HIV. *Science* **2014**, *343*, 428–432. [CrossRef] [PubMed]
63. Dembowski, J.A.; Deluca, N.A. Temporal viral genome-protein interactions define distinct stages of productive herpesviral infection. *MBio* **2018**, *9*, 9. [CrossRef] [PubMed]
64. Kim, B.; Arcos, S.; Rothamel, K.; Jian, J.; Rose, K.L.; McDonald, W.H.; Bian, Y.; Reasoner, S.; Barrows, N.J.; Bradrick, S.; et al. Discovery of Widespread Host Protein Interactions with the Pre-replicated Genome of CHIKV Using VIR-CLASP. *Mol. Cell* **2020**, *78*, 624–640.e7. [CrossRef] [PubMed]
65. Ouellet Lavallée, G.; Pearson, A. Upstream binding factor inhibits herpes simplex virus replication. *Virology* **2015**, *483*, 108–116. [CrossRef]
66. Cabral, J.M.; Oh, H.S.; Knipe, D.M. ATRX promotes maintenance of herpes simplex virus heterochromatin during chromatin stress. *Elife* **2018**, *7*. [CrossRef]

67. Gu, H.; Roizman, B. The Two Functions of Herpes Simplex Virus 1 ICP0, Inhibition of Silencing by the CoREST/REST/HDAC Complex and Degradation of PML, Are Executed in Tandem. *J. Virol.* **2009**, *83*, 181–187. [CrossRef]
68. Fanis, P.; Gillemans, N.; Aghajanirefah, A.; Pourfarzad, F.; Demmers, J.; Esteghamat, F.; Vadlamudi, R.K.; Grosveld, F.; Philipsen, S.; Van Dijk, T.B. Five friends of methylated chromatin target of protein-arginine-methyltransferase[Prmt]-1 (Chtop), a complex linking arginine methylation to desumoylation. *Mol. Cell. Proteom.* **2012**, *11*, 1263–1273. [CrossRef]
69. Wu, J.; Sun, L.; Chen, X.; Du, F.; Shi, H.; Chen, C.; Chen, Z.J. Cyclic GMP-AMP is an endogenous second messenger in innate immune signaling by cytosolic DNA. *Science* **2013**, *339*, 826–830. [CrossRef]
70. Seo, G.J.; Kim, C.; Shin, W.J.; Sklan, E.H.; Eoh, H.; Jung, J.U. TRIM56-mediated monoubiquitination of cGAS for cytosolic DNA sensing. *Nat. Commun.* **2018**, *9*, 1–13. [CrossRef]
71. Almine, J.F.; O'Hare, C.A.J.; Dunphy, G.; Haga, I.R.; Naik, R.J.; Atrih, A.; Connolly, D.J.; Taylor, J.; Kelsall, I.R.; Bowie, A.G.; et al. IFI16 and cGAS cooperate in the activation of STING during DNA sensing in human keratinocytes. *Nat. Commun.* **2017**, *8*, 14392. [CrossRef] [PubMed]
72. Wu, J.J.; Li, W.; Shao, Y.; Avey, D.; Fu, B.; Gillen, J.; Hand, T.; Ma, S.; Liu, X.; Miley, W.; et al. Inhibition of cGAS DNA Sensing by a Herpesvirus Virion Protein. *Cell Host Microbe* **2015**, *18*, 333–344. [CrossRef] [PubMed]
73. Olsen, J.V.; Vermeulen, M.; Santamaria, A.; Kumar, C.; Miller, M.L.; Jensen, L.J.; Gnad, F.; Cox, J.; Jensen, T.S.; Nigg, E.A.; et al. Quantitative phosphoproteomics revealswidespread full phosphorylation site occupancy during mitosis. *Sci. Signal.* **2010**, *3*, ra3. [CrossRef]
74. Hendriks, I.A.; D'Souza, R.C.J.; Yang, B.; Verlaan-De Vries, M.; Mann, M.; Vertegaal, A.C.O. Uncovering global SUMOylation signaling networks in a site-specific manner. *Nat. Struct. Mol. Biol.* **2014**, *21*, 927–936. [CrossRef] [PubMed]
75. Hendriks, I.A.; Lyon, D.; Young, C.; Jensen, L.J.; Vertegaal, A.C.O.; Nielsen, M.L. Site-specific mapping of the human SUMO proteome reveals co-modification with phosphorylation. *Nat. Struct. Mol. Biol.* **2017**, *24*, 325–336. [CrossRef]
76. Hendriks, I.A.; Treffers, L.W.; Verlaan-deVries, M.; Olsen, J.V.; Vertegaal, A.C.O. SUMO-2 Orchestrates Chromatin Modifiers in Response to DNA Damage. *Cell Rep.* **2015**, *10*, 1778–1791. [CrossRef]
77. Xia, P.; Ye, B.; Wang, S.; Zhu, X.; Du, Y.; Xiong, Z.; Tian, Y.; Fan, Z. Glutamylation of the DNA sensor cGAS regulates its binding and synthase activity in antiviral immunity. *Nat. Immunol.* **2016**, *17*, 369–378. [CrossRef]
78. Seo, G.J.; Yang, A.; Tan, B.; Kim, S.; Liang, Q.; Choi, Y.; Yuan, W.; Feng, P.; Park, H.S.; Jung, J.U. Akt Kinase-Mediated Checkpoint of cGAS DNA Sensing Pathway. *Cell Rep.* **2015**, *13*, 440–449. [CrossRef]
79. Hu, M.M.; Yang, Q.; Xie, X.Q.; Liao, C.Y.; Lin, H.; Liu, T.T.; Yin, L.; Shu, H.B. Sumoylation Promotes the Stability of the DNA Sensor cGAS and the Adaptor STING to Regulate the Kinetics of Response to DNA Virus. *Immunity* **2016**, *45*, 555–569. [CrossRef]
80. Van Damme, P.; Lasa, M.; Polevoda, B.; Gazquez, C.; Elosegui-Artola, A.; Kim, D.S.; De Juan-Pardo, E.; Demeyer, K.; Hole, K.; Larrea, E.; et al. N-terminal acetylome analyses and functional insights of the N-terminal acetyltransferase NatB. *Proc. Natl. Acad. Sci. USA* **2012**, *109*, 12449–12454. [CrossRef]
81. Choudhary, C.; Kumar, C.; Gnad, F.; Nielsen, M.L.; Rehman, M.; Walther, T.C.; Olsen, J.V.; Mann, M. Lysine acetylation targets protein complexes and co-regulates major cellular functions. *Science* **2009**, *325*, 834–840. [CrossRef] [PubMed]
82. Guo, A.; Gu, H.; Zhou, J.; Mulhern, D.; Wang, Y.; Lee, K.A.; Yang, V.; Aguiar, M.; Kornhauser, J.; Jia, X.; et al. Immunoaffinity enrichment and mass spectrometry analysis of protein methylation. *Mol. Cell. Proteomics* **2014**, *13*, 372–387. [CrossRef]
83. Zhou, H.; Di Palma, S.; Preisinger, C.; Peng, M.; Polat, A.N.; Heck, A.J.R.; Mohammed, S. Toward a comprehensive characterization of a human cancer cell phosphoproteome. *J. Proteome Res.* **2013**, *12*, 260–271. [CrossRef] [PubMed]
84. Mayya, V.; Lundgren, D.H.; Hwang, S.L.L.; Rezaul, K.; Wu, L.; Eng, J.K.; Rodionov, V.; Han, D.K. Quantitative phosphoproteomic analysis of T Cell receptor signaling reveals system-wide modulation of protein-protein interactions. *Sci. Signal.* **2009**, *2*, ra46. [CrossRef] [PubMed]
85. Rigbolt, K.T.G.; Prokhorova, T.A.; Akimov, V.; Henningsen, J.; Johansen, P.T.; Kratchmarova, I.; Kassem, M.; Mann, M.; Olsen, J.V.; Blagoev, B. System-wide temporal characterization of the proteome and phosphoproteome of human embryonic stem cell differentiation. *Sci. Signal.* **2011**, *4*, rs3. [CrossRef] [PubMed]

86. Bian, Y.; Song, C.; Cheng, K.; Dong, M.; Wang, F.; Huang, J.; Sun, D.; Wang, L.; Ye, M.; Zou, H. An enzyme assisted RP-RPLC approach for in-depth analysis of human liver phosphoproteome. *J. Proteom.* **2014**, *96*, 253–262. [CrossRef]
87. Zhao, J.; Zeng, Y.; Xu, S.; Chen, J.; Shen, G.; Yu, C.; Knipe, D.; Yuan, W.; Peng, J.; Xu, W.; et al. A Viral Deamidase Targets the Helicase Domain of RIG-I to Block RNA-Induced Activation. *Cell Host Microbe* **2016**, *20*, 770–784. [CrossRef]
88. Stingley, S.W.; Ramirez, J.J.G.; Aguilar, S.A.; Simmen, K.; Sandri-Goldin, R.M.; Ghazal, P.; Wagner, E.K. Global Analysis of Herpes Simplex Virus Type 1 Transcription Using an Oligonucleotide-Based DNA Microarray. *J. Virol.* **2000**, *74*, 9916–9927. [CrossRef]
89. Aguilar, J.S.; Devi-Rao, G.V.; Rice, M.K.; Sunabe, J.; Ghazal, P.; Wagner, E.K. Quantitative comparison of the HSV-1 and HSV-2 transcriptomes using DNA microarray analysis. *Virology* **2006**, *348*, 233–241. [CrossRef]
90. Browne, E.P.; Wing, B.; Coleman, D.; Shenk, T. Altered Cellular mRNA Levels in Human Cytomegalovirus-Infected Fibroblasts: Viral Block to the Accumulation of Antiviral mRNAs. *J. Virol.* **2001**, *75*, 12319–12330. [CrossRef]
91. Simmen, K.A.; Singh, J.; Luukkonen, B.G.M.; Lopper, M.; Bittner, A.; Miller, N.E.; Jackson, M.R.; Compton, T.; Früh, K. Global modulation of cellular transcription by human cytomegalovirus is initiated by viral glycoprotein B. *Proc. Natl. Acad. Sci. USA* **2001**, *98*, 7140–7145. [CrossRef]
92. Song, Y.J.; Stinski, M.F. Effect of the human cytomegalovirus IE86 protein on expression of E2F-responsive genes: A DNA microarray analysis. *Proc. Natl. Acad. Sci. USA* **2002**, *99*, 2836–2841. [CrossRef] [PubMed]
93. Challacombe, J.F.; Rechtsteiner, A.; Gottardo, R.; Rocha, L.M.; Browne, E.P.; Shenk, T.; Altherr, M.R.; Brettin, T.S. Evaluation of the host transcriptional response to human cytomegalovirus infection. *Physiol. Genom.* **2004**, *18*, 51–62. [CrossRef] [PubMed]
94. Naranatt, P.P.; Krishnan, H.H.; Svojanovsky, S.R.; Bloomer, C.; Mathur, S.; Chandran, B. Host Gene Induction and Transcriptional Reprogramming in Kaposi's Sarcoma-Associated Herpesvirus (KSHV/HHV-8)-Infected Endothelial, Fibroblast, and B Cells: Insights into Modulation Events Early during Infection. *Cancer Res.* **2004**, *64*, 72–84. [CrossRef] [PubMed]
95. Chandriani, S.; Ganem, D. Host transcript accumulation during lytic KSHV infection reveals several classes of host responses. *PLoS ONE* **2007**, *2*, e811. [CrossRef]
96. Ray, N.; Enquist, L.W. Transcriptional Response of a Common Permissive Cell Type to Infection by Two Diverse Alphaherpesviruses. *J. Virol.* **2004**, *78*, 3489–3501. [CrossRef]
97. Brukman, A.; Enquist, L.W. Suppression of the Interferon-Mediated Innate Immune Response by Pseudorabies Virus. *J. Virol.* **2006**, *80*, 6345–6356. [CrossRef]
98. Wang, Z.; Gerstein, M.; Snyder, M. RNA-Seq: A revolutionary tool for transcriptomics. *Nat. Rev. Genet.* **2009**, *10*, 57–63. [CrossRef]
99. Hu, B.; Li, X.; Huo, Y.; Yu, Y.; Zhang, Q.; Chen, G.; Zhang, Y.; Fraser, N.W.; Wu, D.; Zhou, J. Cellular responses to HSV-1 infection are linked to specific types of alterations in the host transcriptome. *Sci. Rep.* **2016**, *6*, 28075. [CrossRef]
100. Oberstein, A.; Shenk, T. Cellular responses to human cytomegalovirus infection: Induction of a mesenchymal-to-epithelial transition (MET) phenotype. *Proc. Natl. Acad. Sci. USA* **2017**, 114, E8244–E8253. [CrossRef]
101. Lidenge, S.J.; Kossenkov, A.V.; Tso, F.Y.; Wickramasinghe, J.; Privatt, S.R.; Ngalamika, O.; Ngowi, J.R.; Mwaiselage, J.; Lieberman, P.M.; West, J.T.; et al. Comparative transcriptome analysis of endemic and epidemic Kaposi's sarcoma (KS) lesions and the secondary role of HIV-1 in KS pathogenesis. *PLoS Pathog.* **2020**, *16*, e1008681. [CrossRef]
102. Kaakoush, N.O.; Deshpande, N.P.; Man, S.M.; Burgos-Portugal, J.A.; Khattak, F.A.; Raftery, M.J.; Wilkins, M.R.; Mitchell, H.M. Transcriptomic and proteomic analyses reveal key innate immune signatures in the host response to the gastrointestinal pathogen Campylobacter concisus. *Infect. Immun.* **2015**, *83*, 832–845. [CrossRef] [PubMed]
103. Luther, J.; Khan, S.; Gala, M.K.; Kedrin, D.; Sridharan, G.; Goodman, R.P.; Garber, J.J.; Masia, R.; Diagacomo, E.; Adams, D.; et al. Hepatic gap junctions amplify alcohol liver injury by propagating cGAS-mediated IRF3 activation. *Proc. Natl. Acad. Sci. USA* **2020**, *117*, 11667–11673. [CrossRef] [PubMed]

104. Kulej, K.; Avgousti, D.C.; Sidoli, S.; Herrmann, C.; Della Fera, A.N.; Kim, E.T.; Garcia, B.A.; Weitzman, M.D. Time-resolved global and chromatin proteomics during herpes simplex virus type 1 (HSV-1) infection. *Mol. Cell. Proteom.* **2017**, *16*, S92–S107. [CrossRef] [PubMed]
105. Weekes, M.P.; Tomasec, P.; Huttlin, E.L.; Fielding, C.A.; Nusinow, D.; Stanton, R.J.; Wang, E.C.Y.; Aicheler, R.; Murrell, I.; Wilkinson, G.W.G.; et al. Quantitative temporal viromics: An approach to investigate host-pathogen interaction. *Cell* **2014**, *157*, 1460–1472. [CrossRef]
106. Hashimoto, Y.; Sheng, X.; Murray-Nerger, L.A.; Cristea, I.M. Temporal dynamics of protein complex formation and dissociation during human cytomegalovirus infection. *Nat. Commun.* **2020**, *11*, 1–20. [CrossRef]
107. Hartenian, E.; Gilbertson, S.; Federspiel, J.D.; Cristea, I.M.; Glaunsinger, B.A. RNA decay during gammaherpesvirus infection reduces RNA polymerase II occupancy of host promoters but spares viral promoters. *PLoS Pathog.* **2020**, *16*, e1008269. [CrossRef]
108. Gabaev, I.; Williamson, J.C.; Crozier, T.W.M.; Schulz, T.F.; Lehner, P.J. Quantitative Proteomics Analysis of Lytic KSHV Infection in Human Endothelial Cells Reveals Targets of Viral Immune Modulation. *Cell Rep.* **2020**, *33*, 33. [CrossRef]
109. Samaniego, L.A.; Neiderhiser, L.; DeLuca, N.A. Persistence and Expression of the Herpes Simplex Virus Genome in the Absence of Immediate-Early Proteins. *J. Virol.* **1998**, *72*, 3307–3320. [CrossRef]
110. Sanfilippo, C.M.; Blaho, J.A. ICP0 Gene Expression Is a Herpes Simplex Virus Type 1 Apoptotic Trigger. *J. Virol.* **2006**, *80*, 6810–6821. [CrossRef]
111. Rodríguez, M.C.; Dybas, J.M.; Hughes, J.; Weitzman, M.D.; Boutell, C. The HSV-1 ubiquitin ligase ICP0: Modifying the cellular proteome to promote infection. *Virus Res.* **2020**, *285*, 198015. [CrossRef]
112. Ouwendijk, W.J.D.; Dekker, L.J.M.; van den Ham, H.-J.; Lenac Rovis, T.; Haefner, E.S.; Jonjic, S.; Haas, J.; Luider, T.M.; Verjans, G.M.G.M. Analysis of Virus and Host Proteomes During Productive HSV-1 and VZV Infection in Human Epithelial Cells. *Front. Microbiol.* **2020**, *11*, 1179. [CrossRef] [PubMed]
113. Soh, T.K.; Davies, C.T.R.; Muenzner, J.; Hunter, L.M.; Barrow, H.G.; Connor, V.; Bouton, C.R.; Smith, C.; Emmott, E.; Antrobus, R.; et al. Temporal Proteomic Analysis of Herpes Simplex Virus 1 Infection Reveals Cell-Surface Remodeling via pUL56-Mediated GOPC Degradation. *Cell Rep.* **2020**, *33*, 108235. [CrossRef] [PubMed]
114. Antrobus, R.; Grant, K.; Gangadharan, B.; Chittenden, D.; Everett, R.D.; Zitzmann, N.; Boutell, C. Proteomic analysis of cells in the early stages of herpes simplex virus type-1 infection reveals widespread changes in the host cell proteome. *Proteomics* **2009**, *9*, 3913–3927. [CrossRef] [PubMed]
115. Xing, J.; Wang, S.; Li, Y.; Guo, H.; Zhao, L.; Pan, W.; Lin, F.; Zhu, H.; Wang, L.; Li, M.; et al. Characterization of the subcellular localization of herpes simplex virus type 1 proteins in living cells. *Med. Microbiol. Immunol.* **2011**, *200*, 61–68. [CrossRef]
116. Bell, C.; Desjardins, M.; Thibault, P.; Radtke, K. Proteomics analysis of Herpes Simplex Virus type 1-infected cells reveals dynamic changes of viral protein expression, ubiquitylation, and phosphorylation. *J. Proteome Res.* **2013**, *12*, 1820–1829. [CrossRef]
117. Berard, A.R.; Coombs, K.M.; Severini, A. Quantification of the host response proteome after herpes simplex virus type 1 infection. *J. Proteome Res.* **2015**, *14*, 2121–2142. [CrossRef]
118. Drayman, N.; Karin, O.; Mayo, A.; Danon, T.; Shapira, L.; Rafael, D.; Zimmer, A.; Bren, A.; Kobiler, O.; Alon, U. Dynamic Proteomics of Herpes Simplex Virus Infection. *MBio* **2017**, *8*, e01612-17. [CrossRef]
119. Cui, Y.H.; Liu, Q.; Xu, Z.Y.; Li, J.H.; Hu, Z.X.; Li, M.J.; Zheng, W.L.; Li, Z.J.; Pan, H.W. Quantitative proteomic analysis of human corneal epithelial cells infected with HSV-1. *Exp. Eye Res.* **2019**, *185*, 107664. [CrossRef]
120. Liu, H.; Huang, C.X.; He, Q.; Li, D.; Luo, M.H.; Zhao, F.; Lu, W. Proteomics analysis of HSV-1-induced alterations in mouse brain microvascular endothelial cells. *J. Neurovirol.* **2019**, *25*, 525–539. [CrossRef]
121. Wan, W.; Wang, L.; Chen, X.; Zhu, S.; Shang, W.; Xiao, G.; Zhang, L.K. A subcellular quantitative proteomic analysis of herpes simplex virus type 1-infected HEK 293T cells. *Molecules* **2019**, *24*, 4215. [CrossRef]
122. Gatto, L.; Breckels, L.M.; Lilley, K.S. Assessing sub-cellular resolution in spatial proteomics experiments. *Curr. Opin. Chem. Biol.* **2019**, *48*, 123–149. [CrossRef] [PubMed]
123. Jean Beltran, P.M.; Cook, K.C.; Cristea, I.M. Exploring and Exploiting Proteome Organization during Viral Infection. *J. Virol.* **2017**, *91*, 91. [CrossRef] [PubMed]
124. Jean Beltran, P.M.; Mathias, R.A.; Cristea, I.M. A Portrait of the Human Organelle Proteome In Space and Time during Cytomegalovirus Infection. *Cell Syst.* **2016**, *3*, 361–373.e6. [CrossRef] [PubMed]

125. Rathore, D.; Nita-Lazar, A. Phosphoproteome Analysis in Immune Cell Signaling. *Curr. Protoc. Immunol.* **2020**, *130*, 130. [CrossRef]
126. Murray, L.A.; Sheng, X.; Cristea, I.M. Orchestration of protein acetylation as a toggle for cellular defense and virus replication. *Nat. Commun.* **2018**, *9*, 4967. [CrossRef]
127. Sloan, E.; Tatham, M.H.; Groslambert, M.; Glass, M.; Orr, A.; Hay, R.T.; Everett, R.D. Analysis of the SUMO2 Proteome during HSV-1 Infection. *PLoS Pathog.* **2015**, *11*, 130. [CrossRef]
128. Cai, Y.; Su, J.; Wang, N.; Zhao, W.; Zhu, M.; Su, S. Comprehensive analysis of the ubiquitome in rabies virus-infected brain tissue of Mus musculus. *Vet. Microbiol.* **2020**, *241*, 108552. [CrossRef]
129. Keller, A.; Chavez, J.D.; Eng, J.K.; Thornton, Z.; Bruce, J.E. Tools for 3D Interactome Visualization. *J. Proteome Res.* **2019**, *18*, 753–758. [CrossRef]
130. Liu, X.; Chang, C.; Han, M.; Yin, R.; Zhan, Y.; Li, C.; Ge, C.; Yu, M.; Yang, X. PPIExp: A Web-Based Platform for Integration and Visualization of Protein-Protein Interaction Data and Spatiotemporal Proteomics Data. *J. Proteome Res.* **2019**, *18*, 633–641. [CrossRef]
131. Michalak, W.; Tsiamis, V.; Schwämmle, V.; Rogowska-Wrzesińska, A. ComplexBrowser: A tool for identification and quantification of protein complexes in large-scale proteomics datasets. *Mol. Cell. Proteom.* **2019**, *18*, 2324–2334. [CrossRef]
132. Miryala, S.K.; Anbarasu, A.; Ramaiah, S. Discerning molecular interactions: A comprehensive review on biomolecular interaction databases and network analysis tools. *Gene* **2018**, *642*, 84–94. [CrossRef] [PubMed]
133. Rudolph, J.D.; Cox, J. A Network Module for the Perseus Software for Computational Proteomics Facilitates Proteome Interaction Graph Analysis. *J. Proteome Res.* **2019**, *18*, 2052–2064. [CrossRef] [PubMed]
134. Salamon, J.; Goenawan, I.H.; Lynn, D.J. Analysis and Visualization of Dynamic Networks Using the DyNet App for Cytoscape. *Curr. Protoc. Bioinform.* **2018**, *63*, e55. [CrossRef] [PubMed]
135. Choi, M.; Carver, J.; Chiva, C.; Tzouros, M.; Huang, T.; Tsai, T.H.; Pullman, B.; Bernhardt, O.M.; Hüttenhain, R.; Teo, G.C.; et al. MassIVE.quant: A community resource of quantitative mass spectrometry–based proteomics datasets. *Nat. Methods* **2020**, *17*, 981–984. [CrossRef]
136. Federspiel, J.D.; Cook, K.C.; Kennedy, M.A.; Venkatesh, S.S.; Otter, C.J.; Hofstadter, W.A.; Jean Beltran, P.M.; Cristea, I.M. Mitochondria and Peroxisome Remodeling across Cytomegalovirus Infection Time Viewed through the Lens of Inter-ViSTA. *Cell Rep.* **2020**, *32*. [CrossRef]
137. Goodwin, C.M.; Xu, S.; Munger, J. Stealing the Keys to the Kitchen: Viral Manipulation of the Host Cell Metabolic Network. *Trends Microbiol.* **2015**, *23*, 789–798. [CrossRef]
138. Shenk, T.; Alwine, J.C. Human cytomegalovirus: Coordinating cellular stress, signaling, and metabolic pathways. *Annu. Rev. Virol.* **2014**, *1*, 355–374. [CrossRef]
139. Xi, Y.; Harwood, S.; Wise, L.M.; Purdy, J.G. Human Cytomegalovirus pUL37x1 Is Important for Remodeling of Host Lipid Metabolism. *J. Virol.* **2019**, *93*, 843–862. [CrossRef]
140. Taylor, M.W. Interferons. In *Viruses and Man: A History of Interactions*; Springer International Publishing: Cham, Switzerland, 2014; pp. 101–119.
141. Schneider, W.M.; Chevillotte, M.D.; Rice, C.M. Interferon-stimulated genes: A complex web of host defenses. *Annu. Rev. Immunol.* **2014**, *32*, 513–545. [CrossRef]
142. Khan, M.M.; Koppenol-Raab, M.; Kuriakose, M.; Manes, N.P.; Goodlett, D.R.; Nita-Lazar, A. Host-pathogen dynamics through targeted secretome analysis of stimulated macrophages. *J. Proteom.* **2018**, *189*, 34–38. [CrossRef]
143. Bello-Morales, R.; Ripa, I.; López-Guerrero, J.A. Extracellular vesicles in viral spread and antiviral response. *Viruses* **2020**, *12*, 623. [CrossRef] [PubMed]
144. Miettinen, J.J.; Matikainen, S.; Nyman, T.A. Global Secretome Characterization of Herpes Simplex Virus 1-Infected Human Primary Macrophages. *J. Virol.* **2012**, *86*, 12770–12778. [CrossRef]
145. Pocock, J.M.; Storisteanu, D.M.L.; Reeves, M.B.; Juss, J.K.; Wills, M.R.; Cowburn, A.S.; Chilvers, E.R. Human cytomegalovirus delays neutrophil apoptosis and stimulates the release of a prosurvival secretome. *Front. Immunol.* **2017**, *8*, 8. [CrossRef] [PubMed]
146. Ni, X.; Ru, H.; Ma, F.; Zhao, L.; Shaw, N.; Feng, Y.; Ding, W.; Gong, W.; Wang, Q.; Ouyang, S.; et al. New insights into the structural basis of DNA recognition by HINa and HINb domains of IFI16. *J. Mol. Cell Biol.* **2016**, *8*, 51–61. [CrossRef] [PubMed]
147. Tian, Y.; Yin, Q. Structural analysis of the HIN1 domain of interferon-inducible protein 204. *Acta Crystallogr. Sect. F Struct. Biol. Commun.* **2019**, *75*, 455–460. [CrossRef] [PubMed]

148. Brázda, V.; Coufal, J.; Liao, J.C.C.; Arrowsmith, C.H. Preferential binding of IFI16 protein to cruciform structure and superhelical DNA. *Biochem. Biophys. Res. Commun.* **2012**, *422*, 716–720. [CrossRef] [PubMed]
149. Hároníková, L.; Coufal, J.; Kejnovská, I.; Jagelská, E.B.; Fojta, M.; Dvoøáková, P.; Muller, P.; Vojtesek, B.; Brázda, V. IFI16 preferentially binds to DNA with quadruplex structure and enhances DNA quadruplex formation. *PLoS ONE* **2016**, *11*, e0157156. [CrossRef]
150. Morrone, S.R.; Wang, T.; Constantoulakis, L.M.; Hooy, R.M.; Delannoy, M.J.; Sohn, J. Cooperative assembly of IFI16 filaments on dsDNA provides insights into host defense strategy. *Proc. Natl. Acad. Sci. USA* **2014**, *111*, E62–E71. [CrossRef]
151. Civril, F.; Deimling, T.; De Oliveira Mann, C.C.; Ablasser, A.; Moldt, M.; Witte, G.; Hornung, V.; Hopfner, K.P. Structural mechanism of cytosolic DNA sensing by cGAS. *Nature* **2013**, *498*, 332–337. [CrossRef]
152. Gao, P.; Ascano, M.; Wu, Y.; Barchet, W.; Gaffney, B.L.; Zillinger, T.; Serganov, A.A.; Liu, Y.; Jones, R.A.; Hartmann, G.; et al. Cyclic [G(2′,5′)pA(3′,5′)p] is the metazoan second messenger produced by DNA-activated cyclic GMP-AMP synthase. *Cell* **2013**, *153*, 1094–1107. [CrossRef]
153. Zhou, W.; Whiteley, A.T.; de Oliveira Mann, C.C.; Morehouse, B.R.; Nowak, R.P.; Fischer, E.S.; Gray, N.S.; Mekalanos, J.J.; Kranzusch, P.J. Structure of the Human cGAS–DNA Complex Reveals Enhanced Control of Immune Surveillance. *Cell* **2018**, *174*, 300–311.e11. [CrossRef] [PubMed]
154. Li, X.; Shu, C.; Yi, G.; Chaton, C.T.; Shelton, C.L.; Diao, J.; Zuo, X.; Kao, C.C.; Herr, A.B.; Li, P. Cyclic GMP-AMP Synthase Is Activated by Double-Stranded DNA-Induced Oligomerization. *Immunity* **2013**, *39*, 1019–1031. [CrossRef] [PubMed]
155. Dremel, S.E.; Deluca, N.A. Herpes simplex viral nucleoprotein creates a competitive transcriptional environment facilitating robust viral transcription and host shut off. *Elife* **2019**, *8*, 8. [CrossRef] [PubMed]
156. Knipe, D.M.; Lieberman, P.M.; Jung, J.U.; McBride, A.A.; Morris, K.V.; Ott, M.; Margolis, D.; Nieto, A.; Nevels, M.; Parks, R.J.; et al. Snapshots: Chromatin control of viral infection. *Virology* **2013**, *435*, 141–156. [CrossRef]
157. Roy, A.; Ghosh, A.; Kumar, B.; Chandran, B. IfI16, a nuclear innate immune DNA sensor, mediates epigenetic silencing of herpesvirus genomes by its association with H3K9 methyltransferases SUV39H1 and GLP. *Elife* **2019**, *8*, 8. [CrossRef]
158. Abrisch, R.G.; Eidem, T.M.; Yakovchuk, P.; Kugel, J.F.; Goodrich, J.A. Infection by Herpes Simplex Virus 1 Causes Near-Complete Loss of RNA Polymerase II Occupancy on the Host Cell Genome. *J. Virol.* **2016**, *90*, 2503–2513. [CrossRef]
159. Lang, F.; Li, X.; Vladimirova, O.; Hu, B.; Chen, G.; Xiao, Y.; Singh, V.; Lu, D.; Li, L.; Han, H.; et al. CTCF interacts with the lytic HSV-1 genome to promote viral transcription. *Sci. Rep.* **2017**, *7*, 1–15. [CrossRef]
160. Bowman, G.D.; Poirier, M.G. Post-translational modifications of histones that influence nucleosome dynamics. *Chem. Rev.* **2015**, *115*, 2274–2295. [CrossRef]
161. Shivaswamy, S.; Bhinge, A.; Zhao, Y.; Jones, S.; Hirst, M.; Iyer, V.R. Dynamic remodeling of individual nucleosomes across a eukaryotic genome in response to transcriptional perturbation. *PLoS Biol.* **2008**, *6*, e65. [CrossRef]
162. Song, L.; Crawford, G.E. DNase-seq: A high-resolution technique for mapping active gene regulatory elements across the genome from mammalian cells. *Cold Spring Harb. Protoc.* **2010**, *5*. [CrossRef]
163. Giresi, P.G.; Kim, J.; McDaniell, R.M.; Iyer, V.R.; Lieb, J.D. FAIRE (Formaldehyde-Assisted Isolation of Regulatory Elements) isolates active regulatory elements from human chromatin. *Genome Res.* **2007**, *17*, 877–885. [CrossRef] [PubMed]
164. Buenrostro, J.D.; Giresi, P.G.; Zaba, L.C.; Chang, H.Y.; Greenleaf, W.J. Transposition of native chromatin for fast and sensitive epigenomic profiling of open chromatin, DNA-binding proteins and nucleosome position. *Nat. Methods* **2013**, *10*, 1213–1218. [CrossRef] [PubMed]

MDPI
St. Alban-Anlage 66
4052 Basel
Switzerland
Tel. +41 61 683 77 34
Fax +41 61 302 89 18
www.mdpi.com

Biomolecules Editorial Office
E-mail: biomolecules@mdpi.com
www.mdpi.com/journal/biomolecules

www.ingramcontent.com/pod-product-compliance
Lightning Source LLC
LaVergne TN
LVHW070907170726
843515LV00005B/722

* 9 7 8 3 0 3 6 5 2 5 8 2 2 *